MATHS IN PERSPECTIVE 1

Pure Mathematics

Barbara Young
Tarporley County High School, Cheshire

Edward Arnold
A division of Hodder & Stoughton
LONDON   MELBOURNE   AUCKLAND

ACKNOWLEDGEMENTS

Thanks are due to my colleagues B. Law, J. D. Hatton and G. Burrows for their advice and encouragement and, in particular, for their help in trialling the material in this text. Thanks are due also to the many students who have helped with comments, suggestions and error spotting. But, most of all, thanks are due to Lance Ball, Michael Hand, Louisa Hitchin, Adrian Johnson and Sally Peters who have born the brunt of the experimentation needed to produce this book, have constructively criticised throughout and have sparked off many of the new ideas in presentation and content contained therein.

Last, but by no means least, thanks are due to Alan Young who undertook the onerous task of proof reading.

The publishers would like to thank the following examination boards for their permission to include questions from past examination papers: The Associated Examining Board; Joint Matriculation Board and University of London School Examinations Board.

© 1989 Barbara Young

First published in Great Britain 1989

British Library Cataloguing in Publication Data
Young, Barbara
  Pure mathematics : a complete and concise
  course.
  1. Mathematics——1961-
  I. Title
  510   QA39.2
  ISBN 0   7131   7643   1

Phototypesetting by Thomson Press (India) Limited, New Delhi. Printed and bound in Great Britain for Edward Arnold, the educational, academic and medical publishing division of Hodder and Stoughton Limited, Mill Road, Dunton Green, Sevenoaks, Kent by Richard Clay Ltd, Bungay, Suffolk

CONTENTS

Unit 5 CALCULUS II

Unit 6 SERIES

Unit 7 FURTHER TRIGONOMETRY

Unit 8 CALCULUS III

Unit 9 EXPONENTIAL AND LOGARITHMIC FUNCTIONS

Unit 10 ALGEBRAIC TECHNIQUES

INTRODUCTION

This is an A level course in pure mathematics—not an A level text book. It was developed to meet the following objectives.

- To provide a course that would cater for A level classes that include both average and very able students.
- To enable the students to spend as much time as possible *doing* mathematical problems, rather than writing up dictated notes.
- To make each piece of work skill-centered rather than just topic-centerd.
- To make each piece of work include plenty of rote practice of the skills involved but also include work to stretch abler students.
- To develop a 'rolling' course where techniques and topics are met over and over again rather than just in one chapter. In particular to repeat techniques that students find difficult or tend to forget easily.
- To include examples that teach mathematical 'tricks of the trade' as well as basic techniques.
- To introduce students to A level questions as early in the course as possible.
- To include A level questions as part of the course material as often as possible.
- To provide material for abler students (or for any keen students) to do extra work on their own.
- To provide a comprehensive set of reference/revision notes.

To meet these objectives:
- The course consists of a series of sections of work. The time allowed for each section is one (60–70 minute) lesson plus an equal amount of private study time.
- This allocated time includes time to introduce and/or explain the theory involved, to go over examples and to do the work associated with the section.
- Each piece of work includes plenty of practice in the skills and techniques that are to be developed at that point and also includes more difficult questions for abler students. Students do as much as they have time for, but the exercises are so planned that even slow students should have tried all the basic techniques in the time allowed.
- The teacher provides all the theory and explanations required in each case. The only theory provided in the text is that needed for reference and revision.
- The examples in the text cover not only basic skills but also many 'tricks of a mathematician's trade'.
- Certain techniques that students find difficult or tend to forget easily (particularly in calculus) are repeated and revised at regular intervals throughout the course.
- At the end of each unit of work (comprising 3–10 sections) one or two miscellaneous exercises are provided (leading possibly to a test on the section).
- These miscellaneous exercises are of two distinct types. At the beginning of the course each section has a Miscellaneous Exercise A consisting of simple questions similar to those covered in the work just done. As students' skills develop along with the course then there is also a second exercise—B consisting of A level questions. Later still in the course only the second type of miscellaneous exercise is provided.
- **Most Important**   This second type of miscellaneous exercise is composed of A level questions, but not only on the topic just done. Each question *includes* a

skill developed in the section just covered but covers also topics and skills met earlier in the course.

• The sections leading to the second type of miscellaneous exercise are labelled P.Q. + to indicate that they require more time than the other sections. Sufficient questions are included to provide *both* practice on A level questions during the course *and* revision of the course prior to the A level exams.

Notes on teaching the course

1 Although this course has been prepared as a rolling course there are certain units that stand alone and can be taught at any point in the A Level course. These are units 12, 13, 14, 15 and 16, although 12 should precede 15.

2. If a teacher introduces and explains any necessary background and theory, it is quite feasible for students to work through the examples in many of the sections themselves. However, the teaching value of the examples of greatly increased by either the teacher going through the examples with the students, or the students tackling the examples as a group (with guidance), without reference to the model answers.

Also, whilst care has been taken to include one example for each technique required, often students need to go through several examples of the same kind before the idea is assimilated.

Students' guide

All A level Pure Mathematics syllabuses contain an agreed common core syllabus. Thus most of the topics covered in this course are included in all Pure Mathematics syllabuses. The table below lists the sections NOT included in each of the syllabuses listed.

The A level questions included in the Miscellaneous Exercises NB, at the end of most units, each have a bracket after the question which tells the student

(i) the A level board: A = AEB, J = JMB & L = London

(ii) the section(s) which contain the relevent theory used in the question.
This last piece of information has two ourposes:

first—to tell the student not to attempt this question if any one of the sections listed is not in his/her syllabus;

second—to instruct the student where to look, within the course, for help with the question.

Thus it would help the student if he/she used a piece of card as a bookmark upon which were listed the sections not in his/her syllabus, taken from the table on the next page.

Table of Non-Core Sections

An X indicates that the section is *NOT* in the syllabus.
An F indicates that the section is in the Further Mathematics syllabus.

	J	L	A	S	O	C	O&C	W	NI
6.4	F		F	X	X			X	X
6.5	F		F	X	X	X	X	X	X
9.5	F	F	X	①	①			X	
9.6 & 9.7				X	X	X	X	X	X
10.3	F			X	X	X			
10.4						X			
10.10	F		F	X	X	X	X	X	X
11.3 & 11.4		F						X	
11.9				X	X	X	X	X	X
11.10		F	F	X	X	X	X	X	X
12.3				X	X	X	X	X	X
12.6 & 12.7	F		X	X	X	X	X	X	X
13.4			F	X	X	X	X	X	X
13.5	F		F	X	X	X	X	X	X
14.1						X			
14.2						X			
14.3						X			
14.4		F	F		X	X		X	
15.4					X	X			
16.1	X		F	X	X	X		X	
16.2	X		F	X	X	X		X	
16.3	X		F	X	X		X	X	
16.4	X		F	X	X		X	X	
16.5	X		F	X	X		X	X	
16.6	X	F		X	X	②	X		

Note① Series expansions for sin and cos are not included on these syllabuses but the other expansions are included.
Note ②: The Cambridge syllabus includes the Trapezium Rule but not Simpson's rule.

Note for students who wish to tackle extra questions on their own to try and improve their grade potential
The more A level questions you do, the better you will get at doing them. So, at the end of each unit there is a large selection of A level questions which can be tackled at any time after the work in that unit has been completed. But, before you attempt any question, check, by looking at the end of the question and at the above table, to see whether it includes any sections that are not on your syllabus.

Syllabus notes for students and teachers

J = JMB; this course covers the whole of the JMB Pure Mathematics syllabus (PI)

L = London; this course covers all the pure mathematics covered in the Mathematics Syllabus B 371

A = AEB; this course covers the whole of the Pure Mathematics Common Paper for use as 632/1, 636/1 and 646/1

S = SUJB; this course covers the whole of the SUJB Pure Mathematics I syllabus (PM1). However it should be noted that the sections on vectors (13.1, 13.2, 13.3) included in the agreed common core are not included in PM1 but are included in the companion syllabuses AM1 and PM2.

O = Oxford; this course covers all the Basic Work for the Pure Mathematics Paper 9850/1 with the exception of
    (a) solutions of differential equations in the form

$$\frac{d^2y}{dx^2} \pm k^2y = \text{constant}$$

    (b) de Moivre's theorem for integral values of n.

These topics are to be found in the Further Pure Mathematics text in this series.

C = Cambridge; this course covers all the mathematics covered in Paper 1 of Mathematics (9205) with the exception of the work on matrices and transformations which is to be found in the Further Pure Mathematics text in this series.

O&C = Oxford and Cambridge; this course covers all the pure mathematics covered in the Mathematics syllabus (9650) with the exception of
    (a) solutions of differential equations in the form

$$\frac{d^2y}{dx^2} \pm k^2y = \text{constant}$$

    (b) de Moivre's theorem for integral values of n.
    (c) polar coordinates
    (d) proof by contradiction and counter example
    (e) transformations and matrices.

These topics are to be found in the Further Pure Mathematics text in this series.
NI = Northern Ireland; this course covers all the pure mathematics in this syllabus with the exception of the work on matrices determinants and transformations which can be found in the further Pure Mathematics text in this series.

Mathematics notation

N	the set of positive integers and zero, $\{0, 1, 2, \ldots\}$
Z	the set of integers, $\{0, \pm 1, \pm 2, \pm 3, \ldots\}$
Z^+	the set of positive integers, $\{1, 2, 3, \ldots\}$
Q	the set of rational numbers, $\{p/q : p \in Z, q \in Z^+\}$
Q^+	the set of positive rational numbers, $\{x \in Q : x > 0\}$
Q_0^+	the set of positive rational numbers and zero, $\{x \in Q : x \geqslant 0\}$
R	the set of real numbers
R^+	the set of positive real numbers, $\{x \in R : x > 0\}$
R_0^+	the set of positive real numbers and zero, $\{x \in R : x \geqslant 0\}$
C	the set of complex numbers
$[a, b]$	the closed interval $\{x \in R : a \leqslant x \leqslant b\}$
$[a, b[$	the interval $\{x \in R : a \leqslant x < b\}$
$]a, b]$	the interval $\{x \in R : a < x \leqslant b\}$
$]a, b[$	the open interval $\{x \in R : a < x < b\}$
$=$	is equal to
$\neq$	is not equal to
$\equiv$	is identical to or is congruent to
$\approx$	is approximately equal to
$a \times b, ab, a \cdot b$	a multiplied by b
$a \div b, \dfrac{a}{b}, a/b$	a divided by b
$\displaystyle\sum_{i=1}^{n} a_i$	$a_1 + a_2 + \cdots + a_n$
$\propto$	is proportional to
$<$	is less than
$\leqslant, \not>$	is less than or equal to, is not greater than
$>$	is greater than
$\geqslant, \not<$	is greater than or equal to, is not less than
∞	infinity
$\sqrt{a}$	the positive square root of a
$|a|$	the modulus of a
$n!$	n factorial
$\dbinom{n}{r}$	the binomial coefficient $\dfrac{n!}{r!(n-r)!}$ for $n \in Z^+$
	$\dfrac{n(n-1) \cdots (n-r+1)}{r!}$ for $n \in Q$
e	base of natural logarithms
e^x	exponential function of x
$\log_a x$	logarithm to the base a of x
$\ln x, \log_e x$	natural logarithm of x
$\lg x, \log_{10} x$	logarithm of x to base 10

$p \Rightarrow q$	p implies q (if p then q)
$p \Leftarrow q$	p is implied by q (if q then p)
$p \Leftrightarrow q$	p implies and is implied by q (p is equivalent to q)
$\exists$	there exists
$\forall$	for all
$f(x)$	the value of the function f at x
$f : x \mapsto y$	the function f maps the element x to the element y
f^{-1}	the inverse function of the function f
gof, gf	the composite function of f and g which is defined by $(\text{gof})(x)$ or $\text{gf}(x) = \text{g}(f(x))$
$\lim_{x \to a} f(x)$	the limit of $f(x)$ as x tends to a
$\Delta x, \delta x$	an increment of x
$\dfrac{dy}{dx}$	the derivative of y with respect to x
$\dfrac{d^n y}{dx^n}$	the nth derivative of y with respect to x
$f'(x), f''(x), \ldots f^{(n)}(x)$	the first, second, ..., nth derivatives of $f(x)$ with respect to x
$\displaystyle\int y \, dx$	the indefinite integral of y with respect to x
$\displaystyle\int_a^b y \, dx$	the definite integral of y with respect to x between the limits $x = a$ and $x = b$
$\dot{x}, \ddot{x}, \ldots$	the first, second, ... derivatives of x with respect to t
sin, cos, tan cosec, sec, cot	the circular functions
$\sin^{-1}, \cos^{-1}, \tan^{-1},$ $\text{cosec}^{-1}, \sec^{-1}, \cot^{-1}$	the inverse circular functions
arcsin, arccos, arctan, arccosec, arcsec, arccot	the inverse circular functions
i	square root of -1
z	a complex number, $z = x + iy$ $\qquad\qquad\qquad = r(\cos\theta + i\sin\theta)$
Re z	the real part of z, Re $z = x$
Im z	the imaginary part of z, Im $z = y$
$|z|$	the modulus of z, $|z| = \sqrt{(x^2 + y^2)}$
arc z	the argument of z, $\arg z = \theta,\ -\pi < \theta \leqslant \pi$
z^*	the complex conjugate of z, $x - iy$
$\mathbf{a}$	the vector a
$\overrightarrow{AB}$	the vector represented in magnitude and direction by the directed line segment AB
$\hat{\mathbf{a}}$	a unit vector in the direction of $\mathbf{a}$

i, j, k	unit vectors in the directions of the cartesian coordinate axes
$\lvert\mathbf{a}\rvert, a$	the magnitude of a
$\lvert AB \rvert, AB$	the magnitude of $\overrightarrow{AB}$
a·b	the scalar product of a and b

Greek letters

α	alpha	ϕ	phi
β	beta	μ	mu
γ	gamma	ω	omega
δ	delta	χ	ci
ε	epsilon	ψ	psi
Σ	sigma	ρ	rho
λ	lambda		

BASIC TECHNIQUES

1.1 Techniques for Solving Simultaneous Equations

1 Method of Elimination

Example 1   Solve

$$x - 2y + z = 6 \tag{1}$$
$$3x + y - 2z = 4 \tag{2}$$
$$7x - 6y - z = 10 \tag{3}$$

Adding (1) and (3)

$$x - 2y + z = 6$$
$$\underline{7x - 6y - z = 10}$$

Adding $\Rightarrow 8x - 8y = 16$

$$\Rightarrow x - y = 2 \tag{4}$$

$2 \times (1) + (2)$

$$2x - 4y + 2z = 12$$
$$\underline{3x + y - 2z = 4}$$

Adding $\Rightarrow 5x - 3y = 16 \tag{5}$

$(5) - 3 \times (4)$

$$5x - 3y = 16$$
$$\underline{3x - 3y = 6}$$

Subtracting $\Rightarrow \quad 2x = 10$

$$\Rightarrow \quad x = 5$$

$(4) \Rightarrow 5 - y = 2 \quad \Rightarrow \quad y = 3$

$(1) \Rightarrow 5 - 6 + z = 6 \Rightarrow \quad z = 7$

$\Rightarrow$ Solution is $x = 5$, $y = 3$, $z = 7$

2 Method of Substitution

Example 2   Solve

$$x^2 + y^2 = 25 \tag{1}$$
$$y = x - 1 \tag{2}$$

Substituting for y from (2) into (1)

$$\Rightarrow x^2 + (x - 1)^2 = 25$$
$$x^2 + x^2 - 2x + 1 = 25$$
$$2x^2 - 2x - 24 = 0$$
$$x^2 - x - 12 = 0$$
$$(x + 3)(x - 4) = 0$$
$$\Rightarrow x = -3 \text{ and } x = 4$$

when $x = -3$, (2) $\Rightarrow y = -4$
and when $x = 4$, (2) $\Rightarrow y = 3$

$$\Rightarrow \text{Solutions are} \quad \left.\begin{array}{l} x = -3 \\ y = -4 \end{array}\right\} \quad \text{and} \quad \left.\begin{array}{l} x = 4 \\ y = 3 \end{array}\right\}$$

Note   Solutions must be given as two corresponding pairs of values of x and y not $x = -3$ and 4, $y = -4$ and 3.

3 Graphical Method

Example 3   Use a graphical method to solve the simultaneous equations $y = x^2$ and $y + x = 3$.

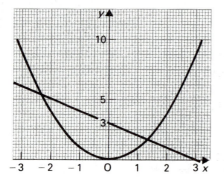

The solutions of the simultaneous equations are given by the values of x and y at the points of intersection.

$$\Rightarrow \text{Solutions are} \quad \left.\begin{array}{l} x = 1.3 \\ y = 1.7 \end{array}\right\}$$

$$\text{and} \quad \left.\begin{array}{l} x = -2.3 \\ y = 5.3 \end{array}\right\}$$

● **Exercise 1.1**   See page 20.

(All quadratic equations in the exercise can be solved by factorisation.)

1.2 Circular Measure

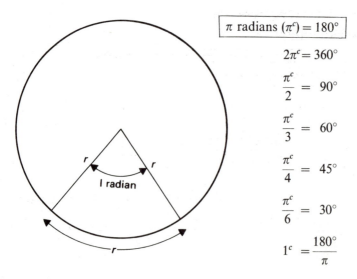

π radians $(\pi^c) = 180°$

$$2\pi^c = 360°$$

$$\frac{\pi^c}{2} = 90°$$

$$\frac{\pi^c}{3} = 60°$$

$$\frac{\pi^c}{4} = 45°$$

$$\frac{\pi^c}{6} = 30°$$

$$1^c = \frac{180°}{\pi}$$

Example 4   Express $\pi^c/10$, $2\pi^c/5$, $0.2\pi^c$ in degrees.

We know that

$$\pi^c = 180° \longleftarrow$$

> Always start with this statement

$$\Rightarrow \frac{\pi^c}{10} = \frac{180°}{10} \qquad = \underline{18°}$$

and

$$\frac{2\pi^c}{5} = \frac{2}{5} \times 180° \quad = \underline{72°}$$

and

$$0.2\pi^c = 0.2 \times 180° = \underline{36°}$$

Example 5   Express $35°$, $112.5°$ in radians.

We know that

$$180° \quad = \pi^c$$

$$\Rightarrow \quad 1° \quad = \frac{\pi^c}{180}$$

$$\Rightarrow \quad 35° \quad = 35 \times \frac{\pi^c}{180} \quad = \underline{\frac{7\pi^c}{36}}$$

and

$$112.5° = \frac{225}{2} \times \frac{\pi^c}{180} = \underline{\frac{5\pi^c}{8}}$$

Example 6   Using a calculator, evaluate $\sin 0.24^c$ and $\cos \pi/6$.

> If an angle is given in terms of π, it is assumed to be in radians and no units are necessary.

Put your calculator into **radian mode** first.

key sequences $\boxed{0.24} \longrightarrow \boxed{\sin} \longrightarrow 0.2377$

$\boxed{\pi} \longrightarrow \boxed{\div} \longrightarrow \boxed{6} \longrightarrow \boxed{=} \longrightarrow \boxed{\cos} \longrightarrow 0.866$

$\Rightarrow \underline{\sin 0.24^c = 0.2377 \text{ and } \cos \pi/6 = 0.866}$

Arc Length

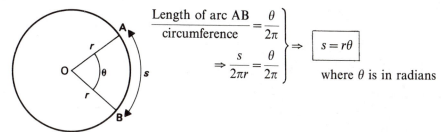

$$\frac{\text{Length of arc AB}}{\text{circumference}} = \frac{\theta}{2\pi}$$

$$\Rightarrow \frac{s}{2\pi r} = \frac{\theta}{2\pi}$$

$$\Rightarrow \boxed{s = r\theta}$$

where θ is in radians

Area of Sector of Circle

$$\frac{\text{area of sector AOB}}{\text{area of circle}} = \frac{\theta}{2\pi}$$

$$\Rightarrow \frac{\text{area}}{\pi r^2} = \frac{\theta}{2\pi}$$

$$\Rightarrow \boxed{\text{Area} = \tfrac{1}{2}r^2\theta}$$

where θ is in radians

Example 7 Find the arc length and area of the sector of a circle, radius 5 cm, which contains an angle of 72°.

First, we need the angle in radians:

$$180° = \pi^c$$

$$1° = \frac{\pi^c}{180}$$

$$\Rightarrow 72° = 72 \times \frac{\pi^c}{180} = \frac{2\pi^c}{5}$$

Arc length $= r\theta = 5 \times \dfrac{2\pi}{5} = 2\pi$ cm

| It is perfectly acceptable to leave answers as multiples of π |

Area of sector $= \tfrac{1}{2}r^2\theta = \tfrac{1}{2}\cdot 25\cdot\dfrac{2\pi}{5} = 5\pi$ cm^2

Area of Segment of Circle

Example 8 Find the area of the minor segment of a circle of radius 10 cm, given that the arc of the segment subtends an angle of $\pi^c/6$ at the centre of the circle.

Area of sector $= \tfrac{1}{2}r^2\theta = \tfrac{1}{2}\cdot 10^2\cdot\dfrac{\pi}{6} = 26.18$

Area of triangle $= \tfrac{1}{2}\cdot r\cdot r\cdot\sin\theta$ ←

| Using area of $\triangle$ $= \tfrac{1}{2}ab\sin C.$ |

$$= \tfrac{1}{2}\cdot 10^2\cdot\sin\frac{\pi}{6} = 25$$

Area of segment $=$ area of sector $-$ area of triangle

$\Rightarrow$ Area of segment $= 26.18 - 25 = 1.18$ cm^2

● **Exercise 1.2.** See page 21.

1.3 Manipulation of Surds

A **surd** is an irrational number of the form $\sqrt{n}$ where n is a positive integer which is not a perfect square.

Addition and Subtraction

$$3\sqrt{7} - \sqrt{7} = 2\sqrt{7}$$
$$4\sqrt{11} + 2\sqrt{11} = 6\sqrt{11}$$

but $\sqrt{7} - \sqrt{5}$ and $\sqrt{2} + \sqrt{3}$ cannot be simplified.

Multiplication and Division

$$\sqrt{2} \times \sqrt{3} = \sqrt{6}$$
$$\sqrt{15} \div \sqrt{3} = \sqrt{5}$$
$$(\sqrt{6})^2 = 6$$
$$2\sqrt{5} \times 3\sqrt{3} = 6\sqrt{15}$$
$$(\sqrt{3} - 2)(\sqrt{2} + 1) = \sqrt{3} \cdot \sqrt{2} - 2\sqrt{2} + \sqrt{3} - 2 = \sqrt{6} - 2\sqrt{2} + \sqrt{3} - 2$$
$$(\sqrt{2} + 1)(3 - \sqrt{2}) = 3\sqrt{2} + 3 - 2 - \sqrt{2} \qquad = 2\sqrt{2} + 1$$
$$(1 + \sqrt{5})^2 = 1 + 5 + 2\sqrt{5} \qquad = 6 + 2\sqrt{5}$$

Reduction to Lowest Terms

$$\sqrt{32} = \sqrt{16 \times 2} = \sqrt{16} \times \sqrt{2} = 4\sqrt{2}$$
$$\sqrt{27} - \sqrt{12} = 3\sqrt{3} - 2\sqrt{3} = \sqrt{3}$$

Rationalising the Denominator of a Fraction

$$\frac{3}{\sqrt{5}} = \frac{3}{\sqrt{5}} \times \frac{\sqrt{5}}{\sqrt{5}} = \frac{3\sqrt{5}}{5}$$

$$\frac{3}{2\sqrt{2} - 1} = \frac{3}{(2\sqrt{2} - 1)} \times \frac{(2\sqrt{2} + 1)}{(2\sqrt{2} + 1)} = \frac{6\sqrt{2} + 3}{(2\sqrt{2})^2 - 1} = \frac{6\sqrt{2} + 3}{7}$$

- Exercise 1.3   See page 22.

1.4 Functions

A **function** is a one–one or many–one mapping.

Notation and Terminology

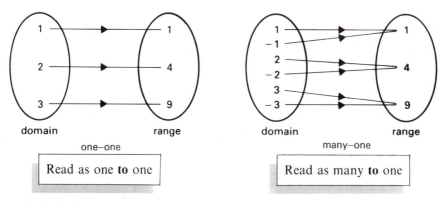

domain range	domain range
one–one	many–one

Read as one **to** one	Read as many **to** one

In both of these cases

1   The function may be described in one of two ways:

(a) $f(x) = x^2$ literally, the image of x under f is x^2

(b) $f:x \to x^2$ literally, f is the function such that x maps onto x^2

2 $f(x)$ is the **image** of x under f.

3 $x \to x^2$ is the **rule** for the function.

4 $\{1, 4, 9\}$ is the **image set** or **range** of the function.

In the first case
$\{1, 2, 3\}$ is the **domain** of the function.

In the second case
$\{\pm 1, \pm 2, \pm 3\}$ is the **domain** of the function.

Example 9   If $f:x \to 2x + 1$ for $x \in \{1, 2, 3, 4\}$ find the range of f.

The domain is $\{1, 2, 3, 4\}$

$$\left.\begin{array}{l} f(1) = 2(1) + 1 = 3 \\ f(2) = 2(2) + 1 = 5 \\ f(3) = 2(3) + 1 = 7 \\ f(4) = 2(4) + 1 = 9 \end{array}\right\} \Rightarrow \text{range} = \{3, 5, 7, 9\}$$

Example 10   If $g(x) = x^2 + 1$, find $g(3)$, $g(y)$, $g(x^2)$, $g(\sin x)$.

$g(x) = x^2 + 1$
$g(3) = 3^2 + 1$ $\quad\Rightarrow\quad$ $g(3) = 10$
$g(y) = y^2 + 1$
$g(x^2) = (x^2)^2 + 1$ $\quad\Rightarrow g(x^2) = x^4 + 1$
$g(\sin x) = (\sin x)^2 + 1 \Rightarrow g(\sin x) = \sin^2 x + 1$

> $\sin^2 x$ is the usual notation for $(\sin x)^2$

Inverse Functions

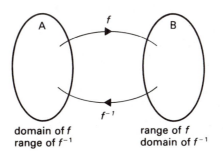

domain of f
range of f^{-1}

range of f
domain of f^{-1}

If f maps A onto B then the **inverse function** f^{-1} maps B onto A.

> **Important**   The inverse function f^{-1} only exists if f is one–one for the given domain.

Example 11   Find the inverse functions of the following one–one functions:
(a) $f:x \to x + 2$ (b) $g:x \to 3x$ (c) $h:x \to \frac{1}{4}x$.

(a) $f(x) = x + 2 \Rightarrow f^{-1}(x) = x - 2$
$\quad\Rightarrow f^{-1}:x \to x - 2$

> The inverse of 'add 2 to a number' is 'take 2 from a number

(b) $g(x) = 3x$ $\quad\Rightarrow g^{-1}(x) = \frac{1}{3}x$
$\quad\Rightarrow g^{-1}:x \to \frac{1}{3}x$

(c) $h(x) = \frac{1}{4}x$ $\quad\Rightarrow h^{-1}(x) = 4x$
$\quad\Rightarrow h^{-1}:x \to 4x$

> **Note**   The notations $f:x \to x + 2$ and $f(x) = x + 2$ are alternative ways of describing the same function. Either may be used but it is usual to give the answer in the same format as the question.

Example 12

(a)   Find the inverse function of $f(x) = x^2$ for the domain $\{0, 1, 2\}$.
(b)   State the domain and range of f^{-1}.
(c)   Why does f^{-1} not exist for f with domain $\{-1, 0, 1\}$?

(a) $f(x) = x^2 \Rightarrow f^{-1}(x) = \sqrt{x}$

(b) $\underline{\text{Domain of } f^{-1}} = \text{range of } f = \{0, 1, 4\}$
      $\underline{\text{Range of } f^{-1}} = \text{domain of } f = \{0, 1, 2\}$

(c) $f(-1) = 1$ and $f(1) = 1$
      $\Rightarrow f$ is many–one for this domain.
      $\therefore f^{-1}$ does not exist.

> where $\sqrt{x}$ is defined as the positive square root of x. $\pm\sqrt{x}$ indicates both positive and negative square roots

Finding More Complex Inverse Functions

Example 13   Find the inverse of f where $f : x \to \dfrac{(3x-1)}{2}$.

Method 1—using flow charts

$$x \longrightarrow \boxed{\times 3} \xrightarrow{3x} \boxed{-1} \xrightarrow{3x-1} \boxed{\div 2} \longrightarrow \frac{3x-1}{2}$$

Now reverse the flow chart operation

$$\frac{2x+1}{3} \xleftarrow{} \boxed{\div 3} \xleftarrow{2x+1} \boxed{+1} \xleftarrow{2x} \boxed{\times 2} \xleftarrow{} x$$

$$\therefore f^{-1} : x \to \frac{2x+1}{3}$$

Method 2—using algebraic techniques

$$y = f(x) \Rightarrow \quad y = \frac{3x-1}{2}$$

$$\Rightarrow 2y \quad = 3x - 1$$

$$\Rightarrow 2y + 1 = 3x$$

$$\Rightarrow x \quad = \frac{2y+1}{3}$$

$$\Rightarrow f^{-1}(y) = \frac{2y+1}{3}$$

$$\Rightarrow f^{-1} : y \to \frac{2y+1}{3}$$

which is equivalent to $f^{-1} : x \to \dfrac{2x+1}{3}$

● Exercise 1.4   See page 23.

1.5 Trigonometric Ratios

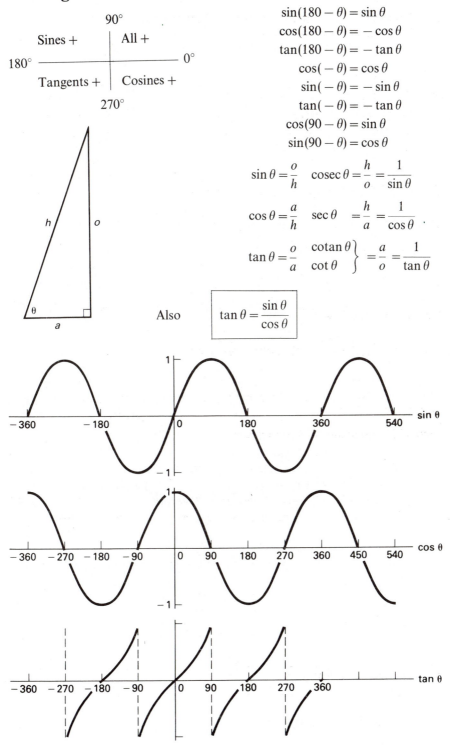

$$\sin(180 - \theta) = \sin \theta$$
$$\cos(180 - \theta) = -\cos \theta$$
$$\tan(180 - \theta) = -\tan \theta$$
$$\cos(-\theta) = \cos \theta$$
$$\sin(-\theta) = -\sin \theta$$
$$\tan(-\theta) = -\tan \theta$$
$$\cos(90 - \theta) = \sin \theta$$
$$\sin(90 - \theta) = \cos \theta$$

$$\sin \theta = \frac{o}{h} \qquad \operatorname{cosec} \theta = \frac{h}{o} = \frac{1}{\sin \theta}$$

$$\cos \theta = \frac{a}{h} \qquad \sec \theta = \frac{h}{a} = \frac{1}{\cos \theta}$$

$$\tan \theta = \frac{o}{a} \qquad \left.\begin{array}{l}\operatorname{cotan} \theta \\ \cot \theta\end{array}\right\} = \frac{a}{o} = \frac{1}{\tan \theta}$$

Also $\qquad \boxed{\tan \theta = \dfrac{\sin \theta}{\cos \theta}}$

Example 14   Solve $\operatorname{cosec} \theta = 1.2$ where $0^c < \theta < \pi^c/2$.

$$\operatorname{cosec} \theta = 1.2 \Rightarrow \sin \theta = \tfrac{1}{1.2} = 0.83°$$
$$\Rightarrow \theta = 0.985^c$$

> Remember that, if you want an answer in radians, you must put your calculator into radian mode first.

Example 15   If $\sin \theta = 0.5$, find all values of θ between $0°$ and $720°$.

S +	S +
S −	S −

$$\theta = 30°, (180 - 30°), (360 + 30°)$$
$$\text{and } (540 - 30°)$$
$$\Rightarrow \theta = 30°, 150°, 390°, 510°$$

> Positive values of sine appear only in the first and second quadrants

Example 16   If $\cos 2\theta = -0.9511$, find all values of θ between $-90°$ and $90°$.

C −	C +
C −	C +

$$-90° < \theta < 90° \Rightarrow -180° < 2\theta < 180°$$

> This calculation must be the first step of the solution to enable us to know what values of 2θ we need to calculate.

The acute angle whose cosine is 0.9511 is $18°$. Hence

$$2\theta = (180 - 18°) \text{ and } -(180 - 18°)$$

$$\Rightarrow 2\theta = \pm 162°$$
$$\Rightarrow \theta = \pm 81°$$

> Negative values of cosine appear only in the second and third quadrants.

Example 17   (A useful technique) Solve $\cos \theta + \sin \theta = 0$, giving all solutions between 0^c and $2\pi^c$.

T −	T +
T +	T −

$$\cos \theta + \sin \theta = 0$$
$$\Rightarrow \sin \theta = -\cos \theta$$
$$\Rightarrow \tan \theta = -1$$
$$\Rightarrow \theta = \frac{3\pi}{4}, \frac{7\pi}{4}$$

● **Exercise 1.5**   See page 25.

1.6 Useful Values and Fundamental Identities

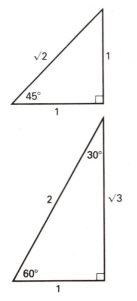

$$\sin 45° = \cos 45° = \frac{1}{\sqrt{2}}$$

$$\tan 45° = 1$$

$$\cos 60° = \sin 30° = \tfrac{1}{2}$$

$$\sin 60° = \cos 30° = \frac{\sqrt{3}}{2}$$

$$\tan 60° = \sqrt{3} \qquad \tan 30° = \frac{1}{\sqrt{3}}$$

$$\begin{aligned}
\sin 0° &= 0 & \cos 0° &= 1 \\
\sin 90° &= 1 & \cos 90° &= 0 \\
\sin 180° &= 0 & \cos 180° &= -1
\end{aligned}$$

Fundamental Identities

$$\boxed{\begin{aligned}
\sin^2 \theta + \cos^2 \theta &\equiv 1 \\
\tan^2 \theta + 1 &\equiv \sec^2 \theta \\
\cot^2 \theta + 1 &\equiv \operatorname{cosec}^2 \theta
\end{aligned}} \tag{1}$$

$\div$ (1) by $\cos^2 \theta \Rightarrow$

or $\div$ (1) by $\sin^2 \theta \Rightarrow$

Example 18   Express $5\cos^2 \theta - 3\sin^2 \theta$ in terms of (a) $\cos^2 \theta$ (b) $\sin^2 \theta$

(a) $\quad 5\cos^2 \theta - 3\sin^2 \theta = 5\cos^2 \theta - 3(1 - \cos^2 \theta)$
$$= 8\cos^2 \theta - 3$$

(b) $\quad 5\cos^2 \theta - 3\sin^2 \theta = 5(1 - \sin^2 \theta) - 3\sin^2 \theta$
$$= 5 - 8\sin^2 \theta$$

Example 19   Solve the equations
(a) $\sec^2 \theta = 2\tan \theta$ and (b) $2\cos^2 \theta + \sin \theta = 1$, for values of θ between 0° and 360°

(a) $\quad \begin{aligned}
\sec^2 \theta &= 2\tan \theta \\
\tan^2 \theta + 1 &= 2\tan \theta \\
\tan^2 \theta - 2\tan \theta + 1 &= 0 \\
(\tan \theta - 1)^2 &= 0 \\
\therefore \tan \theta &= 1
\end{aligned}$

using $\sec^2 \theta = \tan^2 \theta + 1$

$\therefore \theta = 45°, 225°$

(b) $2\cos^2\theta + \sin\theta$ $\qquad = 1$ 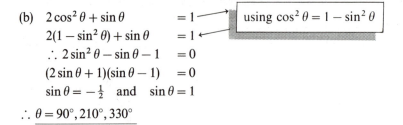 using $\cos^2\theta = 1 - \sin^2\theta$

$\qquad 2(1 - \sin^2\theta) + \sin\theta \quad = 1$

$\qquad \therefore \; 2\sin^2\theta - \sin\theta - 1 \quad = 0$

$\qquad (2\sin\theta + 1)(\sin\theta - 1) \quad = 0$

$\qquad \sin\theta = -\tfrac{1}{2} \quad \text{and} \quad \sin\theta = 1$

$\therefore \; \theta = 90°, 210°, 330°$

Note   The most common error in solving these equations is to say that, since $\cos^2\theta = 1 - \sin^2\theta$
then $\cos\theta = 1 - \sin\theta$, which is not true.

Example 20   Prove that $\tan\theta + \cot\theta \equiv \sec\theta\,\text{cosec}\,\theta$.

$$\text{LHS} = \frac{\sin\theta}{\cos\theta} + \frac{\cos\theta}{\sin\theta}$$

$$= \frac{\sin^2\theta + \cos^2\theta}{\cos\theta\sin\theta}$$

$$= \frac{1}{\cos\theta\sin\theta}$$

$$= \sec\theta\,\text{cosec}\,\theta = \text{RHS} \qquad\qquad \text{Q.E.D.}$$

● Exercise 1.6   See page 26.

1.7 Elementary Curve Sketching

Revision of Straight Lines

The equation of any straight line can be expressed in the form $y = mx + c$ where m is the gradient of the line and c is the intercept on the y-axis.

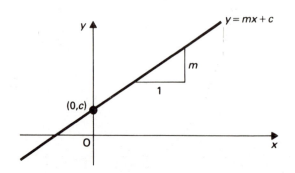

Example 21   Sketch the lines given by $y = 2x + 3$ and $2y - 4x + 1 = 0$.

$y = 2x + 3$ has gradient 2 and y-intercept 3.
$2y - 4x + 1 = 0$ rearranges to become $y = 2x - \frac{1}{2}$ which has gradient 2 and y-intercept $-\frac{1}{2}$.

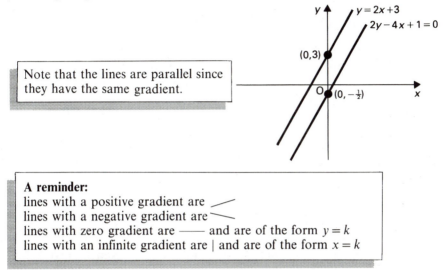

Note that the lines are parallel since they have the same gradient.

A reminder:
lines with a positive gradient are ⟋
lines with a negative gradient are ⟍
lines with zero gradient are ——— and are of the form $y = k$
lines with an infinite gradient are | and are of the form $x = k$

Interval Notation

$\{1, 2, 3\}$ is a set containing just three values. $1 \leqslant x \leqslant 3$ is the set of **all** the values between 1 and 3 inclusive. We now introduce a new form of notation.

$x \in [a, b]$	is equivalent to	$a \leqslant x \leqslant b$
$x \in \,]a, b[$	is equivalent to	$a < x < b$
$x \in \,]a, b]$	is equivalent to	$a < x \leqslant b$
$x \in [a, b[$	is equivalent to	$a \leqslant x < b$
$x \in [0, \infty[$	is equivalent to	$x \geqslant 0$

Graph Notation

Functions may be represented graphically by letting $f(x) = y$.

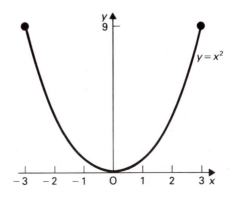

This graph may be labelled $y = x^2$ or $f(x) = x^2$.
This graph of $y = x^2$ is **symmetrical about $x = 0$**.
The **vertex** or **minimum point** occurs at $(0, 0)$.
The **minimum value** of $f(x) = x^2$ is zero.
In this diagram the **domain** of f is $[-3, 3]$ and the **range** of f is $[0, 9]$.
   If, however, the domain of f were to be extended to the set of real numbers $\mathbb{R}$, then the range of f would become $[0, \infty[$.

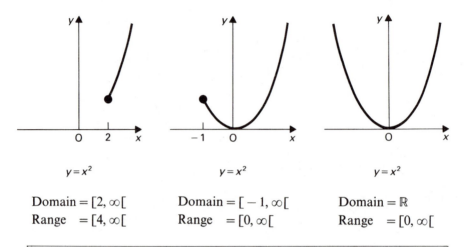

$y = x^2$	$y = x^2$	$y = x^2$
Domain $= [2, \infty[$	Domain $= [-1, \infty[$	Domain $= \mathbb{R}$
Range $= [4, \infty[$	Range $= [0, \infty[$	Range $= [0, \infty[$

> **Note** If the end point of a graph is not obviously labelled it is to be assumed that the graph continues indefinitely.

Related Graphs

Example 22 Sketch the graphs of $y = x^2$, $y = x^2 + 2$, $y = 2x^2$, $y = -x^2$, $y = -x^2 + 2$, $y = (x-2)^2$, $y = (x+2)^2$.

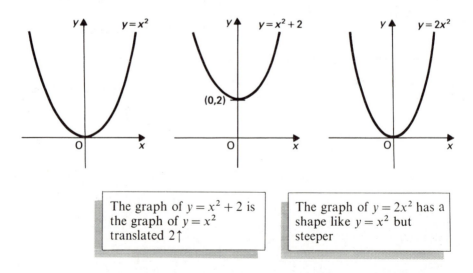

> The graph of $y = x^2 + 2$ is the graph of $y = x^2$ translated $2\uparrow$

> The graph of $y = 2x^2$ has a shape like $y = x^2$ but steeper

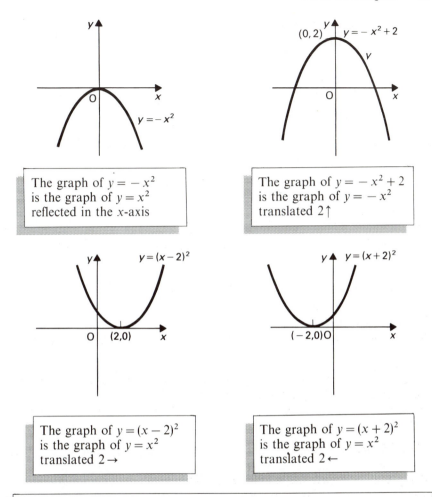

The graph of $y = -x^2$
is the graph of $y = x^2$
reflected in the x-axis

The graph of $y = -x^2 + 2$
is the graph of $y = -x^2$
translated $2\uparrow$

The graph of $y = (x - 2)^2$
is the graph of $y = x^2$
translated $2 \rightarrow$

The graph of $y = (x + 2)^2$
is the graph of $y = x^2$
translated $2 \leftarrow$

Note This technique of curve sketching using translations and scalings is dealt with much more rigorously later in the course

Graphs of Equations of the Form $y = (x - a)(x - b)$

Example 23 Sketch the curve $y = (x + 1)(x - 4)$ and use it to find the solution of $(x + 1)(x - 4) > 0$.

When $y = 0$, $(x + 1)(x - 4) = 0$.
Hence the curve crosses $y = 0$
when $x = -1$ and 4.
Since the coefficient of x^2 is positive,
the graph is $\cup$.

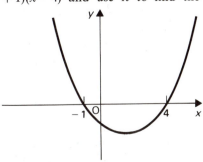

$(x + 1)(x - 4) > 0 \Rightarrow y > 0$

and this is true for $x < -1$ and $x > 4$.

$$\therefore (x+1)(x-4) > 0 \quad \text{for} \quad \underline{x < -1 \text{ and } x > 4}$$

$$\text{or} \quad \underline{x \in]-\infty, -1[\cup]4, \infty[}$$

> alternative notation for answers

Using Simultaneous Equations to Find the Points of Intersection

In 1.1 we found, graphically, the intersection of two lines to solve a pair of simultaneous equations.

Conversely: if we solve the pair of simultaneous equations it will give us the point(s) of intersection of the two lines.

●   **Exercise 1.7**   See page 27. (All quadratics used in this exercise can be factorised.)

1.8 Quadratic Functions and Equations

A **quadratic function** is a function of the form

$$f(x) = ax^2 + bx + c \qquad a \neq 0$$

Thus, the graph of a quadratic function has an equation of the form

$$y = ax^2 + bx + c$$

Quadratic Equations

Any equation of the form $ax^2 + bx + c = 0$ is called a **quadratic equation**. There are three main methods of solving quadratic equations.

1   *Factorisation*
If the equation can be expressed in the form

$$(x - \alpha)(x - \beta) = 0$$

then α and β are the **roots of the equation.**

2   *Completing the square*

Example 24   Solve $2x^2 - 5x + 1 = 0$ by completing the square.

$$2x^2 - 5x + 1 = 0$$

$$\Rightarrow \quad x^2 - \frac{5x}{2} + \frac{1}{2} = 0$$

> The first step is always to make the coefficient of x^2 into 1

$$\Rightarrow \quad x^2 - \frac{5x}{2} = -\frac{1}{2}$$

> Adding $(\frac{1}{2} \times \frac{5}{2})^2$ to both sides to 'complete the square' on the LHS

$$\Rightarrow x^2 - \tfrac{5}{2}x + (\tfrac{5}{4})^2 = -\tfrac{1}{2} + (\tfrac{5}{4})^2$$

$$\Rightarrow (x - \tfrac{5}{4})^2 = \tfrac{17}{16}$$

$$\Rightarrow x - \tfrac{5}{4} = \pm \tfrac{\sqrt{17}}{4}$$

$$\Rightarrow x \approx \tfrac{5}{4} \pm \tfrac{4.123}{4}$$

$$\Rightarrow x \approx 2.28 \text{ or } 0.22$$

3 Using 'the formula'

The formula is derived from $ax^2 + bx + c = 0$ using the method of completing the square and is

$$x = \frac{-b \pm \sqrt{b^2 - 4ac}}{2a}$$

Example 25 Use the formula to solve $2x^2 + 3x - 4 = 0$.

$$a = 2, b = 3, c = -4$$

$$\Rightarrow x = \frac{-3 \pm \sqrt{9 - 4.2(-4)}}{2.2}$$

$$\Rightarrow x = \frac{-3 \pm \sqrt{9 + 32}}{4}$$

$$\Rightarrow x = \frac{-3 + \sqrt{41}}{4} \quad \text{and} \quad x = \frac{-3 - \sqrt{41}}{4}$$

$$\Rightarrow x \approx 0.85 \text{ and } -2.35$$

The Nature of the Roots of a Quadratic Equation

The equation $ax^2 + bx + c = 0$

has **two real distinct roots**   if $b^2 - 4ac > 0$
has **two real equal roots**   if $b^2 - 4ac = 0$

has $\begin{cases} \textbf{no real roots} \\ \textbf{two imaginary roots} \end{cases}$ if $b^2 - 4ac < 0$

and, if $b^2 - 4ac = $ a perfect square, then the equation has two distinct, real, rational roots.

Intersection of a Line and Curve

To find the intersection of a line and a curve, their two equations are solved simultaneously (see 1.7). Often this results in a quadratic expression in x which will either have two real distinct roots, two real equal roots or two imaginary roots.

Two real distinct roots ⇔ two distinct points of intersection

Two real equal roots ⇔ the line is a tangent to the curve

Two imaginary roots ⇔ the line and curve do not intersect

Example 26   Show that $y = 4x - 2$ is a tangent to $y = x^2 + 2$.

To find the points of intersection we solve the simultaneous equations

$$y = 4x - 2 \quad \text{and} \quad y = x^2 + 2$$

$$\Rightarrow x^2 + 2 \qquad = 4x - 2$$

$$\Rightarrow x^2 - 4x + 4 = 0 \quad \longleftarrow \quad \boxed{\textbf{or,} \text{ you could show that } b^2 = 4ac}$$

$$\Rightarrow (x - 2)^2 \quad = 0$$

Since the equation has two equal roots, the line is a tangent to the curve

● Exercise 1.8   See page 28.

1.9 Principle of Undetermined Coefficients

Example 27   Find the values of a and b in the identity $(2x + a)^2 \equiv 4x^2 + 20x + b$.

This is equivalent to

$$4x^2 + 4ax + a^2 \equiv 4x^2 + 20x + b$$

Since this expression is true for all x then the two terms in x must be the same.

$$\therefore 4ax = 20x$$

$$\therefore a \quad = 5$$

Similarly, the constant terms must be the same

$$\therefore a^2 = b$$

$$\therefore b = 25$$

Example 28   Express the function $4x^2 + x - 1$ in the form $A + B(x + 1) + Cx(x + 1)$ where A, B and C are constants.

Let $\qquad\qquad 4x^2 + x - 1 \equiv A + B(x + 1) + Cx(x + 1)$

Equating coefficients of $x^2 : C = 4$

Equating coefficients of x $\ : B + C = 1$ $\qquad \therefore B = -3$

Equating coefficients of $x^0 : A + B = -1$ $\quad \therefore A = 2.$

$\therefore 4x^2 + x - 1 \equiv 2 - 3(x + 1) + 4x(x + 1)$

Example 29   Find l and m so that $9x^4 - 6x^3 + 13x^2 + lx + m$ may be a perfect square

Let $\qquad 9x^4 - 6x^3 + 13x^2 + lx + m \equiv (3x^2 + Ax + B)(3x^2 + Ax + B).$

Equating coefficients of $x^3 : 6A = -6$ $\quad \therefore A = -1$

Equating coefficients of $x^2 : 6B + A^2 = 13 \therefore B = 2.$

Equating coefficients of x $\ : 2AB = l$ $\qquad \therefore l = -4$

Equating coefficients of $x^0 : B^2 = m$ $\qquad \therefore m = 4.$

$\therefore l = -4$ and $m = 4$

Example 30
Given that
$$\frac{4x - 13}{2x^2 + x - 6} \equiv \frac{A}{x + 2} + \frac{B}{2x - 3} \quad \text{find } A \text{ and } B.$$
Let
$$\frac{4x - 13}{2x^2 + x - 6} \equiv \frac{A}{x + 2} + \frac{B}{2x - 3}$$
$$= \frac{A(2x - 3) + B(x + 2)}{2x^2 + x - 6}$$

Method 1
$$4x - 13 \equiv A(2x - 3) + B(x + 2)$$

Equating coefficients of x $\ : \quad 2A + B = 4$

Equating coefficients of $x^0 : -3A + 2B = -13$

Solving these equations simultaneously $\Rightarrow A = 3, B = -2$

Method 2
$$4x - 13 \equiv A(2x - 3) + B(x + 2)$$

Let $x = -2$, then $-8 - 13 = A(-7) \therefore A = 3$

Let $x = \frac{3}{2}$, then $6 - 13 = B(\frac{7}{2})$ $\quad \therefore B = -2$

Factorising $(x^3 \pm y^3)$

$$\boxed{\begin{array}{l} x^3 - y^3 \equiv (x - y)(x^2 + xy + y^2) \\ x^3 + y^3 \equiv (x + y)(x^2 - xy + y^2) \end{array}}$$

● Exercise 1.9   See page 29.

1.10 Revision of Unit

- Miscellaneous Exercise 1A   See page 30.

Unit 1 EXERCISES

Exercise 1.1

Solve:

1 $x + y + z = 4$
    $2x + 3y - z = 7$
    $x - y = -1$

2 $2x + y + z = 10$
    $3x - y + z = 12$
    $x - y - z = 2$

3 $3x - y + 4z = 19$
    $5x + y - 2z = 3$
    $2x + 2y + 3z = 11$

4 $x + 2y + 3z = -1$
    $-x + y + 3z = 4$
    $2x - y - z = -4$

5 $3u + 2v - 4w = 15$
    $u + v + 2w = 2$
    $2u + v - 3w = 10$

6 $3x + 2y - z = -1$
    $2x + 5y = 16$
    $z = 3$

7 $y = x^2$
    $y = x + 6$

8 $y = x^2$
    $y = 3 - 2x$

9 $y^2 + x^2 = 20$
    $2y - x = 0$

10 $xy = 4$
    $y = 2x + 2$

11 $y^2 = 4x$
    $2x + y = 4$

12 $4y^2 - 3x^2 = 1$
    $x - 2y = 1$

13 Use this graph to solve the following pairs of simultaneous equations:

(a) $y = x^2 - 4x + 3$
    $x + y = 3$

(b) $x + y = 3$
    $y = x$

(c) $y = x^2 - 4x + 3$
    $y = 0$

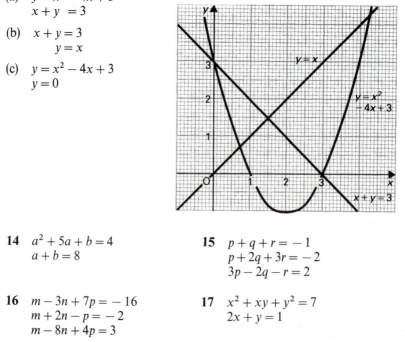

14 $a^2 + 5a + b = 4$
    $a + b = 8$

15 $p + q + r = -1$
    $p + 2q + 3r = -2$
    $3p - 2q - r = 2$

16 $m - 3n + 7p = -16$
    $m + 2n - p = -2$
    $m - 8n + 4p = 3$

17 $x^2 + xy + y^2 = 7$
    $2x + y = 1$

18 $x^2 + y^2 + 4x + 6y - 40 = 0$
   $x - y = 10$

19 $5x^2 - 6xy + 4y^2 = 3$
   $3x - 2y = 2$

20 $2x + 5y = 1$
   $x^2 + 5xy - 4y^2 = -10$

21 $0.5x + 0.3y + 0.2z = 46$
   $0.2x - 0.5y + 0.4z = 0$
   $0.1x + 0.8y - 0.6z = 26$

Exercise 1.2

1 Express the following angles in degrees:

$$\frac{\pi^c}{5}, \frac{\pi^c}{10}, \frac{3\pi^c}{4}, \frac{\pi^c}{9}, \frac{\pi^c}{8}, \frac{3\pi^c}{8}, \frac{3\pi^c}{10}, \frac{2\pi^c}{3}, \frac{3\pi^c}{2}, \frac{3^c}{2}$$

2 Express the following angles in radians, leaving your answers as multiples of π:
   $18°, 72°, 162°, 240°, 360°, 85°, 50°, 15°, 20°, 720°$

3 Use your calculator to evaluate each of the following, giving your answers correct to three decimal places:

$$\sin\frac{\pi}{6}, \sin\frac{3\pi}{8}, \cos\frac{\pi}{7}, \tan\frac{5\pi}{4}, \sin\frac{5\pi}{6}, \cos\frac{2\pi}{9}, \tan\pi,$$

$$\cos\frac{3\pi}{5}, \sin\frac{7\pi}{6}, \tan\frac{\pi}{4}$$

4 Find the arc lengths of the following sectors:
   (a) radius 6 cm, angle at centre = $36°$
   (b) radius 4 cm, angle at centre = $90°$
   (c) radius 10 cm, angle at centre = 0.2^c
   (d) radius 15 cm, angle at centre = 0.5^c
   (e) radius 20 cm, angle at centre = $144°$

5 Find the areas of each of the sectors in question 4

6 Find the angle, in radians, subtended by an arc of length 10 cm, at the centre of a circle of radius 8 cm

7 The area of a sector is 7.079 cm^2 and its angle is $30°$. Find the radius

8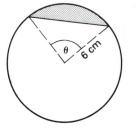

Find the area of the minor segment (shaded), to 2 decimal places, given that

   (a) $\theta = \dfrac{\pi}{2}$ (b) $\theta = \dfrac{2\pi}{3}$

9 Find the area of a sector of a circle, radius 5 cm, bounded by an arc of length 8 cm

10 Find the areas of the two segments into which a circle, of radius 8 cm, is divided by a chord of 8 cm, giving the areas to 2 d.p.

11 The radius of a cycle wheel is 35 cm. How many radians does the wheel turn through in travelling 1 m?

12 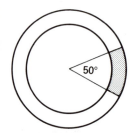 Find the area enclosed by two concentric circles of radii 12 cm and 9 cm, and two radii inclined at an angle of 50°, giving the answer as a multiple of π

13 The radius of a circle is 15 cm. Find the length of an arc of this circle if the length of the chord of the arc is 15 cm

*14 A cylindrical log of radius 30 cm, length 10 m, is floating with its axis horizontal and its highest point 5 cm above the level of the water. Find the volume of the log below the water

Exercise 1.3

1 Express in terms of the simplest possible surds:

(a) $\sqrt{8}$ (b) $\sqrt{12}$ (c) $\sqrt{20}$
(d) $\sqrt{50}$ (e) $\sqrt{80}$ (f) $\sqrt{18}$
(g) $\sqrt{72}$ (h) $\sqrt{125}$ (i) $\sqrt{200}$
(j) $\sqrt{450}$ (k) $\sqrt{2000}$ (l) $\sqrt{75}$

2 Simplify, where possible:

(a) $2\sqrt{3}+5\sqrt{3}$ (b) $3\sqrt{2}+2\sqrt{3}$ (c) $5\sqrt{11}-\sqrt{11}$
(d) $\sqrt{2}(3-\sqrt{2})$ (e) $\sqrt{2}(3-2\sqrt{2})$ (f) $\sqrt{5}(\sqrt{5}-1)$

3 Simplify:

(a) $\sqrt{6}\times\sqrt{8}$ (b) $3\sqrt{2}\times5\sqrt{3}$ (c) $\sqrt{3}\times\sqrt{24}$
(d) $\sqrt{112}\div\sqrt{28}$ (e) $7\sqrt{20}\div3\sqrt{5}$ (f) $\sqrt{120}\div\sqrt{24}$

4 Simplify:

(a) $5\sqrt{3}+\sqrt{48}$ (b) $\sqrt{32}-2\sqrt{8}$
(c) $\sqrt{3}(3\sqrt{3}-\sqrt{12})$ (d) $\sqrt{3}(\sqrt{27}-1)$
(e) $\sqrt{5}(\sqrt{5}-\sqrt{20})$ (f) $\sqrt{72}+\sqrt{8}-\sqrt{98}+\sqrt{50}$

5 Simplify:

(a) $(\sqrt{2}-1)(\sqrt{2}+1)$ (b) $(\sqrt{3}-2)(\sqrt{3}-1)$
(c) $(\sqrt{5}+2)(2\sqrt{5}-1)$ (d) $(3-\sqrt{2})(2+\sqrt{3})$

(e) $(x + \sqrt{2})(x - \sqrt{2})$ (f) $(y + 3\sqrt{3})(y + \sqrt{3})$

6 Simplify:

(a) $(\sqrt{2} - 1)^2$ (b) $(\sqrt{2} + 1)^2$

(c) $(\sqrt{x} - 1)(\sqrt{x} + 1)$ (d) $(2\sqrt{x} - 1)^2$

(e) $(\sqrt{x} + \sqrt{y})(\sqrt{x} - \sqrt{y})$ (f) $(\sqrt{a} - \sqrt{b})^2$

7 Rationalise the denominators and simplify:

(a) $\dfrac{1}{\sqrt{3}}$ (b) $\dfrac{1}{\sqrt{8}}$ (c) $\dfrac{2}{\sqrt{32}}$ (d) $\dfrac{3}{\sqrt{2}}$ (e) $\dfrac{9}{\sqrt{3}}$

8 Rationalise the denominators and simplify:

(a) $\dfrac{1}{\sqrt{2} + 1}$ (b) $\dfrac{1}{\sqrt{3} - 1}$ (c) $\dfrac{1}{2 + \sqrt{3}}$

(d) $\dfrac{\sqrt{5}}{2 + \sqrt{5}}$ (e) $\dfrac{2}{2\sqrt{3} - 3}$ (f) $\dfrac{3}{2 - \sqrt{3}}$

(g) $\dfrac{1}{2\sqrt{3} + \sqrt{2}}$ (h) $\dfrac{\sqrt{7}}{2 - \sqrt{7}}$

9 Rationalise the denominators and simplify:

(a) $\dfrac{2 + \sqrt{3}}{2 - \sqrt{3}}$ (b) $\dfrac{\sqrt{2} - 1}{2\sqrt{2} + 1}$ (c) $\dfrac{\sqrt{6} - \sqrt{2}}{\sqrt{6} + \sqrt{2}}$ (d) $\dfrac{\sqrt{5} - 1}{3 + \sqrt{5}}$

*10 Simplify, leaving the denominators in rational form:

(a) $\dfrac{1}{\sqrt{2} + 1} + \dfrac{1}{\sqrt{2} - 1}$ (b) $\dfrac{1}{\sqrt{x} + 2} - \dfrac{1}{\sqrt{x} - 2}$

(c) $\dfrac{\sqrt{x}}{\sqrt{x} + 1}$ (d) $\dfrac{x - y}{\sqrt{x} + \sqrt{y}}$

(e) $\dfrac{\sqrt{x} + 2}{\sqrt{x}}$ (f) $\left(\sqrt{x} + \dfrac{1}{\sqrt{x}}\right)\left(\sqrt{x} - \dfrac{1}{\sqrt{x}}\right)$

(g) $\dfrac{1}{1 + \sqrt{x}} + \dfrac{1}{1 - \sqrt{x}} + \dfrac{1}{1 - x}$

Exercise 1.4

1 If $g : x \to x + 2$ find $g(3)$, $g(0)$, $g(p^3)$, $g(\cos x)$

2 If $h(x) = 12/x$ find $h(1)$, $h(2)$, $h(6)$, $h(y)$

3 If $f : y \to 4 - 2y$ find $f(0)$, $f(1)$, $f(2)$, $f(3)$, $f(-3)$, $f(x)$, $f(2x)$, $f(3y)$

4 If $k:x \rightarrow \sqrt{[(x-1)(x-3)]}$

    (a) find $k(1)$, $k(3)$, $k(-3)$

    (b) is k one–one or many–one?

5 $f(x) = x^2 - 3$

    (a) Find $f(3)$, $f(-3)$, $f(1)$

    (b) Find the range if the domain is $\{1, 2, 3\}$

    (c) Find the range if the domain is $\{-1, 0, 1\}$

    (d) If the domain is $\{1, 2, 3\}$ is f one–one or many–one?

    (e) If the domain is $\{-1, 0, 1\}$ is f one–one or many–one?

6 State the inverse functions of:

    (a) $f:x \rightarrow x - 3$

    (b) $g:x \rightarrow x/2$

    (c) $h:x \rightarrow 100x$

7 If $k:x \rightarrow x + 1$ for the domain $\{1, 2, 3\}$

    (a) find the range of k

    (b) find the inverse function k^{-1}

    (c) state the domain and range of k^{-1}

8 Find the inverses of the following functions:

    (a) $f:x \rightarrow 5x - 3$

    (b) $f:x \rightarrow 5(x - 3)$

    (c) $g:x \rightarrow 3(2x + 1)$

    (d) $g:x \rightarrow \frac{1}{3}(2x + 1)$

    (e) $h:x \rightarrow 3 - x$

    (f) $h:x \rightarrow \dfrac{6}{x}$

*9 If $h:x \rightarrow 2^x + 1$ find:

    (a) $h(2)$, $h(3)$, $h(-1)$

    (b) the value of x if $h(x) = 17$

*10 $f(x) = \sin x$ and $g(x) = x + 2$

    (a) Are f and g one–one or many–one, if the domain for both is

$$0 \leqslant x \leqslant \frac{\pi}{2}$$

    (b) Find $f(0)$, $f\left(\dfrac{\pi}{2}\right)$, $f\left(\dfrac{\pi}{6}\right)$

    (c) Find $f(x^2)$, $f(x - 1)$, $f(x + 3)$, $f(g(x))$

    (d) Find $g(1)$, $g(y)$, $g(p^2)$, $g(\cos x)$, $g(f(x))$

Exercise 1.5

Give all answers to the nearest 0.1 of a degree or as a multiple of π^c.

1 Solve $\sec \theta = 2$ where $0° < \theta < 90°$

2 Solve $\operatorname{cosec} \theta = 4$ where $0° < \theta < 180°$

3 Solve $\cot \theta = 1$ where $0^c < \theta < 2\pi^c$

4 Solve $\sin \theta = -0.5$ where $0° < \theta < 360°$

5 Find all the angles between $0°$ and $360°$ whose cosine is 0.766

6 Solve, for $-180° < \theta < 180°$
    (a) $\cos \theta = 0.636$ (b) $\cos \theta = -0.636$ (c) $\tan \theta = 1.732$
    (d) $\tan \theta = -1.732$ (e) $\sin \theta = -0.5$

7 Solve, for $0° \leqslant \theta \leqslant 360°$
    (a) $\tan \theta = 1$ (b) $\tan \theta = -1$ (c) $\sin \theta = -0.3$
    (d) $\cos \theta = -0.3$ (e) $\tan \theta = 0$

8 If $\cos 2\theta = 0.777$ find all values of θ between $0°$ and $360°$

9 If $\tan 2\theta = 1$ find all values of θ between 0^c and π^c

10 If $\sin 3\theta = \frac{1}{2}$ find all values of θ between $0°$ and $180°$

11 If $\cos \theta/2 = \frac{1}{2}$ find all values of θ between $-180°$ and $+180°$

12 On one diagram, sketch the graphs of $\sin x$, $\sin 2x$, $\sin x/2$ for values of x from $-360°$ to $+360°$

13 On one diagram, sketch the graphs of $\cos x$, $\cos 2x$, $\cos x/2$, for values of x from $-360°$ to $+360°$

14 Solve $3 \sin \theta = 2 \cos \theta$ giving all solutions between $0°$ and $360°$

15 Solve $\sin \theta + 2 \cos \theta = 0$ giving all solutions between $-180°$ and $180°$

16 Solve, for $0° \leqslant \theta \leqslant 360°$:
    (a) $\sin 2\theta = -0.3636$ (b) $\tan 2\theta = 1.9878$
    (c) $\cos \theta/2 = -0.35$ (d) $\tan \theta/3 = -3.078$
    (e) $\sin 3\theta/2 = 0.5$ (f) $\sin \theta/4 = 0$

Exercise 1.6

1 Without using a calculator, complete this table:

Angle	sin x	cos x	tan x	cosec x	sec x	cot x
0^c	0	1		∞		
$\pi/6^c$	$\frac{1}{2}$	$\sqrt{3}/2$				
$\pi/4^c$		$1/\sqrt{2}$	1			
$\pi/3^c$						
$\pi/2^c$	1		∞			
$2\pi/3^c$						
$3\pi/4^c$						
$5\pi/6^c$						
π^c	0	-1				

2 Express in terms of sin θ:
   (a) $\cos^2 \theta - \sin^2 \theta$ (b) $2\cos^2 \theta - 3\sin \theta$

3 Simplify:

   (a) $\dfrac{\sin^2 \theta + \cos^2 \theta}{\cos^2 \theta}$ (b) $\dfrac{\sin^2 \theta \cos \theta + \cos^3 \theta}{\sin \theta}$

   (c) $\dfrac{\sec^2 \theta - 1}{\cosec^2 \theta - 1}$ (d) $\dfrac{\sin \theta}{\sqrt{1 - \cos^2 \theta}}$

   (e) $\dfrac{\sqrt{(1 + \tan^2 \theta)}}{\sqrt{(1 - \sin^2 \theta)}}$ (f) $\dfrac{1}{\cos \theta \sqrt{(1 + \cot^2 \theta)}}$

4 If $\tan^2 \theta + 3\sec^2 \theta = 5$ find the possible values of $\tan \theta$

Solve the following equations for $0° \leqslant \theta \leqslant 180°$

5 $4\cos^2 \theta + 5\sin \theta = 5$

6 $2\sec^2 \theta = 3\tan^2 \theta$

7 $2\sin^2 \theta + \cos \theta = 1$

8 $2\cosec^2 \theta = 3\cot \theta + 1$

9 $\sin^2 \theta = \cos^2 \theta + \cos \theta$

10 $\tan^2 \theta = \sec \theta + 1$

11 $1 + \sin \theta \cdot \cos^2 \theta = \sin \theta$

Prove the following identities:

12 $\tan^2 \theta + \cot^2 \theta \equiv \sec^2 \theta + \text{cosec}^2 \theta - 2$

13 $(\sin \theta + \cos \theta)^2 \equiv 1 + 2 \sin \theta \cdot \cos \theta$

*14 $(1 + \sin \theta + \cos \theta)^2 \equiv 2(1 + \sin \theta)(1 + \cos \theta)$

*15 $(\text{cosec } \theta - \sin \theta)(\sec \theta - \cos \theta) \equiv \cos \theta \cdot \sin \theta$

Exercise 1.7

1 On one diagram sketch (do not plot!) and label each of the following lines:
$y = x$, $y = 3x$, $y = \frac{1}{3}x$, $y + x = 0$, $y + 2x = 0$, $y = 0$, $x = 0$

2 On one diagram sketch and label each of the following lines:
$y = x$, $y - x = 2$, $2y + 7 = 2x$

3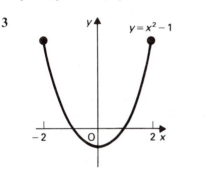

(a) What is the equation of the axis of symmetry of $y = x^2 - 1$?
(b) What are the coordinates of the vertex?
(c) What is the minimum value of $x^2 - 1$?
(d) The domain of $x^2 - 1$ is $[-2, 2]$. What is the range?

4 (a) What is the equation of the axis of symmetry of $f(x) = (x + 1)(5 - x)$?
(b) What are the coordinates of the vertex?
(c) What is the maximum value of $(x + 1)(5 - x)$?
(d) What is the domain of f? What is the range?

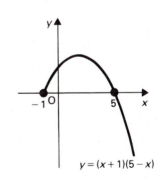

5 On separate diagrams, sketch each of the curves given by the following equations. Mark on each sketch the coordinates of the vertex

(a) $y = x^2$ (b) $y = x^2 - 1$ (c) $y = (x - 1)^2$
(d) $y = (x + 3)^2$ (e) $y = 2(x - 1)^2$ (f) $y = 3 - x^2$

6 (a) For what values of x does $y = (x - 1)(x + 2)$ cross the x-axis?
(b) Sketch the graph of $y = (x - 1)(x + 2)$
(c) Use the graph to solve $(x - 1)(x + 2) < 0$
(d) What is the axis of symmetry?

7 Without drawing graphs, find the points of intersection of the following pairs of straight lines and curves:

(a) $y = x$
$y + x = 3$

(b) $y = x^2 - 1$
$y = 1 - x$

(c) $y + x = 4$
$x = 1$

*8 Use sketch graphs to find the solutions of each of the following:

(a) $x^2 - 2x < 0$

(b) $4 - x^2 \geqslant 0$

(c) $x^2 + 8 \leqslant 6x + 3$

(d) $4x + 1 < x^2 + 4$

(e) $3x^2 + 2x - 1 < 0$

(f) $2x^2 + 7x + 3 \geqslant 0$

*9 This is the graph of $y = \sin x$, with x in radians

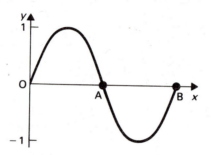

(a) What are the coordinates of A and B?

(b) Sketch the graph of $y = 1 + \sin x$ for $x \in [0, 2\pi]$
What are the coordinates of the point where it touches the x-axis?

(c) Sketch the graph of $y = \sin(x - \pi)$ for $x \in [0, 2\pi]$

Exercise 1.8

Solve the following by completing the square:

1 $x^2 - 2x - 1 = 0$

2 $x^2 + 4x - 8 = 0$

3 $x^2 - 5x + 3 = 0$

4 $x^2 + 7x - 5 = 0$

5 $2x^2 - 6x - 1 = 0$

6 $2x^2 + 7x + 3 = 0$

7 $4x^2 - 6x + 1 = 0$

8 $3x^2 + 4x + 1 = 0$

9 $x^2 - 2ax + b = 0$

Solve the following using the formula:

10 $x^2 - 5x + 1 = 0$

11 $x^2 + 4x - 6 = 0$

12 $x^2 - 2x + 1 = 0$

13 $2x^2 - 3x - 2 = 0$

14 $3x^2 + 5x + 2 = 0$

15 $2x^2 + 10x + 3 = 0$

16 $3x^2 - 8x - 4 = 0$

17 $5x^2 + 3x - 2 = 0$

18 $5x^2 - 17x + 3 = 0$

Determine the nature of the roots of each of the following equations: (Do not solve the equations)

19 $2x^2 + 3x + 1 = 0$

20 $3x^2 + 5x + 5 = 0$

21 $3x^2 + 5x - 5 = 0$

22 $4x^2 + 4x + 1 = 0$

22 $4x^2 - 3x + 2 = 0$

24 $31x^2 - 57x - 101 = 0$

25 Determine the nature of the points of intersection of the curve $y^2 - x^2 + 2x + 3 = 0$ and the straight line $y = 2x - 1$

26 Prove that the straight line $4y = 8x + 5$ touches the curve $y^2 = 10x$

27 The roots of $3x^2 + 12x + k = 0$ are equal. Find k

28 If $y^2 + py + 25 = 0$ has equal roots, find p

29 For what values of k is $16x^2 + kx + 25$ a perfect square?

*30 Prove that $kx^2 + 6x - (k - 6) = 0$ has real roots for all values of k

*31 By considering the value of '$b^2 - 4ac$' show that the roots of $px^2 + (p + q)x + q = 0$ are real for all values of p and q

*32 If x is real and $t = \dfrac{5x^2 - 1}{2x - 1}$, show that $t^2 - 5t + 5 \geqslant 0$

Exercise 1.9

1 Factorise:
   (a) $a^3 - b^3$ (b) $x^3 - 27$ (c) $p^3 + 8$ (d) $1 + y^3$

2 Find the values of the constants A and B in each of the following:
   (a) $2(x - 1)^2 \equiv A(x^2 + 1) + Bx$
   (b) $3x - 1 \equiv A(x - 1) + B(x + 1)$

3 Express $6 - 8x + 6x^2$ in the form $A(1 - x)^2 + B(1 + x)^2$ where A and B are constants

4 If $x^4 + 40x - 96 \equiv (x^2 - ax + 12)(x^2 + 2x - b)$ find a and b

5 If $x^4 + x^3 - 3x^2 + 7x - 6 \equiv (x^2 - x + 2)(x^2 + ax + b)$ find a and b and hence completely factorise the expression

6 Express $5x^2 - 18x + 17$ in the form
$$a(x - 1)(x - 2) + b(x - 3)(x - 1) + c(x - 2)(x - 3)$$

7 If $4x^4 - 4x^3 + 5x^2 + ax + b$ is a perfect square, find a and b

8 In the following identities find the values of A and B:
   (a) $\dfrac{x + 1}{(x - 1)(x - 3)} \equiv \dfrac{A}{x - 1} + \dfrac{B}{x - 3}$

   (b) $\dfrac{6x}{(2x - 1)(x + 1)} \equiv \dfrac{A}{2x - 1} + \dfrac{B}{x + 1}$

   (c) $\dfrac{2}{1 - x^2} \equiv \dfrac{A}{1 - x} + \dfrac{B}{1 + x}$

9 Find the constants p and q where:

(a) $\dfrac{x+2}{x^2-1} \equiv \dfrac{p}{x-1} + \dfrac{q}{x+1}$

(b) $\dfrac{10x}{(3x-1)(x+3)} \equiv \dfrac{p}{x+3} + \dfrac{q}{3x-1}$

(c) $\dfrac{x}{(1+x)^2} \equiv \dfrac{p}{(1+x)^2} + \dfrac{q}{1+x}$

*10 If $(lx+my)^3 \equiv 8x^3 - 12x^2y + pxy^2 - y^3$, find l, m and p

*11 Express the fraction $\dfrac{x^3}{x^3+1}$ in the form $A + \dfrac{B}{x+1} + \dfrac{Cx+D}{x^2-x+1}$

*12 Express the fraction $\dfrac{4}{x(x+2)^2}$ in the form $\dfrac{A}{x} + \dfrac{B}{x+2} + \dfrac{C}{(x+2)^2}$

*13 Express the fraction $\dfrac{x^2}{(x+2)^3}$ as the sum of three fractions of the form $\dfrac{A}{x+2}$,
$\dfrac{B}{(x+2)^2}$ and $\dfrac{C}{(x+2)^3}$

Miscellaneous Exercise 1A

1 Solve $2x+y+2z=4$
$x-2y+4z=6$
$3x+3y-2z=2$

2 Solve $x^2+y^2=25$ and $x+2y=5$

3 Express in degrees: $(\pi^c/2)\ 2\pi^c,\ \pi^c/6,\ \pi^c/3,\ \pi^c/10,\ \pi^c/4,\ 3\pi^c/4$

4 Express in radians: $18°,\ 72°,\ 180°,\ 270°,\ 150°$

5 Find the total length of the perimeter of a sector of a circle, of radius 7.4 cm and angle $45°$ to 2 d.p.

6 Find the area of a sector of a circle, radius 10 cm bounded by an arc of length 10 cm

7 What is the ratio of the radii of two circles, at the centres of which, two arcs of the same length subtend angles of $45°$ and $60°$?

8 Simplify:

(a) $\sqrt{3}(\sqrt{5}-\sqrt{3})$

(b) $\sqrt{75}+\sqrt{12}-\sqrt{48}$

(c) $(\sqrt{3}+\sqrt{2})(\sqrt{3}-\sqrt{2})$

(d) $(3\sqrt{2}-1)(3-\sqrt{2})$

(e) $\dfrac{5}{\sqrt{10}}$

(f) $\dfrac{\sqrt{3}+1}{3\sqrt{3}-1}$

9 Solve, for $0° \leqslant \theta \leqslant 180°$, giving θ correct to 1 d.p.:

(a) $\sec \theta = \sqrt{2}$ (b) $\cot \theta = -0.6$ (c) $\mathrm{cosec}\, \theta = -2$

(d) $\cos 2\theta = 0.5358$ (e) $\sin 3\theta = -0.3795$ (f) $\tan \dfrac{\theta}{2} = 0.2217$

10 On one diagram sketch the graphs of $\tan \theta$ and $\tan 2\theta$, for values of θ from $-180°$ to $+180°$

11 Solve $\sin 2x - 3\cos 2x = 0$ for values of x between $0°$ and $360°$

12 Express in terms of $\cos^2 \theta$:

(a) $\cos^2 \theta - \sin^2 \theta$ (b) $3\sin^2 \theta - 4\cos^2 \theta$

13 State whether the following are true or false:

(a) $\sin \pi/4 = 1/\sqrt{2}$ (b) $\sec 45° = \sqrt{2}$ (c) $\sec 135° = \sqrt{2}$
(d) $\cot \pi/6 = \sqrt{3}$ (e) $\tan 5\pi/6 = -1/\sqrt{3}$ (f) $\tan \pi/2 = 0$

14 Solve for $0° \leqslant \theta \leqslant 360°$

(a) $2\cos^2 \theta = \sin \theta + 1$ (b) $2\tan^2 \theta + 5 = 5\sec \theta$

15 Solve for $0° \leqslant x \leqslant 180°$, $2\,\mathrm{cosec}^2\, 3x = 3\cot 3x + 1$

16 Factorise:

(a) $125 - r^3$ (b) $27a^3 + 64b^3$

17 Solve by completing the square

(a) $x^2 - 4x + 2 = 0$ (b) $2x^2 - 6x + 3 = 0$

18 Solve using 'the formula'

(a) $x^2 + 5x + 1 = 0$ (b) $2x^2 - 5x - 1 = 0$

19 If $f : x \rightarrow 2x - 5$ find

(a) $f(5)$ (b) $f(x^2)$ (c) $f^{-1}(x)$ (d) $f^{-1}(1)$

20 (a) Sketch the graph of $y = (x - 2)(x + 3)$
(b) Solve $(x - 2)(x + 3) \geqslant 0$
(c) Find the coordinates of the minimum point of this graph

21 Find the constants A, B, C, D in the following identities:

(a) $\dfrac{3}{(x^2 + 1)(x - 2)} \equiv \dfrac{A}{x - 2} + \dfrac{Bx + C}{x^2 + 1}$

(b) $\dfrac{3}{(1 - x)(1 + x)^2} \equiv \dfrac{A}{1 - x} + \dfrac{B}{1 + x} + \dfrac{C}{(1 + x)^2}$

(c) $\dfrac{x^2 + 5}{(x^2 + 1)(x - 1)^2} \equiv \dfrac{Ax + B}{x^2 + 1} + \dfrac{C}{x - 1} + \dfrac{D}{(x - 1)^2}$

2 INTRODUCTION TO COORDINATE GEOMETRY

2.1 Some Properties of Points and Lines

Distance between two points (x_1, y_1) **and** (x_2, y_2)

$$d^2 = (x_2 - x_1)^2 + (y_2 - y_1)^2$$

Gradient of line joining two points (x_1, y_1) **and** (x_2, y_2)

$$m = \frac{y_2 - y_1}{x_2 - x_1} = \frac{\text{difference in } ys}{\text{difference in } xs}$$

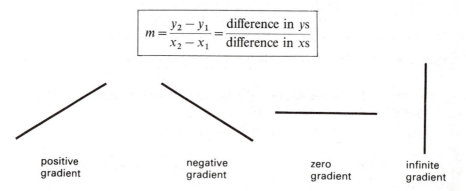

positive gradient   negative gradient   zero gradient   infinite gradient

Coordinates of midpoint of line joining (x_1, y_1) **and** (x_2, y_2)

$$(x_m, y_m) = \left(\frac{x_1 + x_2}{2}, \frac{y_1 + y_2}{2} \right)$$

Example 1   Find the distance between the points $(1, 2)$ and $(3, -1)$.

$$d^2 = (1 - 3)^2 + (2 - (-1))^2 = 2^2 + 3^2 = 13$$

$\therefore$ distance $= \sqrt{13}$

Example 2   Find the gradients of the lines joining the points (a) $(1, 2)$ and $(3, -1)$ (b) $(1, 2)$ and $(1, 4)$.

(a) $m = \dfrac{2 - (-1)}{1 - 3} = \dfrac{3}{-2} = -\dfrac{3}{2}$

(b) The formula gives us $m = \dfrac{4-2}{1-1} = \dfrac{2}{0}$ and we cannot divide by zero. Look at the coordinates again. The two x-coordinates are the same, so the line is $|$. $\therefore$ the gradient is infinite

Parallel lines

Parallel lines have the same gradient.

e.g. $y = 2x + 1$ and $y = 2x - 3$ both have gradient 2 and are therefore parallel.

Perpendicular lines

$y = m_1 x + c_1$ is perpendicular to $y = m_2 x + c_2$

if
$$\boxed{m_1 \times m_2 = -1}$$

Example 3 Determine whether $y - 7x = 3$ and $7y + x = 5$ are parallel, perpendicular, or neither.

$y - 7x = 3$ can be written as $y = 7x + 3$. Its gradient is 7.
$7y + x = 5$ can be written as $y = -\frac{1}{7}x + \frac{5}{7}$. Its gradient is $-\frac{1}{7}$.
$m_1 \times m_2 = 7 \times (-\frac{1}{7}) = -1$. $\therefore$ the lines are perpendicular

Example 4 Find the gradient of the line which is inclined at an angle of 120° to the positive direction of the x-axis.

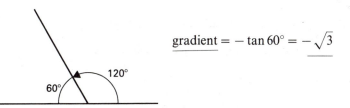

$$\text{gradient} = -\tan 60° = -\sqrt{3}$$

- Exercise 2.1 See page 35.

2.2 Equations of Straight Lines

Equation of a straight line passing through a given point (h, k) with a given gradient m.

$$\boxed{\dfrac{y-k}{x-h} = m}$$

Example 5   Find the equation of the straight line passing through the point $(2, -1)$ with gradients 3

$$\frac{y+1}{x-2} = 3$$

$$\Rightarrow y + 1 = 3x - 6$$

$$\Rightarrow \quad y = 3x - 7$$

Equation of a straight line passing through the points (x_1, y_1) and (x_2, y_2)

$$\boxed{\frac{y - y_1}{x - x_1} = \frac{y_2 - y_1}{x_2 - x_1}}$$

Example 6   Find the equation of the straight line passing through the points $(3, -1), (-2, 1)$.

$$\frac{y+1}{x-3} = \frac{-1-1}{3-(-2)} = -\frac{2}{5}$$

Cross multiplying, we get

$$5y + 5 = -2x + 6$$

i.e. $$5y + 2x = 1$$

- Exercise 2.2. See page 36.

2.3 Points and Lines: Revision of Unit

1 Perpendicular Distance of a Point from a Line

If the point is (h, k)
and the line is $ax + by + c = 0$
then the perpendicular distance is

$$\frac{|ah + bk + c|}{\sqrt{(a^2 + b^2)}}$$

Example 7   Find the length of the perpendicular from the point $(1, -2)$ onto the line $4y - 3x + 2 = 0$.

First the equation must be rearranged into the form $ax + by + c = 0$. The equation becomes $3x - 4y - 2 = 0$ so $a = 3$, $b = -4$, $c = -2$ and the perpendicular distance is

$$\frac{|3.1 - 4(-2) - 2|}{\sqrt{(3^2 + 4^2)}} = \frac{|3 + 8 - 2|}{5} = \frac{9}{5}$$

2   *Dividing a Line Between Two Points in a Given Ratio*

Example 8   If A is (2, 3) and B is (6, 11) find the coordinates of the point P for which AP:PB = 1:3.

X must divide AZ in the ratio 1:3
∴ AX = 1
⇒ x coordinate of P = 2 + 1 = 3.

Y must divide BZ in the ratio 1:3
∴ ZY = 2
⇒ y coordinate of P = 3 + 2 = 5
∴ P is (3, 5)

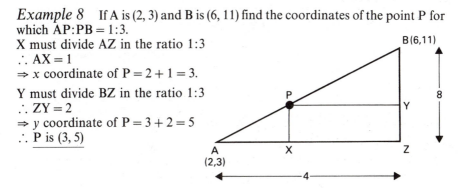

● Miscellaneous Exercise 2A   See page 37.

Unit 2 EXERCISES

Exercise 2.1

1   Find the gradients of the lines joining the following pairs of points:
   (a)   (1, 2), (3, 7)   (b)   (2, − 5), (3, − 1)   (c)   (3, 5), (− 3, 5)
   (d)   (− 4, − 2), (− 1, 2)   (e)   (− 1, 0), (− 1, 4)   (f)   (0, a), (a, 0)

2   Find the coordinates of the midpoints of the lines joining the pairs of points in question 1

3   Calculate the lengths of the sides of △ ABC and determine whether the triangle is right angled:
   (a)   A(0, 0)   B(2, 3)   C(1, 4)
   (b)   A(3, − 1)   B(1, 7)   C(− 7, 5)
   (c)   A(2, − 1)   B(− 1, 1)   C(1, 5)

4   Find the gradients of each of the following lines:
   (a) $5x − 2y = 3$ (b) $x = 3y − 4$ (c) $(x/2) − (y/3) = 1$
   (d) $7x − 4y = 8$ (e) $ax − by = c$

5   Determine whether the following pairs of lines are parallel, perpendicular or neither:
   (a) $2y − 3x + 5 = 0$ $4y = 6x + 1$
   (b) $y + 5x = 0$ $5y = x − 3$
   (c) $3y − 4x = 2$ $4x − 3y = 4$
   (d) $7y − 11x + 31 = 0$ $21y = 33x + 4$
   (e) $y + ax = 0$ $a^2x − ay = 1/a$
   (f) $py − p^2x = 0$ $p^2y + px = 0$

6   Find the gradient of the line joining the points on the curve $y = x^3$, whose x-coordinates are 1 and 3

7 Prove that the point (4, 4) is equidistant from the points (1, 0) and (− 1, 4)

8 Find the gradients of the sides of the triangle whose vertices are the points A(5, 4), B(2, − 2), C(− 6, 2). Determine whether the triangle is right angled

9 Prove that the triangle, whose vertices are the points (1, 2), (3, 4) and (− 1, 6), is isosceles

10 Write down the gradients of the lines that are inclined at the following angles to the positive *x*-axis:
   (a) 45°   (b) 135°   (c) 90°

11 The distance between the points $(\alpha, 0)$ and $(0, \alpha)$ is equal to the distance between the points (1, 2) and (− 1, 3). Find α

12 The distance of the point (a, b) from the origin is equal to its distance from the point (− 2, 3). Find an equation connecting a and b

13 Prove that the lines OA and OB are perpendicular where A is (4, 3) and B is (3, − 4)

14 Find the gradient of the straight line joining the points of intersection of the curve $y = x^2$ and the line $y = x + 2$

*15 A, B, C are the points (7, 3), (− 4, 1) and (− 3, − 2) respectively.
   (a) Show that $\triangle$ ABC is isosceles
   (b) Find the midpoint of the base of $\triangle$ ABC and hence find the area of triangle ABC

*16 The straight line $(x/3) + (y/4) = 1$ meets the *x*- and *y*-axes at the points A and B respectively. Find the length and gradient of AB.

*17 M is the midpoint of AB. A is (3, 5). M is (0, 2). Find the coordinates of B

Exercise 2.2

1 Find the equations of the following straight lines:
   (a) gradient 2, passing through (2, 5)
   (b) gradient − 3, passing through (0, 0)
   (c) gradient $\frac{1}{4}$, passing through (1, − 3)
   (d) gradient − *p*, passing through (*p*, − *p*)

2 Find the equations of the straight lines joining the following pairs of points:
   (a) (1, 4), (2, 7)   (b) (− 1, 1), (1, − 3)
   (c) (2, 3), (− 3, 4)   (d) (*p*, 0), (0, *p*)
   (e) (*p*, *q*), (*q*, *p*)

3 Find the equation of the line parallel to the line $3x − y = 4$ which passes through the point (1, − 2)

4   Find the equation of the line perpendicular to the line $y + 2x = 3$ which passes through the point $(3, 0)$

5   What is the equation of the straight line which passes through the point of intersection of the lines $y = x$ and $y = 2x - 3$ and the point $(0, -2)$?

6   Find the equation of the chord joining the points on the curve $y = x^3 - 2$ whose x-coordinates are 0 and 2

7   Find the equation of the chord joining the points on the curve $y = x^3$ whose y-coordinates are -1 and 1

8   What is the equation of the perpendicular bisector of the line joining the points $(0, 0)$ and $(4, 4)$?

*9   Find the equation of the perpendicular to the line $2x - 3y = 5$ drawn from the point $(0, 2)$. Hence find the coordinates of the foot of the perpendicular

*10   The vertices of the quadrilateral PQRS are $P(0, 1)$, $Q(3, 2)$, $R(2, 7)$ $S(-2, 3)$. Find the equations of the diagonals PR and QS and hence find the point of intersection of the diagonals

*11 $A(0, 0)$, $B(1, 2)$, $C(2, 3)$ are three vertices of a parallelogram ABCD. Find:
   (a)   the gradients of AB and BC
   (b)   the equation of the line through A parallel to BC
   (c)   the equation of the line through C parallel to AB
   (d)   the coordinates of D

*12   The points $A(-2, -1)$, $C(-7, 4)$, $D(-1, 1)$ are three vertices of a rectangle ABCD. Find:
   (a)   the equation of the line through C perpendicular to CD.
   (b)   the equation of the line through A perpendicular to AD.
   (c)   the coordinates of B.

*13

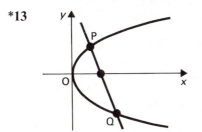

The straight line joining the point P on the curve $y^2 = 16x$ whose y-coordinate is 4, to the point $(4, 0)$ meets the curve again at Q. Find the equation of PQ and the coordinates of Q

Miscellaneous Exercise 2A

1   Find the lengths of the perpendiculars from $(3, 2)$ onto the lines:
   (a) $3y - 2x = 4$ (b) $5x - 2y + 7 = 0$
   (c) $5y = 2x$ (d) $y = 3$

2   Find the distance of the point $(3, -5)$ from the line $3x - 4y + 1 = 0$

3 Find the distance of the point $(1, -2)$ from the line $2y = 5x$

4 Find the coordinates of the point P which divides the line AB in the ratio $1:2$ if

(a) A is $(-1, 1)$, B is $(5, 10)$ (b) A is $(-3, 2)$, B is $(9, -1)$

5 Find the equation of the straight line passing through the point of intersection of the lines $3y - x + 2 = 0$ and $y - 5x = 4$ and perpendicular to the line $y - 2x = 3$

6 Determine by comparing gradients, whether the three points given are collinear (i.e. lie on the same straight line)

(a) $(0, -1)$, $(1, 1)$, $(2, 3)$
(b) $(1, -3)$, $(0, -1)$, $(-1, 2)$
(c) $(-6, -5)$, $(1, -3)$, $(8, -1)$

7 Determine whether PQ is parallel or perpendicular to RS, or neither where:

(a) P is $(1, 0)$ Q is $(5, 2)$, R is $(6, -2)$, S is $(10, 0)$
(b) P is $(1, 0)$, Q is $(6, 1)$, R is $(5, -1)$, S is $(4, 3)$
(c) P is $(1, 0)$, Q is $(-2, 2)$, R is $(8, 2)$, S is $(10, 5)$

8 The points $A(2, 2)$, $B(7, 4)$, $C(0, 7)$ form a triangle ABC. Prove that the triangle is right angled and find its area

9 Prove that the quadrilateral with vertices $(2, 1)$, $(2, 3)$, $(5, 6)$, $(5, 4)$ is a parallelogram and find the point of intersection of the diagonals

10 A point (p, q) is equidistant from the y-axis and the point $(3, 0)$. Find a relationship between p and q

11 Show that the triangle whose vertices are $(0, 4)$, $(2, 0)$, $(4, 2)$ is isosceles

12 Find the equation of the line which goes through the point $(7, 3)$ and which is inclined at an angle of $45°$ to the positive x-axis.?

13 Find the coordinates of the points of intersection of $y = x^2 - 4$ and $y = x - 2$. Find the length of the line joining these points

*14 Find the equations of the perpendicular bisectors of the lines joining the points $(2, 1)$, $(6, 3)$ and $(6, 3)$, $(8, 1)$. Show that the point of intersection of these bisectors is on the x-axis.

*15 The coordinates of A, B, C are $(0, 2)$, $(-2, 0)$, $(1, -3)$ respectively. ABCD is a rectangle. Find the coordinates of D

*16 The coordinates of the points A, B, C are $(-3, -1)$, $(11, 13)$, $(-1, -3)$. Find the equations of the medians of the triangle ABC and the coordinates of their point of intersection. (A median joins one vertex of a triangle to the midpoint of the opposite side.)

3.1 Differentiation from first principles

Gradient of a curve

The gradient of a curve at any point is defined as the gradient of the tangent at that point.

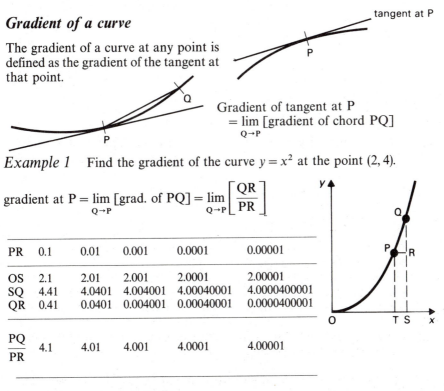

tangent at P

Gradient of tangent at P
$= \lim_{Q \to P} [\text{gradient of chord PQ}]$

Example 1 Find the gradient of the curve $y = x^2$ at the point $(2, 4)$.

gradient at $P = \lim_{Q \to P} [\text{grad. of PQ}] = \lim_{Q \to P} \left[\dfrac{QR}{PR} \right]$

PR	0.1	0.01	0.001	0.0001	0.00001
OS	2.1	2.01	2.001	2.0001	2.00001
SQ	4.41	4.0401	4.004001	4.00040001	4.0000400001
QR	0.41	0.0401	0.004001	0.00040001	0.0000400001

$\dfrac{PQ}{PR}$	4.1	4.01	4.001	4.0001	4.00001

$\therefore$ gradient at $P = \lim_{Q \to P} \left[\dfrac{QR}{PR} \right] = 4$

- Exercise 3.1 See page 49. Question 1 only.

Gradient of a curve at any point, from first principles

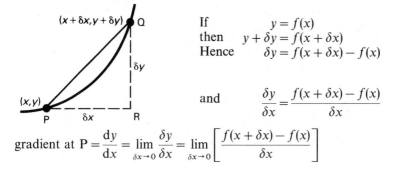

If $\qquad y = f(x)$
then $\qquad y + \delta y = f(x + \delta x)$
Hence $\qquad \delta y = f(x + \delta x) - f(x)$

and $\qquad \dfrac{\delta y}{\delta x} = \dfrac{f(x + \delta x) - f(x)}{\delta x}$

gradient at $P = \dfrac{dy}{dx} = \lim_{\delta x \to 0} \dfrac{\delta y}{\delta x} = \lim_{\delta x \to 0} \left[\dfrac{f(x + \delta x) - f(x)}{\delta x} \right]$

Example 2   Find the gradient of $y = x^2$ at any point, from first principles.

If
$$y = x^2$$

then
$$y + \delta y = (x + \delta x)^2$$
$$\delta y = (x + \delta x)^2 - x^2$$
$$\therefore \delta y = x^2 + 2x \cdot \delta x + (\delta x)^2 - x^2$$
$$\therefore \delta y = 2x \cdot \delta x + (\delta x)^2$$
$$\frac{\delta y}{\delta x} = 2x + \delta x$$

$$\text{gradient} = \frac{dy}{dx} = \lim_{\delta x \to 0} \frac{\delta y}{\delta x} = \lim_{\delta x \to 0} (2x + \delta x) = \underline{2x}$$

Example 3   Find the gradient of $y = 1/x^2$ at any point, from first principles.

If
$$y = \frac{1}{x^2}$$

then
$$y + \delta y = \frac{1}{(x + \delta x)^2}$$

$$\delta y = \frac{1}{(x + \delta x)^2} - \frac{1}{x^2}$$

$$\delta y = \frac{x^2 - (x + \delta x)^2}{(x + \delta x)^2 x^2} = \frac{x^2 - (x^2 + 2x\delta x + (\delta x)^2)}{(x + \delta x)^2 x^2}$$

$$\delta y = \frac{-2x\delta x - (\delta x)^2}{(x + \delta x)^2 x^2}$$

$$\frac{\delta y}{\delta x} = \frac{-2x - \delta x}{(x + \delta x)^2 x^2}$$

$$\text{gradient} = \frac{dy}{dx} = \lim_{\delta x \to 0} \left[\frac{-2x - \delta x}{(x + \delta x)^2 x^2} \right] = \frac{-2x}{x^4} = \underline{-\frac{2}{x^3}}$$

- Exercise 3.1   See page 49.
  Question 2 onwards.

3.2 Differentiation by Rule

Terminology

$2x$ is the **derived function** or **derivative** of x^2

The process of obtaining the derivative is called **differentiation**.

dy/dx is the **derivative of y** with respect to x or the **rate of change of y w.r.t. x**

Hence the derivative of a function is **the gradient function**. If you are asked to differentiate a function, you may be asked in one of several ways.

1 *Example 4* find $\dfrac{dy}{dx}$ if $y = x^2$

Answer: $\dfrac{dy}{dx} = 2x$

2 *Example 5* Find $\dfrac{d}{dx}(x^2)$.

Answer: $\dfrac{d}{dx}(x^2) = 2x$　　　　| Literally: the derivative of x^2 is $2x$ |

3 *Example 6* If $f(x) = x^2$, find $f'(x)$.

Answer: $f'(x) = 2x$

Differentiation by Rule

1 Derivative of a constant: $\dfrac{d}{dx}(c) = 0$

2 Derivative of ax 　　: $\dfrac{d}{dx}(ax) = a$

3 Derivative of x^n 　　: $\dfrac{d}{dx}(x^n) = nx^{n-1}$

4 Derivative of ax^n 　　: $\dfrac{d}{dx}(ax^n) = a\dfrac{d}{dx}(x^n) = anx^{n-1}$

5 Derivative of a sum 　: $\dfrac{d}{dx}(f(x) + g(x)) = \dfrac{d}{dx}(f(x)) + \dfrac{d}{dx}(g(x))$

Example 7 Find (a) $\dfrac{d}{dx}(x^9)$ 　(b) $\dfrac{d}{dx}\left(\dfrac{1}{x^4}\right)$ 　(c) $\dfrac{d}{dx}(\sqrt[3]{x})$

(a) $\dfrac{d}{dx}(x^9) = 9x^8$

(b) $\dfrac{d}{dx}\left(\dfrac{1}{x^4}\right) = \dfrac{d}{dx}(x^{-4}) = -4x^{-5} = -\dfrac{4}{x^5}$

(c) $\dfrac{d}{dx}(\sqrt[3]{x}) = \dfrac{d}{dx}(x^{1/3}) = \tfrac{1}{3}x^{-2/3}$　| Before differentiating, the function must be put in the form x^n. |

Example 8 Find dy/dx if $y = 4x^5$.

$$\dfrac{dy}{dx} = 4\cdot(5x^4) = 20x^4$$

Example 9   Find $f'(x)$ if (a) $f(x) = x^2 + 7x$ (b) $f(x) = 3x^4 - (1/x) + 2$

(a) $f'(x) = 2x + 7$

(b) $f'(x) = \dfrac{d}{dx}(3x^4 - x^{-1} + 2) = 12x^3 + x^{-2}$

| must be put in the form x^n first |

Example 10   Differentiate $\dfrac{2x^3 + 3x - 1}{x}$

$$\frac{d}{dx}\left(\frac{2x^3 + 3x - 1}{x}\right) = \frac{d}{dx}(2x^2 + 3 - x^{-1}) = 4x + x^{-2}$$

Example 11   Find the gradient of the curve $y = (x + 1)(x + 2)$ at the point where $x = 3$.

$$y = x^2 + 3x + 2$$

$$\frac{dy}{dx} = 2x + 3$$

| gradient = value of $\dfrac{dy}{dx}$ at the point |

when $x = 3$, gradient $= 9$

Example 12   Find the coordinates of the point at which $y = 2/x^2$ has gradient $\frac{1}{2}$.

$$\frac{dy}{dx} = \frac{d}{dx}(2x^{-2}) = -4x^{-3} = -\frac{4}{x^3}$$

when $\dfrac{dy}{dx} = \dfrac{1}{2}, \dfrac{-4}{x^3} = \dfrac{1}{2} \Rightarrow x^3 = -8 \Rightarrow x = -2$

and when $x = -2$, $y = \frac{1}{2}$ ∴ point is $(-2, \frac{1}{2})$

● Exercise 3.2   See page 50.

3.3 Equations of Tangents and Normals

Example 13   Find the equation of the tangent to the curve $y = 2x^2 - 3x - 2$ at the point where it cuts the positive x-axis.

| First we must find where it cuts the x-axis |

$y = 2x^2 - 3x - 2$ cuts the x-axis where $y = 0$

i.e. $2x^2 - 3x - 2 = 0$

$\therefore (2x + 1)(x - 2) = 0$

$\therefore x = -\frac{1}{2}$ and $x = 2$

∴ it cuts the positive x-axis where $x = 2$, $y = 0$.

The gradient of the tangent at $(2, 0)$ is the value of dy/dx when $x = 2$.

As
$$y = 2x^2 - 3x - 2$$

$$\frac{dy}{dx} = 4x - 3$$

when $x = 2$, the gradient is 5.
∴ the tangent has gradient 5 and passes through $(2, 0)$.

Equation of tangent:

$$\frac{y}{x - 2} = 5$$

i.e.
$$y = 5x - 10$$

Example 14   Find the equation of the normal to the curve $xy = 2$ at the point where $x = 1$. Find the coordinates of the point where this normal intersects the curve again.

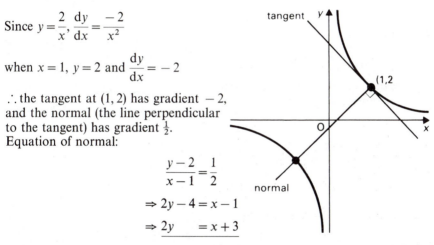

Since $y = \dfrac{2}{x}$, $\dfrac{dy}{dx} = \dfrac{-2}{x^2}$

when $x = 1$, $y = 2$ and $\dfrac{dy}{dx} = -2$

∴ the tangent at $(1, 2)$ has gradient -2, and the normal (the line perpendicular to the tangent) has gradient $\frac{1}{2}$.
Equation of normal:

$$\frac{y - 2}{x - 1} = \frac{1}{2}$$

$$\Rightarrow 2y - 4 = x - 1$$

$$\Rightarrow 2y \quad = x + 3$$

The points of intersection of the normal and the curve are found by solving simultaneously

$$2y = x + 3$$

and
$$y = \frac{2}{x}$$

$$\Rightarrow \qquad \frac{4}{x} = x + 3$$

$$\Rightarrow \quad x^2 + 3x - 4 = 0$$

$$(x - 1)(x + 4) = 0$$

∴ the normal cuts the curve at $x = 1$ and $x = -4$. $x = 1$ is the given point
∴ it cuts the curve again where $x = -4$. i.e. at $(-4, -\frac{1}{2})$

● **Exercise 3.3**   See page 51.

3.4 Stationary Values and Turning Points: Maxima and Minima

Terminology

If $f(x)$ is increasing w.r.t. x then $\dfrac{d}{dx} f(x) > 0$

If $f(x)$ is decreasing w.r.t. x then $\dfrac{d}{dx} f(x) < 0$

If $f(x)$ is neither increasing nor decreasing, then $\dfrac{d}{dx}(f(x)) = 0$, and at this point $f(x)$ has a **stationary value.**

Consider the graph of $y = f(x)$. At A and B, $f(x)$ has stationary values. A and B are also called **turning points** of the curve $y = f(x)$.

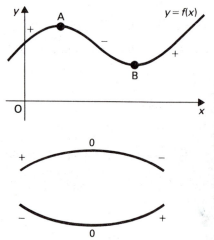

Turning Points

At A: If $\dfrac{dy}{dx} = 0$ and $\dfrac{d^2y}{dx^2} < 0$

then there is a **maximum turning point**

At B: If $\dfrac{dy}{dx} = 0$ and $\dfrac{d^2y}{dx^2} > 0$

then there is a **minimum turning point**

Example 15   Find the values of x at which the function $x^4 - 4x^3$ has stationary values and find the stationary values.

Let
$$f(x) = x^4 - 4x^3$$
$$f'(x) = 4x^3 - 12x^2$$

For stationary values

$$f'(x) = 0 \Rightarrow 4x^2(x - 3) = 0$$
$$\Rightarrow x = 0 \text{ and } x = 3$$

$\therefore$ stationary values occur when $x = 0$ and $x = 3$

stationary values are $f(0)$ and $f(3)$ i.e. 0 and -27

Example 16   Find and classify the $\left.\begin{array}{c}\text{turning}\\\text{stationary}\end{array}\right\}$ points of the curve $y = x^3 - 3x$. Sketch the curve.

Method 1 (Using the second derivative test)

$$y = x^3 - 3x$$ Turning points occur when $\dfrac{dy}{dx} = 0$

$$\frac{dy}{dx} = 3x^2 - 3$$ $$\Rightarrow 3x^2 - 3 = 0$$

$$\frac{d^2y}{dx^2} = 6x$$ $$\Rightarrow 3(x^2 - 1) = 0$$

$$\Rightarrow x = \pm 1$$

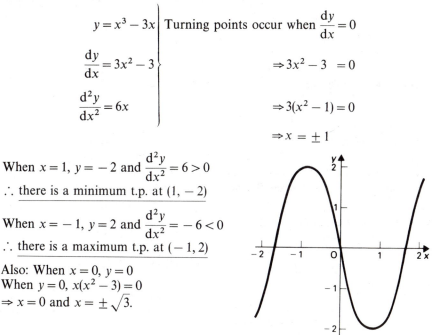

When $x = 1$, $y = -2$ and $\dfrac{d^2y}{dx^2} = 6 > 0$

$\therefore$ there is a minimum t.p. at $(1, -2)$

When $x = -1$, $y = 2$ and $\dfrac{d^2y}{dx^2} = -6 < 0$

$\therefore$ there is a maximum t.p. at $(-1, 2)$

Also: When $x = 0$, $y = 0$

When $y = 0$, $x(x^2 - 3) = 0$

$\Rightarrow x = 0$ and $x = \pm\sqrt{3}$.

Example 17   Determine the turning point(s) of $y = 2x - x^2$ and sketch the curve.

Method 2 (Using the gradients before and after the t.p.)

$$y = 2x - x^2$$ Turning points occur when $\dfrac{dy}{dx} = 0$

$$\frac{dy}{dx} = 2 - 2x$$ $$\Rightarrow 2 - 2x = 0$$

$$\Rightarrow x = 1$$

When $x = 0.9$, $\dfrac{dy}{dx} = 2 - 1.8 > 0$

When $x = 1.1$, $\dfrac{dy}{dx} = 2 - 2.2 < 0$

$\Rightarrow + \diagup \overline{} \diagdown -$

When $x = 1$, $y = 1$ and we have a maximum turning point at $(1, 1)$

Also: $y = 0 \Rightarrow x = 0$ and $x = 2$.

- Exercise 3.4   See page 52.

3.5 Points of Inflexion: Further Curve Sketching

Points of Inflexion

At a point of inflexion the curve changes direction

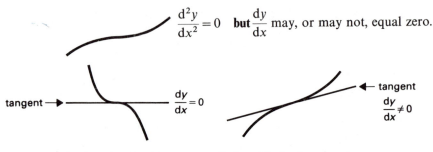

$\dfrac{d^2y}{dx^2} = 0$ **but** $\dfrac{dy}{dx}$ may, or may not, equal zero.

tangent → $\dfrac{dy}{dx} = 0$

← tangent

$\dfrac{dy}{dx} \neq 0$

Point of inflexion that is also a
stationary point

Point of inflexion that is **not**
a stationary point

In all cases, the **tangent crosses the curve** at a point of inflexion.

The Second Derivative Test

$$\frac{d^2y}{dx^2} = \frac{d}{dx}\left(\frac{dy}{dx}\right) \text{ i.e. } y \text{ is differentiated twice}$$

At all points for which $\dfrac{dy}{dx} = 0$ then:

if $\dfrac{d^2y}{dx^2} > 0$ there is a minimum turning point

if $\dfrac{d^2y}{dx^2} < 0$ there is a maximum turning point.

But, if $\dfrac{d^2y}{dx^2} = 0$ there may exist a maximum, a minimum or a point of inflexion.

If $\dfrac{d^2y}{dx^2} = 0$ you should consider the gradients on either side of the turning point.

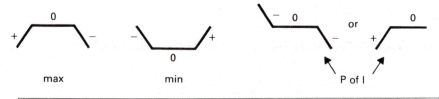

max   min   P of I

Whether you are looking for stationary points in general or points of
inflexion in particular, whenever you consider $\dfrac{d^2y}{dx^2} = 0$ you must
consider the gradients on either side of the point.

Example 18   Find and classify the turning points on the following curves:
(a) $y = x^4$ (b) $y = 2x^3 + 1$.

Sketch each curve.

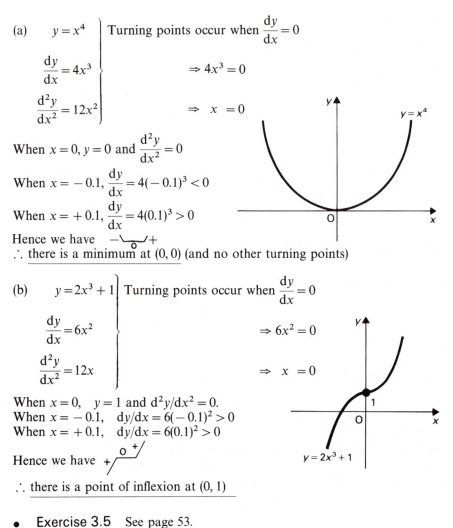

(a) $y = x^4$ | Turning points occur when $\dfrac{dy}{dx} = 0$

$\dfrac{dy}{dx} = 4x^3$ $\Rightarrow 4x^3 = 0$

$\dfrac{d^2y}{dx^2} = 12x^2$ $\Rightarrow x = 0$

When $x = 0$, $y = 0$ and $\dfrac{d^2y}{dx^2} = 0$

When $x = -0.1$, $\dfrac{dy}{dx} = 4(-0.1)^3 < 0$

When $x = +0.1$, $\dfrac{dy}{dx} = 4(0.1)^3 > 0$

Hence we have $-\underset{\scriptstyle 0}{\searrow}+$

$\therefore$ there is a minimum at $(0, 0)$ (and no other turning points)

(b) $y = 2x^3 + 1$ | Turning points occur when $\dfrac{dy}{dx} = 0$

$\dfrac{dy}{dx} = 6x^2$ $\Rightarrow 6x^2 = 0$

$\dfrac{d^2y}{dx^2} = 12x$ $\Rightarrow x = 0$

When $x = 0$, $y = 1$ and $d^2y/dx^2 = 0$.
When $x = -0.1$, $dy/dx = 6(-0.1)^2 > 0$
When $x = +0.1$, $dy/dx = 6(0.1)^2 > 0$

Hence we have $+\underset{\scriptstyle 0}{\nearrow}+$

$\therefore$ there is a point of inflexion at $(0, 1)$

• **Exercise 3.5**   See page 53.

3.6 Applications of the Differential Calculus

$\dfrac{dy}{dx} =$ the rate of change of y w.r.t. x

In general, $\dfrac{dP}{dq} =$ the rate of change of P w.r.t. q.

If $\dfrac{dP}{dq}$ is positive, then it represents the increase in P.

If $\dfrac{dP}{dq}$ is negative, then it represents the decrease in P.

In particular, velocity = rate of change of displacement w.r.t. time

$$\Rightarrow v = \dfrac{ds}{dt}$$

and acceleration = rate of change of velocity w.r.t. time

$$\Rightarrow a = \frac{dv}{dt} = \frac{d^2s}{dt^2}$$

Example 19   If $s = 5t^3 - t^2$ where s m is the distance moved in t seconds, find the velocity and acceleration after $2s$.

$$s = 5t^3 - t^2$$

$$\left.\begin{array}{l} v = \dfrac{ds}{dt} = 15t^2 - 2t \\[3mm] a = \dfrac{d^2s}{dt^2} = 30t - 2 \end{array}\right\} \quad \begin{array}{l} \text{when } t = 2, \underline{\text{velocity}} = (15 \times 4) - 2.2 = \underline{56\,\text{m/s}} \\[3mm] \underline{\text{acceleration}} = (30 \times 2) - 2 = \underline{58\,\text{m/s}^2} \end{array}$$

Example 20   The area A cm^2 of a blot of ink is growing so that, after t s, $A = 3t^2 + \frac{1}{3}t$. Find the rate at which the area is increasing after 1 s.

The rate of increase of A w.r.t time is given by dA/dt.

$$A = 3t^2 + \tfrac{1}{3}t$$

$$\frac{dA}{dt} = 6t + \tfrac{1}{3}$$

$\therefore$ rate of increase of area after $1\,\text{s} = 6\frac{1}{3}\,\text{cm}^2/\text{s}$

Example 21   The rate of working, H, of an engine is given by the expression $H = 10v + (4000/v)$ where v is the speed of the engine. Find the speed at which the rate of working is least.

We want to find the minimum value of H. This occurs when $dH/dv = 0$.

$$H = 10v + \frac{4000}{v} \Rightarrow \frac{dH}{dv} = 10 - \frac{4000}{v^2}$$

$$\frac{dH}{dv} = 0 \Rightarrow v^2 = 400 \Rightarrow v = \pm 20.$$

Also
$$\frac{d^2H}{dv^2} = \frac{8000}{v^3}$$

when $v = 20$, $dH/dv = 0$ and $d^2H/dv^2 > 0$
$\therefore$ least value of H occurs when $v = 20$

Example 22   If $x + y = 20$ find the maximum value of xy.

xy is a function of **two** variables which we cannot differentiate. So we must reduce it to a function of **one** variable.
    Let

$$z = xy \tag{1}$$

$$x + y = 20 \Rightarrow y = 20 - x \tag{2}$$

Substituting for y (from (2)) in (1) gives us

$$z = x(20 - x) \longleftarrow \boxed{\text{a function in one variable.}}$$

Differentiating:

$$\frac{dz}{dx} = 20 - 2x$$

$$\Rightarrow \frac{dz}{dx} = 0 \quad \text{when } x = 10$$

Also

$$\frac{d^2z}{dx^2} = -2 \qquad \text{(for all values of } x\text{)}$$

$\therefore$ when $x = 10$, $\dfrac{dz}{dx} = 0$ and $\dfrac{d^2z}{dx^2} < 0$.

$\Rightarrow z$ is a maximum when $x = 10$
i.e. xy is a maximum when $x = 10$.
When x is 10, $y = 10$ (from (2)).
$\therefore$ maximum value of $xy = 100$

● **Exercise 3.6**   See page 53.

3.7 Revision of Unit

● Miscellaneous Exercise 3A   See page 54.

Unit 3 EXERCISES

Exercise 3.1

1   P is the point $(1, 3)$ on the curve $y = 3x^2$ and Q is a nearby point on the curve. Complete the table below and deduce the gradient of the curve $y = 3x^2$ at the point $(1, 3)$.

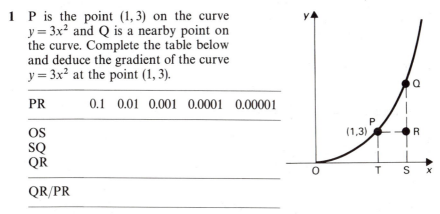

PR	0.1	0.01	0.001	0.0001	0.00001
OS					
SQ					
QR					

QR/PR

Differentiate from first principles:

2 $y = 2x^2$

3 $y = x^2 + 2$

4 $y = 4x - 1$

5 $y = x^3$

6 $y = 1/x$

7 $y = 6x - x^2$

8 $y = x^4$

9 If $s = 1 - t + 3t^2$ find $\dfrac{ds}{dt}$ from first principles

10 If $v = 3r^2 + 2r$ find $\dfrac{dv}{dr}$ from first principles

Exercise 3.2

In this exercise all differentiation is to be done by rule.

1 Find:

(a) $\dfrac{d}{dx}(x^5)$ (b) $\dfrac{d}{dx}(x^{11})$ (c) $\dfrac{d}{dx}(x^{-3})$ (d) $\dfrac{d}{dx}(x^{1/3})$

(e) $\dfrac{d}{dx}\left(\dfrac{1}{x^4}\right)$

2 Find $\dfrac{dy}{dx}$ if:

(a) $y = 2x$ (b) $y = \dfrac{1}{2x^6}$ (c) $y = \sqrt{x^5}$ (d) $y = 6$

(e) $y = -3x$

3 Find $f'(x)$ if $f(x) =:$

(a) $3x^4$ (b) $\dfrac{x^5}{2}$ (c) $\dfrac{3}{x}$ (d) $7x^{3/2}$ (e) $\sqrt[4]{x}$

(f) $\dfrac{1}{\sqrt{x}}$

4 Differentiate with respect to x:

(a) $2x - x^3$ (b) $2 + 3x - 5x^2$ (c) $2 + \dfrac{1}{x}$

(d) $(1 - 3x)^2$ (e) $3x^4 - \dfrac{1}{x^2}$ (f) $\left(x - \dfrac{1}{x}\right)^2$

(g) $\dfrac{(x^3 + 4)}{x}$ (h) $\dfrac{x^2 - 7x + 1}{x^3}$ (i) $\sqrt{x}(x + 1)$

(j) $\dfrac{x-1}{\sqrt{x}}$ (k) $(x-1)(x+3)$ (l) $(2x-5)(2x+5)$

5 Find $\dfrac{ds}{dt}$ where:

(a) $s = 3t - 5t^2 + 1$ (b) $s = 2t^3 - t^2$ (c) $s = 2t - \dfrac{1}{t}$

In each of questions 6–12 find the gradient of the curve at the given point:

6 $y = x^5 - 3$ where $x = 1$

7 $y = 4x^3 - 1$ where $x = 2$

8 $y = \sqrt{x}$ where $x = 3$

9 $y = 1/x$ where $x = 4$

10 $y = (x-2)(x+1)$ where $x = -2$

11 $y = (2x-1)/x$ where $x = \frac{1}{2}$

12 $y = 3\sqrt{x}(2 - \sqrt{x})$ where $x = 4$

In each of questions 13–18 find the coordinates of the point(s) on the curve at which the gradient has its given value:

13 $y = 2x^2 - 3x + 4$, gradient $= 5$

14 $y = 5 - 3x^3$, gradient $= -9$

15 $y = (x+2)(x-2)$, gradient $= 6$

16 $y = x^3 + x^2$, gradient $= 1$

17 $y = 1/\sqrt{x}$, gradient $= -\frac{1}{2}$

18 $y = 3 + 1/x$, gradient $= -\frac{1}{4}$

*19 Find the gradients of the curve $y = 3x^2 + 1$ at the ends of the chord whose equation is $y = 3x + 1$

*20 Find the coordinates of the points on the curve $y = \frac{1}{3}x^3 - \frac{1}{2}x^2 - 11x + 7$ where the tangent makes an angle of $45°$ with the positive direction of the x-axis

Exercise 3.3

1 Find the equation of the tangent to each curve at the given point:

(a) $y = 3x^2$ at $x = 2$
(b) $y = 5x - 3x^2 + 1$ at $x = 0$

(c) $y = 1/x$ at $x = -1$
(d) $y = (x - 1)(3x^2 + 1)$ at $x = 1$
(e) $y = (2x - 1)^2$ at $x = 3$

2   Find the equation of the normal to each curve given in question 1, at the given points

3   Find the equation of the tangent to the curve $y = 3x^2 + 12x - 1$ at the point where the curve cuts the y-axis

4   Find the equation of the tangent to the curve $y = 8x - 3x^2$ at the point where $x = 1$ and find where this tangent meets the line $x = y$

5   Find the coordinates of the point of intersection of the normal to the curve $y = x^3 - 4x^2$ at the point $x = 1$, and the line $y = x + 4$

6   The tangents to the curve $y = 5 - 3x - 2x^2$ at the points whose x-coordinates are 1 and -1 meet at P. Find the coordinates of P

7   Find the equations of the normals to the curve $y = (x - 1)(x - 4)$ at the points where the curve cuts the x-axis. Find the coordinates of the point of intersection of the normals

8   Find the coordinates of the point on $y = x^2$ at which the gradient is 4. Hence find the equation of the tangent to $y = x^2$ with gradient 4

9   Find the equation of the normal to $y = x^2 - 5x + 3$ which has a gradient of $\frac{1}{2}$

10   Find the equation of the tangent to $y = (x - 3)(3x + 1)$ which is parallel to the x-axis

*11   Find the values of k for which $y = 3x + k$ is normal to $y = 1 - x^3$

*12   The tangent and the normal at the point P(3, 0) on the curve $y = x(3 - x)$ meet the axis of y at the points T and N respectively. Find the lengths of PT and PN

Exercise 3.4

1   Find the value(s) of x at which the following functions have stationary values:
   (a) $x^2 - 1$ (b) $x^2 - 6x + 3$ (c) $y = x^3 - 12x - 2$

2   Find the stationary values of the following functions:
   (a) $x^2 - 10x$ (b) $x + 1/x$ (c) $3x^3 - 4x + 1$

In questions 3–8 find and classify the turning points on the curves. Make a rough sketch of each curve.

3   $y = 3x - x^2$ 4 $x^2 - 4x + 3$

5 $y = (3 - x)^2$ **6** $y = (2x - 1)(x + 1)$

7 $y = (x^3/3) - 9x$ ***8** $y = x^2 + 16/x$

9 Find the maximum value of $3x(4 - x)$

*10 Find the value of p which makes $\dfrac{p^2 + 1}{p}$ a minimum

Exercise 3.5

In questions 1–5 find the points on the curves where $dy/dx = 0$. By considering the signs of dy/dx before and after these points, determine whether the points are maximum, minimum or inflexion points.

1 $y = 2x^3$ **2** $y = 3x^6$ **3** $y = 5 - x^3$

4 $y = 3x^8 - 2$ **5** $y = 4x^5 + 1$

6 Find the greatest value of $(1 - 2x)(1 + x)$

7 What is the minimum value of $(2 - 3x)^2$?

8 Find the points of inflexion of each of the following curves and state, in each case, whether the point of inflexion is a stationary point:

    (a) $y = 2x^3 - 3x^2 + 1$ (b) $y = x^4 - 6x^2 + 8x + 12$

In questions 9–14 find and classify the stationary points on each curve and sketch the curve.

9 $y = x^3 - 12x + 2$ **10** $y = 1 - 3x - 3x^2 - x^3$

11 $y = 2x^3 - 9x^2 + 12x - 4$ **12** $y = x^3 - 6x^2 + 12x - 11$

13 $y = 28 - 27x + 9x^2 - x^3$ ***14** $y = x + (1/x) + 1$

Exercise 3.6

1 If a body travels s m in t s where $s = 3t - t^3$ find the velocity and acceleration of the body when $t = \frac{1}{2}$ s. At what time is the body at rest?

2 The velocity v m/s of a particle is equal to $(1 - 5t)^2$ where t is the time taken in seconds. Find the acceleration of the body after t s. When is the body at rest? What is its acceleration at this time?

3 A pebble thrown into the air rises s m in t s where $s = 10t - 4.9t^2$. Find the velocity of the stone when $t = 1$ s and $t = 2$ s. What is the meaning of the negative velocity?

4 The angle turned through by a rotating body, θ, at time t is given by $\theta = 5t^3 - 2t^2$ radians. Find the rate of increase of θ when $t = 2$

5   The acceleration of a particle is equal to $5t(t-2)\,\text{m/s}^2$ where t is the time in s. Find when the acceleration is a minimum and find its minimum value.

6   If $x+y=15$ find the greatest value of $2xy$

7   If $V=\pi r^2 h$ and $r+h=10$ find the greatest and least values of V

8   If $p+q=10$ find the maximum and minimum values of p^3q^2

*9   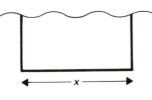 One side of a rectangular pen is formed by a hedge. The total length of fencing available for the other three sides is 400 m. Obtain an expression for the area of the pen $A\,\text{m}^2$ in terms of x. Find the maximum area of the pen

*10   If the volume of a circular, cylindrical block is $800\,\text{cm}^3$, prove that the total surface area is equal to $2\pi x^2 + (1600/x)\,\text{cm}^2$ where x cm is the radius of the base

Find the value of x which makes the surface area a minimum.

*11   The volume of a right, circular cone is $\frac{1}{3}\pi r^2 h$ where $r=$ radius, $h=$ height. If $r+h=8$ find the maximum volume of the cone

*12   A closed rectangular box is made of thin metal. The length of the box is twice its width. If the box has volume $243\,\text{cm}^3$ and its width is x cm show that its surface area is equal to $4x^2 + (729/x)\,\text{cm}^2$.

Find the dimensions of the box with least surface area

*13   A stained glass window is in the shape of a rectangle surmounted by a semicircle. The diameter of the semicircle equals the width of the rectangle. If the perimeter of the window is 30 m find the radius of the semicircle if the window is to let in as much light as possible

*14   A piece of wire of given length is bent to form the perimeter of the sector of a circle. If the area of the sector is a maximum find the angle between the bounding radii in radians

Miscellaneous Exercise 3A

1   Express the following statements in symbols:
   (a)   the rate of increase of y w.r.t. x is 2
   (b)   the rate of increase of x w.r.t. t is -1
   (c)   the rate of decrease of V w.r.t. r is 4

2   Differentiate from first principles:
   (a) $(1-3x)^2$ (b) $\dfrac{1}{1-x}$

3 Differentiate the following w.r.t. p

(a) $5p^4 - 4p^2 + \dfrac{4}{p}$ (b) $(2p^2 - 1)\sqrt{p}$

(c) $\left(p^2 - \dfrac{1}{p^2}\right)^2$ (d) $\dfrac{1 - 3p^2}{4p^3}$

4 Find the gradient of the curve $y = x(x - 1)(x - 4)$ at the points where it cuts the x-axis. Also find the x-coordinates of the points on the curve where the tangent is parallel to the x-axis. Sketch the curve

5 If $x^3 y = 7$ find the value of d^2y/dx^2 when $x = 2$

6. Find the equations of the tangent and the normal to the curve $y = x(3 - x)$ at the point whose x coordinate is p

7 Find the gradient of the curve $y = (x + (1/x))^2$ at the point where $x = -1$

8 Find the gradients of the curves $y = x^2 + 2$, $y = 4 - x^2$ at their points of intersection

9 Find and classify the turning points of $y = 6 + 12x - 3x^2 - 2x^3$. Give a rough sketch of the curve

*10 Find the point of intersection of the normals to the curve $x^2 = 25y$ at the ends of the chord $x = 2y + 2$

*11 The tangent at the point $x = p$ on the curve $y = x^3 - 2x$ meet the x- and y-axes at P and Q respectively. Find the coordinates of P and Q and the length PQ

*12 The path traced out by a shell is given by the equation $y = 4x - x^2/4.8$ where the x-axis is horizontal, the y-axis vertical and the origin is the point of projection. Find the angle of projection and also the directon of motion when $x = 2$

*13 If the function $ax^2 + bx + c$ has gradient function $3x - 1$ and a minimum value of 1, find a, b and c

ADDITION FORMULAE: SOLUTIONS OF TRIANGLES

4.1 Addition Formulae

$$\sin(A + B) = \sin A \cdot \cos B + \cos A \cdot \sin B$$
$$\sin(A - B) = \sin A \cdot \cos B - \cos A \cdot \sin B$$
$$\cos(A + B) = \cos A \cdot \cos B - \sin A \cdot \sin B$$
$$\cos(A - B) = \cos A \cdot \cos B + \sin A \cdot \sin B$$

$$\tan(A + B) = \frac{\tan A + \tan B}{1 - \tan A \cdot \tan B}$$

$$\tan(A - B) = \frac{\tan A - \tan B}{1 + \tan A \cdot \tan B}$$

Example 1   Simplify $\sin(A + B) - \sin(A - B)$

$$\sin(A + B) = \sin A \cdot \cos B + \cos A \cdot \sin B$$
$$\sin(A - B) = \sin A \cdot \cos B - \cos A \cdot \sin B$$

$$\therefore\ \sin(A + B) - \sin(A - B) = 2\cos A \sin B$$

Example 2   Given that $\sin x = \frac{3}{5}$ and $\cos y = \frac{5}{13}$ and x and y are acute, without using tables or calculators, find the values of $\cos x$, $\sin y$, $\sin(x + y)$, $\cos(x - y)$, $\tan(x + y)$.

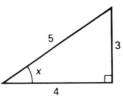

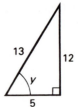

Using the right angled $\triangle$s, it follows that
$\cos x = \frac{4}{5}$ and $\sin y = \frac{12}{13}$

$$\sin(x + y) = \sin x \cdot \cos y + \cos x \cdot \sin y = \frac{3}{5} \cdot \frac{5}{13} + \frac{4}{5} \cdot \frac{12}{13} = \frac{63}{65}$$

$$\cos(x - y) = \cos x \cdot \cos y + \sin x \cdot \sin y = \frac{4}{5} \cdot \frac{5}{13} + \frac{3}{5} \cdot \frac{12}{13} = \frac{56}{65}$$

$$\tan(x + y) = \frac{\tan x + \tan y}{1 - \tan x \cdot \tan y} = \frac{\frac{3}{4} + \frac{12}{5}}{1 - \frac{3}{4} \cdot \frac{12}{5}} = -\frac{63}{16}$$

● Exercise 4.1   See page 66.

4.2 Applications of the Addition Formulae

1 To find the angles between two lines of gradients m_1 and m_2

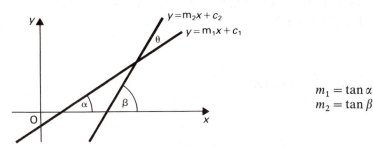

$$m_1 = \tan\alpha$$
$$m_2 = \tan\beta$$

Angle between two lines $= \theta = \beta - \alpha$

$$\tan\theta = \tan(\beta - \alpha) = \frac{\tan\beta - \tan\alpha}{1 + \tan\alpha\cdot\tan\beta} = \frac{m_2 - m_1}{1 + m_1 m_2}$$

$$\boxed{\tan\theta = \frac{m_1 - m_2}{1 + m_1 m_2}}$$

Note: $m_1 - m_2$ gives one angle between the two lines; $m_2 - m_1$ gives the other angle.

2 Example 3　Use $\sin 45° = \cos 45° = 1/\sqrt{2}$ and the addition formulae to show that $\sin 90° = 1$ and $\cos 0° = 1$.

$$\underline{\sin 90°} = \sin(45° + 45°) = \sin 45°\cdot\cos 45° + \cos 45°\cdot\sin 45°$$

$$= \frac{1}{\sqrt{2}}\cdot\frac{1}{\sqrt{2}} + \frac{1}{\sqrt{2}}\cdot\frac{1}{\sqrt{2}} = \underline{1}$$

$$\underline{\cos 0°} = \cos(45° - 45°) = \cos 45°\cdot\cos 45° + \sin 45°\cdot\sin 45°$$

$$= \frac{1}{\sqrt{2}}\cdot\frac{1}{\sqrt{2}} + \frac{1}{\sqrt{2}}\cdot\frac{1}{\sqrt{2}} = \underline{1}$$

3 Example 4　Solve $\sin(x + 60°) = \cos x$ for $0° \leqslant x \leqslant 360°$

$$\sin(x + 60°) = \cos x$$
$$\sin x\cdot\cos 60° + \cos x\cdot\sin 60° = \cos x$$
$$\sin x\cdot\frac{1}{2} + \cos x\cdot\frac{\sqrt{3}}{2} = \cos x$$
$$\sin x + \sqrt{3}\cos x = 2\cos x$$
$$\sin x = (2 - \sqrt{3})\cos x$$
$$\tan x = 2 - \sqrt{3} \longleftarrow \boxed{\tan x = \frac{\sin x}{\cos x}}$$
$$\Rightarrow \underline{x = 15° \text{ and } 195°}$$

4 Example 5　Prove that $\tan A + \tan B \equiv \sin(A + B)\cdot\sec A\cdot\sec B$.

$$\text{LHS} = \frac{\sin A}{\cos A} + \frac{\sin B}{\cos B}$$

$$= \frac{\sin A \cos B + \sin B \cos A}{\cos A \cdot \cos B}$$

$$= \frac{\sin (A + B)}{\cos A \cdot \cos B}$$

$$= \sin (A + B) \cdot \sec A \cdot \sec B = \text{RHS} \qquad \text{Q.E.D.}$$

● Exercise 4.2   See page 67.

4.3 Double Angle Formulae

$$\left. \begin{aligned} \sin 2A &= 2 \sin A \cdot \cos A \\ \cos 2A &= \cos^2 A - \sin^2 A \\ &= 2 \cos^2 A - 1 \\ &= 1 - 2 \sin^2 A \\ \tan 2A &= \frac{2 \tan A}{1 - \tan^2 A} \end{aligned} \right\} (1)$$

Similarly
$$\sin \theta = 2 \sin (\theta/2) \cdot \cos (\theta/2)$$
$$\cos \theta = \cos^2 (\theta/2) - \sin^2 (\theta/2)$$
$$= 2 \cos^2 (\theta/2) - 1$$
$$= 1 - 2 \sin^2 (\theta/2)$$
$$\tan \theta = \frac{2 \tan (\theta/2)}{1 - \tan^2 (\theta/2)}$$

Sometimes we need to write $\sin^2 x$ or $\cos^2 x$ in terms of $\cos 2x$.
Rearranging (1) $\qquad \Rightarrow \cos^2 x = \frac{1}{2}(1 + \cos 2x)$

$$\sin^2 x = \frac{1}{2}(1 - \cos 2x)$$

Note   It is better to learn how to produce these, than to learn the formulae themselves.

Example 6   If $\cos \theta = \frac{1}{3}$, find, without using tables or calculators, the value of $\cos (\theta/2)$, given that θ is acute

$$\cos \theta = 2 \cos^2 (\theta/2) - 1$$
$$\Rightarrow 2 \cos^2 (\theta/2) = \cos \theta + 1$$
$$\therefore \cos^2 (\theta/2) = \frac{1}{2}(\cos \theta + 1)$$

If $\cos \theta = \frac{1}{3}$, $\cos^2 (\theta/2) = \frac{1}{2}(\frac{1}{3} + 1) = \frac{2}{3}$

$$\therefore \cos (\theta/2) = \pm \sqrt{\frac{2}{3}}$$

θ is acute, $\therefore$ $\theta/2$ is also acute $\therefore$ $\cos (\theta/2) = + \sqrt{\frac{2}{3}}$

Example 7   Given that $\tan 30° = 1/\sqrt{3}$, find, without using tables or calculators, the value of $\tan 15°$.

$$\tan 30° = \frac{1}{\sqrt{3}} = \frac{2 \tan 15°}{1 - \tan^2 15°}$$

$$\Rightarrow 1 - \tan^2 15° = 2\sqrt{3}\tan 15°$$

$$\tan^2 15° + 2\sqrt{3}\tan 15° - 1 = 0 \qquad \text{which is a quadratic}$$
$$\text{in } \tan 15°$$

$$\therefore \tan 15° = \frac{-2\sqrt{3} \pm \sqrt{12+4}}{2} = -\sqrt{3} \pm 2$$

but, as $\tan 15° > 0$ we know that $\underline{\tan 15° = 2 - \sqrt{3}}$

- **Exercise 4.3** See page 68.

4.4 Applications of Double Angle Formulae

Example 8   Solve the equation $\sin 2\theta = \sin \theta$, giving all solutions between $0°$ and $360°$ inclusive.

i.e.
$$\sin 2\theta = \sin \theta$$
$$2\sin\theta\cdot\cos\theta = \sin\theta$$
$$2\sin\theta\cdot\cos\theta - \sin\theta = 0$$
$$\sin\theta(2\cos\theta - 1) = 0$$
$$\sin\theta = 0 \text{ or } \cos\theta = \tfrac{1}{2}$$

cancelling the factor $\sin\theta$ would lead to the loss of some of the roots of this equation.

$$\therefore \ \underline{\theta = 0°, 180°, 360°, 60°, 300°}$$

Example 9   Express $\sin 3x$ in terms of $\sin x$.

$$\sin 3x = \sin (2x + x) = \sin 2x \cdot \cos x + \cos 2x \cdot \sin x$$

Note   There is only one way of expanding $\sin 2x$ but three ways of expanding $\cos 2x$. Since we want the answer in terms of $\sin x$ we choose $\cos 2x = 1 - 2\sin^2 x$.

$$\therefore \ \sin 3x = \sin 2x \cdot \cos x + \cos 2x \cdot \sin x$$
$$= [2\sin x \cdot \cos x]\cos x + \sin x [1 - 2\sin^2 x]$$
$$= 2\sin x \cdot \cos^2 x + \sin x - 2\sin^3 x$$

Now, we need to get rid of the $\cos^2 x$ by using $\cos^2 x = 1 - \sin^2 x$.

$$\therefore \ \sin 3x = 2\sin x(1 - \sin^2 x) + \sin x - 2\sin^3 x$$
$$\therefore \ \underline{\sin 3x = 3\sin x - 4\sin^3 x}$$

Example 10   Eliminate θ from the equations $x = \cos 2\theta$, $y = \sec \theta$.

$$x = \cos 2\theta = 2\cos^2 \theta - 1 \quad \text{and} \quad y = \sec \theta = \frac{1}{\cos \theta}$$

$$\Rightarrow x = 2\left(\frac{1}{y}\right)^2 - 1$$

$$\Rightarrow (x+1)y^2 = 2$$

> **Note**   This is a **Cartesian Equation** which we have obtained by eliminating the **parameter** θ from the pair of **parametric equations**.

● Exercise 4.4   See page 70.

4.5 The Sine and Cosine Rules

The Sine Rule

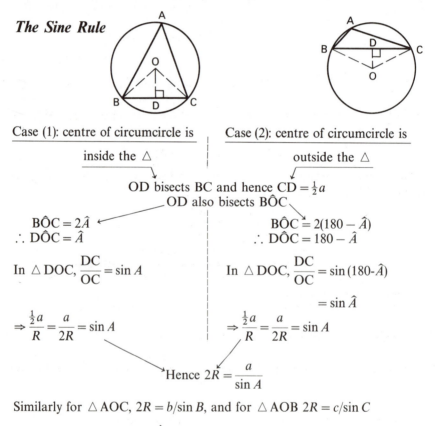

Case (1): centre of circumcircle is inside the △

Case (2): centre of circumcircle is outside the △

OD bisects BC and hence $CD = \frac{1}{2}a$

OD also bisects $B\hat{O}C$

$B\hat{O}C = 2\hat{A}$
$\therefore D\hat{O}C = \hat{A}$

In △ DOC, $\dfrac{DC}{OC} = \sin A$

$\Rightarrow \dfrac{\frac{1}{2}a}{R} = \dfrac{a}{2R} = \sin A$

$B\hat{O}C = 2(180 - \hat{A})$
$\therefore D\hat{O}C = 180 - \hat{A}$

In △ DOC, $\dfrac{DC}{OC} = \sin(180 - \hat{A})$

$= \sin \hat{A}$

$\Rightarrow \dfrac{\frac{1}{2}a}{R} = \dfrac{a}{2R} = \sin A$

Hence $2R = \dfrac{a}{\sin A}$

Similarly for △ AOC, $2R = b/\sin B$, and for △ AOB $2R = c/\sin C$

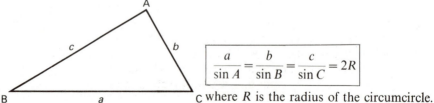

$$\frac{a}{\sin A} = \frac{b}{\sin B} = \frac{c}{\sin C} = 2R$$

where R is the radius of the circumcircle.

The Cosine Rule

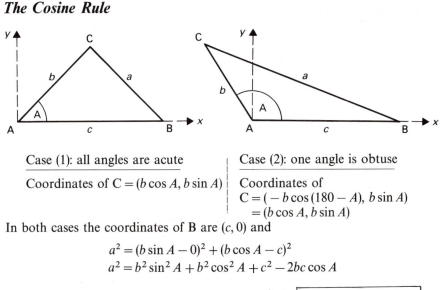

Case (1): all angles are acute	Case (2): one angle is obtuse
Coordinates of $C = (b \cos A, b \sin A)$	Coordinates of $C = (-b \cos(180 - A), b \sin A)$ $= (b \cos A, b \sin A)$

In both cases the coordinates of B are $(c, 0)$ and

$$a^2 = (b \sin A - 0)^2 + (b \cos A - c)^2$$
$$a^2 = b^2 \sin^2 A + b^2 \cos^2 A + c^2 - 2bc \cos A$$

$$\boxed{\therefore\ a^2 = b^2 + c^2 - 2bc \cos A}$$

$$\left. \begin{array}{c} \text{and} \\ \text{rearranging} \end{array} \right\} \Rightarrow \boxed{\cos A = \dfrac{b^2 + c^2 - a^2}{2bc}}$$

Example 11   Find the longest side of $\triangle ABC$ given that $\hat{A} = 102°$, $\hat{B} = 51°$, and $c = 4.71$ cm.

$$\hat{C} = 180 - (102 + 51) = 27°$$

The largest side is opposite the largest angle, so the largest side is a. Using the sine rule:

$$\frac{a}{\sin 102°} = \frac{4.71}{\sin 27°}$$

$$\therefore\ a = 4.71 \times \frac{\sin 102°}{\sin 27°} = 10.14 \text{ cm}$$

$\therefore$ the longest side is 10.14 cm

The Ambiguous Case

Example 12   Find $\hat{B}$ and $\hat{C}$ in the $\triangle$ where $\hat{A} = 24°$, $c = 2.6$ cm and $a = 1.1$ cm.

Using the sine rule: $\dfrac{2.6}{\sin \hat{C}} = \dfrac{1.1}{\sin 24°} \Rightarrow \sin C = 0.9614$

but $\arcsin 0.9614 = 74°$ or $106° = \hat{C}$.
(a)   If $\hat{C} = 74°$, $B = 180 - (74 + 24) = 82°$
(b)   If $\hat{C} = 106°$, $B = 180 - (106 + 24) = 50°$
$\therefore$ two triangle can be formed (a)   where $\hat{B} = 82°$ and $\hat{C} = 74°$
                                    and (b)   where $\hat{B} = 50°$ and $\hat{C} = 106°$

Example 13   If $\hat{A} = 37°$, $a = 4.59\,\text{cm}$ and $c = 2.1\,\text{cm}$ show that there is only one $\triangle$ that can be formed and find the remaining angles.

Using the sine rule: $\dfrac{4.59}{\sin 37°} = \dfrac{2.1}{\sin C} \Rightarrow \hat{C} = \arcsin 0.2753 = 16°$ or $164°$

but if $\hat{C} = 164°$, $\hat{A} + \hat{C} = 164° + 37° > 180$ $\therefore$ the $\triangle$ does not exist.
$\therefore$ only one $\triangle$ can be formed and for this $\hat{C} = 16°$ and $B = 127°$

Example 14   Find the largest angle in $\triangle ABC$ where $a = 17.5\,\text{cm}$, $b = 8.4\,\text{cm}$ and $c = 11.9\,\text{cm}$.

The longest side is opposite the largest angle so we must find $\hat{A}$.

$$\cos A = \frac{8.4^2 + 11.9^2 - 17.5^2}{2(8.4)(11.9)} = -0.4706 \longleftarrow \boxed{\begin{array}{l}\text{show this amount of}\\\text{working in all answers!}\end{array}}$$

$$\therefore \hat{A} = 118.1°$$

> **Note:** Some problems can be solved using either the sine rule or the cosine rule. The sine rule gives an easier calculation but you must always beware of the ambiguous case. The cosine rule, though more complex, gives rise to no ambiguities.

● Exercise 4.5   See page 70.

4.6 Areas of Triangles and Applications of Sine and Cosine Rules

> Area of Triangle $= \frac{1}{2} \times \text{base} \times \text{height}$
> $$= \frac{1}{2} ab \sin \hat{C}$$
> $$= \sqrt{s(s-a)(s-b)(s-c)} \quad \text{where } s = \frac{a+b+c}{2}$$
> Area of Trapezium $= \frac{1}{2}(a+b)h$

Example 15   Find the area of the triangle with sides $13\,\text{cm}$, $5\,\text{cm}$, and $10\,\text{cm}$

$$s = \frac{13 + 5 + 10}{2} = 14$$

$$\text{Area} = \sqrt{14(14-13)(14-5)(14-10)} = \sqrt{14 \times 1 \times 9 \times 4} = 22.45\,\text{cm}^2$$

Example 16   Find the area of the triangle where $a = 5\,\text{cm}$, $b = 3\,\text{cm}$, and $\hat{C} = 40°$

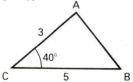

$$\text{Area} = \frac{1}{2} \cdot 3 \cdot 5 \cdot \sin 40° = 4.82\,\text{cm}^2$$

Example 17 In $\triangle$ ABC, $\hat{A} = 90°$, $a = 12$, $b = 3$. Find the length of the altitude through $\hat{A}$.

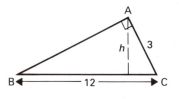

$$AB^2 = 12^2 - 3^2 = 135$$
$$\therefore AB = \sqrt{135}$$

Area of $\triangle = \frac{1}{2} \cdot AC \cdot AB = \frac{1}{2} \cdot 3\sqrt{135} = \frac{9}{2}\sqrt{15}$

Also area of $\triangle = \frac{1}{2} \cdot 12 \cdot h$

$$\therefore \frac{1}{2} \cdot 12 \cdot h = \frac{1}{2} \cdot 9 \cdot \sqrt{15} \qquad \therefore h = \frac{9}{12}\sqrt{15} = 2.9$$

$\therefore$ altitude through $\hat{A}$ is 2.9

> The altitude is the *height* of the triangle

- **Exercise 4.6** See page 71

4.7 Three-Dimensional Problems

Hints on Drawing Clear Three-Dimensional Diagrams

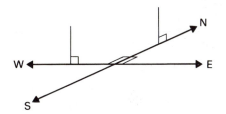

1 Vertical lines should be drawn vertically on the page.

2 East-west lines should be drawn horizontally on the page.

3 North-south lines should be drawn at an acute angle to east-west lines.

4 Parallel lines should be drawn parallel in the diagram.

5 90° angles should be marked as right angles in the diagram, particularly those angles that do not look like right angles.

Angle Between Two Planes

Take any two lines, OA and OB, one in each plane, that are both perpendicular to the lines of intersection of the planes. Angle AOB is the angle between the two planes.

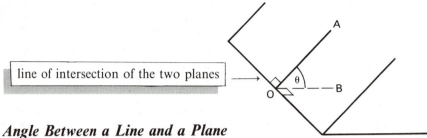

line of intersection of the two planes

Angle Between a Line and a Plane

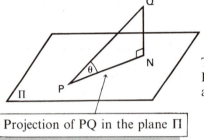

The angle between a line PQ and a plane Π is defined as the angle between PQ and its projection in the plane, PN.

Projection of PQ in the plane Π

Example 18

ABCDPQRS is a cuboid with measurements 8 cm, 6 cm, 5 cm. Find:

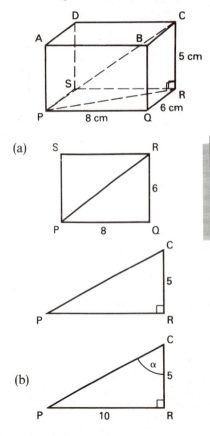

(a) the length of the diagonal PC
(b) the angle between PC and the edge CR
(c) the angle between PC and the face BCRQ.

(a)

1   Draw the 3-D diagram and put on to it all known lengths and angles.

2   For each part of the question draw at least one two-dimensional diagram.

$$PR^2 = 8^2 + 6^2 \Rightarrow PR = 10$$
$$PC^2 = PR^2 + 5^2 = 10^2 + 5^2 = 125$$
$$\Rightarrow PC = 11.2\,\text{cm}$$

(b)

Angle between PC and CR is α.
$$\tan \alpha = \tfrac{10}{5} = 2$$
$$\Rightarrow \text{angle } \alpha = 63.4°$$

(c)

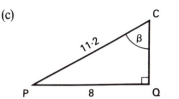

CQ is the projection of PC in BCRQ

Angle between PC and BCRQ is β.
$$\sin \beta = \frac{8}{11.2}$$
$$\Rightarrow \text{angle } \beta = 45.6°$$

Example 19 From a point A due west of a tower, the angle of elevation of the top of the tower is 25°. From a point B due south of the tower, the angle of elevation of the top of the tower is 33°. The tower is 25 m tall. Find the distance AB.

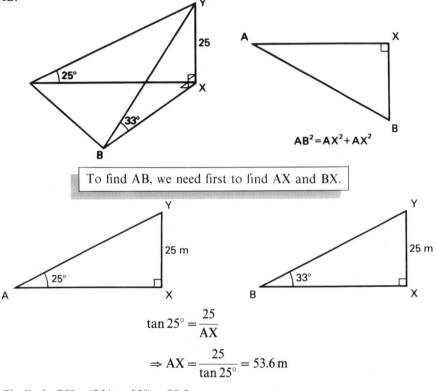

$$AB^2 = AX^2 + AX^2$$

To find AB, we need first to find AX and BX.

$$\tan 25° = \frac{25}{AX}$$

$$\Rightarrow AX = \frac{25}{\tan 25°} = 53.6 \text{ m}$$

Similarly $BX = (25/\tan 33°) = 38.5 \text{ m}$
Hence $AB^2 = 53.6^2 + 38.5^2 \Rightarrow AB = 66 \text{ m}$

- Exercise 4.7 See page 73.

4.8 Miscellaneous Techniques in Trigonometry: Revision of Unit

Triple Angles

$$\cos 3x = 4\cos^3 x - 3\cos x$$
$$\sin 3x = 3\sin x - 4\sin^3 x$$

To express tan x, sin x, cos x in terms of $\tan \frac{1}{2}x$.

Let $t = \tan\frac{1}{2}x$

then $\tan x = \dfrac{2\tan\frac{1}{2}x}{1-\tan^2\frac{1}{2}x}$ $\Rightarrow$ $\boxed{\tan x = \dfrac{2t}{1-t^2}}$

By Pythagoras' Theorem,
the other side of the $\therefore$ $\sin x = \dfrac{2t}{1+t^2}$
$\triangle$ is $1+t^2$

> This technique will be
> met again in Unit 7

and $\cos x = \dfrac{1-t^2}{1+t^2}$

2t $1+t^2$

x

$1-t^2$

> **Note**: these results cannot be quoted but must
> be derived when you need them.

Example 20   By letting $\tan(\theta/2) = t$, solve the equation
$3\cos\theta - 8\sin\theta = -2$, giving solutions between $0°$ and $360°$.

Using the above formulae, the equation becomes

$$3\left(\frac{1-t^2}{1+t^2}\right) - 8\left(\frac{2t}{1+t^2}\right) = -2$$

$\therefore$ $3 - 3t^2 - 16t = -2 - 2t^2$

$\therefore$ $t^2 + 16t - 5 = 0$

$\therefore$ $t = \dfrac{-16 \pm \sqrt{256 + 20}}{2} = 0.31 \text{ or } -16.31$

$\Rightarrow \tan(\theta/2) = 0.31$ and $\tan(\theta/2) = -16.31$

Since $0° < \theta < 360°$ then $0° < \theta/2 < 180°$
$\Rightarrow \theta/2 = 17.2°$ and $93.5° \Rightarrow \theta = 34.4°$ and $187°$

- Miscellaneous Exercise 4A   See page 74.

4.9 + A Level Questions

- Miscellaneous Exercise 4B   See page 75.

Unit 4 EXERCISES

Exercise 4.1

1   Express as single sines or cosines:
   (a) $\sin 43° \cdot \cos 61° + \cos 43° \sin 61°$
   (b) $\sin 22° \cdot \cos 18° - \cos 22° \sin 18°$

(c) $\cos 63°\cdot\cos 11° + \sin 63°\cdot\sin 11°$
(d) $\sin 41°\cdot\sin 22° - \cos 41°\cdot\cos 22°$
(e) $\sin 2x\cdot\cos x + \cos 2x\cdot\sin x$
(f) $\cos (\pi/2)\cdot\cos (\pi/6) + \sin (\pi/2)\cdot\sin (\pi/6)$
(g) $\cos^2 x - \sin^2 x$ (i.e. $\cos x\cdot\cos x - \sin x\cdot\sin x$)

2   Use the addition formulae to simplify the following:

$\sin(180° + A), \cos(180° + A), \sin(90° + A), \cos(90° + A)$

3   Use the addition formulae to prove that:

(a) $\sin(\pi - \theta) = \sin \theta$ (b) $\cos(\pi - \theta) = -\cos \theta$
(c) $\cos(2\pi - \theta) = \cos \theta$ (d) $\sin(2\pi - \theta) = -\sin \theta$

NB: angles are given in radians.

4   Simplify:

(a) $\cos x\cdot\cos y + \sin x\cdot\sin y$ (g) $\cos 4\theta\cdot\cos \theta + \sin 4\theta \sin \theta$
(b) $\sin 3\theta \cos \phi - \cos 3\theta \sin \phi$ (h) $\sin(\theta + \phi)\cdot\cos(\theta - \phi)$
(c) $\sin \phi\cdot\cos 20° + \cos \phi \sin 20°$ $+ \cos(\theta + \phi)\sin(\theta - \phi)$
(d) $\sin A\cdot\cos 2B - \cos A \sin 2B$ (i) $\sin 75°\cdot\cos 15° - \cos 75°\cdot\sin 15°$
(e) $\cos \theta\cdot\cos 4\theta - \sin 4\theta \sin \theta$
(f) $\sin C \sin D - \cos C\cdot\cos D$ (j) $\dfrac{\tan 75° + \tan 15°}{1 - \tan 75°\cdot\tan 15°}$

5   Simplify:

(a) $\sin(\theta + \phi) + \sin(\theta - \phi)$ (b) $\cos(C + D) + \cos(C - D)$
(c) $\cos(\alpha + \beta) - \cos(\alpha - \beta)$

6   If $\sin A = \frac{4}{5}$ and $\tan B = \frac{5}{12}$ and A and B are acute angles, find:

(a) $\cos A$ (b) $\tan A$ (c) $\sin B$ (d) $\cos B$
(e) $\sin(A + B)$ (f) $\sin(A - B)$ (g) $\cos(A + B)$ (h) $\cos(A - B)$
(i) $\tan(A + B)$ (j) $\tan(A - B)$

7   If $\cos x = \frac{3}{5}$, $\sin y = \frac{12}{13}$ and x and y are acute, find $\sin(x + y)$ and $\cos(x + y)$.

Copy and complete:

$\sin(2x + y) = \sin[(x + y) + x] = \sin(x + y)\cdot\cos x + \cos(x + y)\cdot\sin x$
$= (\cdots \times \frac{3}{5}) + (\cdots \times \cdots)$
$=$

Similarly find $\cos(2x + y)$, $\sin(x + 2y)$ and $\cos(x + 2y)$

8   If $\sin \alpha = -\frac{4}{5}$ where $\pi < \alpha < 3\pi/2$ and $\cos \beta = -\frac{24}{25}$ where $(\pi/2) < \beta < \pi$, without using tables, calculate:

(a) $\sin(\alpha - \beta)$ (b) $\cos(\alpha + \beta)$ (c) $\cos 2\alpha$
(d) $\tan(\alpha - \beta)$ (e) $\tan 2\alpha$ (f) $\tan 3\alpha$

Exercise 4.2

1   Find the acute angles between the following pairs of straight lines:
(a) $y = x$ and $y = 2x$

(b) $y = 2x + 1$ and $y = 3x - 2$
(c) $2y - x = 1$ and $3y = x - 1$
(d) $y + 2x = 0$ and $y + 3x = 0$
(e) $2y + x - 3 = 0$ and $y - 3x - 4 = 0$

2   By putting $A = B = \theta$ in the formula for $\cos(A - B)$, deduce that $\cos^2 \theta + \sin^2 \theta = 1$

3   Use $\sin 30° = \frac{1}{2}$ and $\cos 30° = \sqrt{3}/2$ and the addition formulae to prove that $\sin 60° = \sqrt{3}/2$ and $\cos 60° = \frac{1}{2}$.

Hence, use these values to show that $\sin 90° = 1$ and $\cos 90° = 0$

4   Using the values given and found in question 3, and the addition formulae, prove that:
(a) $\sin(x + 60°) = \cos(x - 30°)$
(b) $\sin(x + 30°) = \sin(150° - x)$

5   Solve the following equations, giving angles from $0°$ to $360°$:
(a) $\cos(45° - \theta) = \sin(30° + \theta)$
(b) $3 \sin x = \cos(x + 60°)$
(c) $\tan(A - \theta) = \frac{2}{3}$ and $\tan A = 3$

6   Prove that $\sin\left(\dfrac{\pi}{4} + A\right) + \sin\left(\dfrac{\pi}{4} - A\right) \equiv \sqrt{2} \cos A$

7   Prove that $(\sin A + \cos A)(\sin B + \cos B) \equiv \sin(A + B) + \cos(A - B)$

8   Prove that $\dfrac{\sin(A + B)}{\cos A \cdot \cos B} \equiv \tan A + \tan B$

9   Prove that $\tan(A + B + C) \equiv \dfrac{\tan A + \tan B + \tan C - \tan A \cdot \tan B \cdot \tan C}{1 - \tan B \cdot \tan C - \tan A \cdot \tan B - \tan A \cdot \tan C}$

Exercise 4.3

1   Express as single trig. ratios:
(a) $2 \sin 14° \cdot \cos 14°$

(b) $\dfrac{2 \tan 35°}{1 - \tan^2 35°}$

(c) $\cos^2 26° - \sin^2 26°$

(d) $2 \cos^2 \dfrac{\pi}{6} - 1$

(e) $1 - 2 \sin^2 4\theta$

(f) $\dfrac{2 \tan 3\theta}{1 - \tan^2 3\theta}$

    (g) $2\sin 2\theta \cdot \cos 2\theta$

    (h) $\sin \theta \cdot \cos \theta$

2   Express the following in the terms shown:

    (a) $\sin 2\theta$ in terms of θ

    (b) $\tan 8\theta$ in terms of 4θ

    (c) $\cos 4\theta$ in terms of $\cos 2\theta$

    (d) $2\sin\dfrac{\alpha}{2}\cdot\cos\dfrac{\alpha}{2}$ in terms of α

    (e) $\sin\dfrac{\phi}{2}\cdot\cos\dfrac{\phi}{2}$ in terms of ϕ

    (f) $\dfrac{2\tan 22\frac{1}{2}°}{1-\tan^2 22\frac{1}{2}°}$ in terms of $45°$

    (g) $\cos^2\dfrac{\pi}{12} - \sin^2\dfrac{\pi}{12}$ in terms of $\dfrac{\pi}{6}$

    (h) $\sin^2\phi - \cos^2\phi$ in terms of 2ϕ

    (i) $(\sin A + \cos A)^2$ in terms of $2A$

    (j) $\sin\dfrac{3\theta}{2}\cdot\cos\dfrac{3\theta}{2}$ in terms of 3θ

3   If θ is acute, find the values of $\sin 2\theta$ and $\cos 2\theta$ given that:

    (a) $\cos \theta = 3/5$ (b) $\sin \theta = 7/25$ (c) $\tan \theta = 12/5$

4   If $\tan \theta = \frac{1}{4}$ find the value of $\tan 2\theta$

5   If $\sin x = \frac{3}{5}$ and x is acute find the value of $\tan x$ and hence that of $\tan 2x$

6 $\cos 2A = \cos^2 A - \sin^2 A$
$$= 2\cos^2 A - 1$$
$$= 1 - 2\sin^2 A$$

Rearrange these formulae to express:

    (a) $\cos^2 A$ in terms of $\cos 2A$

    (b) $\sin^2 A$ in terms of $\cos 2A$

    (c) $1 - \cos 2A$ in terms of A

    (d) $1 + \cos 2A$ in terms of A

Similarly express

    (e) $1 - \cos B$ in terms of $B/2$

    (f) $1 + \cos B$ in terms of $B/2$

7   If $\cos 2\theta = \frac{3}{5}$, find the value of $\sin \theta$, without using tables or calculators

8   If $\cos \theta = \frac{1}{2}$, find the value of $\sin \theta/2$, without using tables or calculators

9   Given that $\tan 45° = 1$, find the value of $\tan 22\frac{1}{2}°$ in surd form

10   If $\tan \theta = 5/12$ find the value of $\cos \theta$, without using tables or calculators. Hence find the values of $\cos 2\theta$ and $\cos \theta/2$

Exercise 4.4

In questions 1–4 solve the following equations, giving values of θ between $0°$ and $180°$ inclusive:

1 $\sin\theta(2\sin\theta - 1) = 0$ **2** $2\sin\theta\cdot\cos\theta = \cos\theta$

3 $4\sin 2\theta = \sin\theta$ **4** $\cos 2\theta = \cos\theta$

5 Show that $\cos 3x = 4\cos^3 x - 3\cos x$

6 By eliminating θ from the following pairs of parametric equations, find the corresponding Cartesian equations:
   (a) $x = \tan 2\theta,\ y = \tan\theta$ (b) $x = \cos 2\theta,\ y = \cos\theta$
   (c) $x = \cos 2\theta,\ y = \operatorname{cosec}\theta$ (d) $x = \sin 2\theta,\ y = \cos 4\theta$

7 (a) Express $\cos 4A$ in terms of $\cos 2A$
   (b) Express $\cos 2A$ in terms of $\cos A$
   Hence express $\cos 4A$ in terms of $\cos A$ only

In questions 8–12 solve the following equations for $0° \leqslant x \leqslant 360°$

8 $\cos 2x = \sin x$ **9** $\sin 2x + \cos x = 0$

10 $4 - 5\cos x = 2\sin^2 x$ **11** $\tan x\cdot\tan 2x = 2$

12 $\sin 2x - 1 = \cos 2x$

13 Show that $\tan 3x = \dfrac{3\tan x - \tan^3 x}{1 - 3\tan^2 x}$

Exercise 4.5

Draw a labelled sketch for **each** question. Give all answers correct to 2 d.p. or 0.1 of a degree.

Use the Cosine Rule to find the following:

1 $a = 5,\ b = 9,\ c = 10$. Find $\hat{A}$

2 $x = 5.1,\ y = 2.1,\ z = 4.1$. Find $\hat{Z}$

3 $p = 45,\ q = 33,\ r = 21$. Find $\hat{Q}$

4 $a = 11\,\text{cm},\ b = 4\,\text{cm},\ \hat{C} = 36°$. Find c

5 $x = 1.7\,\text{cm},\ z = 3.2\,\text{cm},\ \hat{Y} = 24°$. Find y

6 $q = 21.2\,\text{cm},\ r = 43\,\text{cm},\ \hat{P} = 119°$. Find p

Use the Sine Rule to find the following:

7 $\hat{A} = 29°,\ \hat{B} = 35°,\ b = 9.3\,\text{cm}$. Find a

8 In $\triangle$ XYZ, $\hat{X} = 34°$, $\hat{Y} = 47°$, $z = 10.5\,\text{cm}$. Find x

9 In $\triangle$ CDE, $\hat{C} = 36°$, $\hat{D} = 29°$, $d = 14\,\text{cm}$. Find e

10 In $\triangle$ WXY, $w = 13.6\,\text{cm}$, $x = 7.1\,\text{cm}$, $\hat{W} = 79°$. Find $\hat{Y}$

11 Find the largest angle of the triangle whose sides are 2 cm, 4 cm, 5 cm

12 $a = 3\,\text{cm}$, $b = 4\,\text{cm}$, $\hat{A} = 20°$. Show that two triangles can be drawn and solve the triangles

13 In $\triangle$ FGH, $g = 3.8\,\text{cm}$, $\hat{F} = 25°$, $f = 1.8\,\text{cm}$. Find two possible values of h

In questions 14–18 solve completely (i.e. find all the missing sides and angles) the $\triangle$s in which:

14 $\hat{A} = 78°$, $\hat{B} = 43°$, $b = 29.4\,\text{cm}$

15 $\hat{A} = 78.2°$, $b = 8.75\,\text{cm}$, $c = 6.36\,\text{cm}$

16 $\hat{X} = 22°$, $\hat{Y} = 126.3°$, $y = 8.63\,\text{cm}$

17 $p = 2.1\,\text{m}$, $q = 3.2\,\text{m}$, $r = 1.2\,\text{m}$

18 $\hat{B} = 109.2°$, $a = 149\,\text{m}$, $b = 163\,\text{m}$

*19 Given that $\hat{B} = 30°$, $b = 150$, $c = 150\sqrt{3}$ prove that, of the two triangles which can be drawn, one will be isosceles and the other right-angled

*20 Two sides of a triangle are $\sqrt{3} + 1$ and $\sqrt{3} - 1$ and the included angle is 60°. Show that the other side is $\sqrt{6}$

*21 From a lighthouse on the quay one buoy is 5 km away on a bearing N50°E and another is 2 km away on a bearing N60°W. How far apart are the buoys?

*22 A parallelogram has sides 4 cm and 7 cm. One of the angles is 65°. Find the length of each diagonal

Exercise 4.6

Draw a labelled sketch for each question. Give all answers correct to 2 d.p. or 0.1 of a degree.

1 Find the areas of the triangles in which:
   (a) $p = 10\,\text{cm}$, $q = 11\,\text{cm}$, $r = 15\,\text{cm}$
   (b) $x = 50\,\text{m}$, $z = 43\,\text{m}$, $\hat{Y} = 88°$
   (c) $q = 2\sqrt{3}$, $r = 6$, $\hat{P} = 30°$

2 Find the shortest altitude of the $\triangle$ ABC in which $a = 6$, $b = 7$, $c = 8$

3   Find the other two sides of $\triangle$ ABC, in which $\hat{A} = 31°$, $\hat{B} = 14°$. $c = 7$ cm. Also find the area of the triangle and its largest altitude

4   Find the area of the trapezium ABCD in which AB is parallel to CD, AB $= 8$ cm, BC $= 11$ cm, CD $= 17.6$ cm, B $= 102.3°$

5   The two sides of a stepladder are of length 2 m and 1.85 m. When fully opened the longer side is inclined at 65° to the horizontal. Find the distance apart of the feet of the ladder

6   A golfer attempting a 150 m shot plays at 18° to the correct line and drives the ball 90 m. How far does the ball land from the hole?

7   To measure the height of a tower, two sightings are made of the top of the tower, from two points on the same level as the base of the tower and 100 m in line with the base. Find the length YZ and the height of the tower

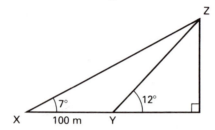

8   Two ships leave port at the same time, one steaming at 5 km/h on a bearing of 046° and the other at 9 km/h on a bearing of 127°. How far apart are the ships after 3 hours?

9   The sides of a triangle are $\sqrt{2}$ m, $\sqrt{8}$ m and 3 m. Find the sizes of all the angles.

*10   The sides of a parallelogram are 7.21 cm and 9.32 cm respectively and one diagonal measures 8.3 cm. Calculate:

(a)   the angles of the parallelogram
(b)   the length of the other diagonal
(c)   the angles between the diagonals

*11   Find the area of the quadrilateral ABCD given AB $= 7$ cm, BC $= 4.5$ cm, CD $= 14.5$ cm, DA $= 12$ cm and B $= 110.5°$

*12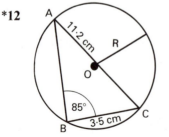

(a)   Find AB
(b)   Find the radius of the circumcircle, R

*13   In the quadrilateral ABCD, $\hat{A} = \hat{C} = 90°$ and $\hat{D} = 123°$. AD $= 16$ cm, CD $= 10$ cm. Find the length of BD

Exercise 4.7

1 A cube has side 12 cm. Find:
  (a) the length of a diagonal
  (b) the angle between a diagonal and an adjacent face
  (c) the angle between a diagonal and an adjacent edge

2

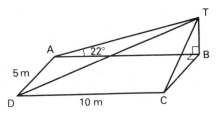

ABCD is a rectangle and TB is a vertical pole. AB = 10 m, AD = 5 m and the angle of elevation of T from A is 22°. Find:
  (a) the height of the pole
  (b) the angle of elevation of T from C
  (c) the length of a diagonal of the rectangle ABCD
  (d) the angle of elevation of T from D

3

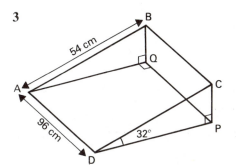

In the wedge shown, ABCD and ADPQ are rectangles. AB = 54 cm, AD = PQ = 96 cm and angle CDP is 32°. Find the inclination of AC to the plane ADPQ, to the nearest degree.

4

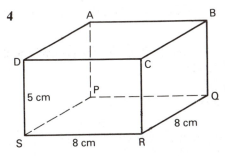

ABCDPQRS is a cuboid where BQ = 5 cm and PQRS is a square of side 8 cm. Find:
  (a) the lengths of QS and BS
  (b) the angle between BS and the plane PQRS
  (c) the angle between the planes AQRD and PQRS

5 The angle of elevation of the top of a tower is 27° from a point P due south of it. From a point Q due east of it, the angle of elevation is 11°. Find the distance PQ if the height of the tower is 7.26 m

6

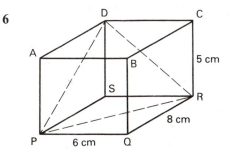

ABCDPQRS is a cuboid whose dimensions are shown in the diagram. Find:
  (a) PR   (b) DR   (c) DP
  (d) the angle PDR

7

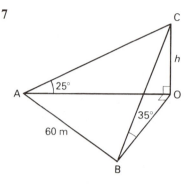

A, B, O are in the same horizontal plane and C is vertically above 0. AB = 60 m. The angle of elevation of C from A is 25° and the angle of elevation of C from B is 35°. Angle AOB is 90°.

(a) Find the length of AO in terms of h

(b) Find the length of BO in terms of h

(c) Find the value of h

8 A triangle ABC lies in a horizontal plane. A point P is 2 m vertically above B. Angle ABC = 150°, angle PAB = 40° and angle PCB = 30°.

(a) Calculate the length of AC

(b) Find the acute angle between the plane PAC and the horizontal, giving your answer to the nearest degree

9 ABC is a straight horizontal road where BC = 2AB = 2x. OH is a vertical tower of height h and O is in the same horizontal plane as ABC. The angles of elevation of H from A, B, C are α, β, γ respectively. Prove that $2x^2 = h^2 (\cot^2 \alpha - \cot^2 \beta)$

For $\alpha = 10°, \beta = 25°$ and AB = 150 m, find the height of the tower

Miscellaneous Exercise 4A

1 Using $t \equiv \tan \theta/2$, solve the following, giving values of θ from $-180°$ to $180°$:

(a) $3 \cos \theta + 2 \sin \theta = 3$ (b) $5 \cos \theta - \sin \theta + 4 = 0$

(c) $\cos \theta + 7 \sin \theta = 5$

2 Find the acute angle between the lines $2y = 3x - 7$ and $2x + 4y = 3$

3 If $\tan \theta = 1.5$ and θ is acute, without using tables or calculators, obtain the values of $\tan 2\theta$, $\sin 2\theta$ and $\cos 2\theta$

4 If $\sin \theta = 20/29$, find the values, without use of tables or calculators, of $\cos \theta$, $\sin 2\theta$, $\cos 2\theta$, $\sin \theta/2$, $\cos \theta/2$

5 If $\tan \theta = -7/24$ and θ is obtuse, find the value of $\tan \theta/2$ and hence find $\sin \theta/2$ and $\cos \theta/2$

6 Find the values of $\sin 15°$ and $\cos 15°$ in surd form. (Use $\cos 30° = \sqrt{3}/2$.)

7 Express $\sin 4A$ in terms of $\sin 2A$ and $\cos 2A$. Hence show that

$$\frac{\sin 4A}{\sin A} = 8 \cos^3 A - 4 \cos A$$

8   Prove that $\dfrac{1 - \cos 2A}{\sin 2A} \equiv \tan A$

9   Prove that $\tan \theta + \cot \theta \equiv 2 \operatorname{cosec} 2\theta$

10   Prove that $\sin 2\theta \equiv \dfrac{2 \tan \theta}{1 + \tan^2 \theta}$

Miscellaneous Exercise 4B

1   Find all values of θ between 0 and 360 which satisfy the equation

$$\cos(60 + \theta)^\circ + 2 \sin(30 + \theta)^\circ = 0 \qquad \text{(L: 4.2)}$$

2   Find the solutions in the range $0^\circ \leqslant \theta \leqslant 180^\circ$ of each of the following equations:

(a) $\sin 2\theta = \cos \theta$ (b) $2 \cos 2\theta = 1 + 4 \cos \theta$ (J: 4.4)

3   The points A, B and C are on a horizontal plane, where $AB = 5a$, $AC = 3a$ and the angle $CAB = 120^\circ$. The bisector of angle CAB meets BC at the point P.

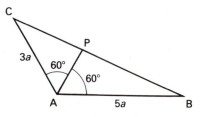

(a)   Prove that $BC = 7a$ (b)   Prove that $CP = 21a/8$

A vertical pole CV is placed at C. The angle of elevation of V from A is 30°

(c)   Calculate the angle of elevation of V from P   (A: 4.5, 4.7)

4   Given that $\theta + \phi = \pi/4$, express $\tan \phi$ in terms of $\tan \theta$ and hence prove that

$$(1 + \tan \theta)(1 + \tan \phi) = 2$$

Deduce that $\tan \pi/8 = \sqrt{2} - 1$ (A: 4.2)

5   In the triangle ABC, $BC = a$ and angle $CAB = \theta$. Given also that $B\hat{C}D = 4\theta$ where D is on AC produced prove that

(a) $AB = 4a \cos \theta \cdot \cos 2\theta$
(b) $AC = a(1 + 2 \cos 2\theta)$ (J: 4.3, 4.5)

6   By using identifies for $\cos(A - B)$ and $\cos(A + B)$, prove that

$$1 + \cos(A - B) \cos(A + B) = \cos^2 A + \cos^2 B$$

Hence prove that

(a) $1 + \cos 2\theta = 2 \cos^2 \theta$
(b) $3 + \cos(P - Q) \cos(P + Q) + \cos(Q - R) \cos(Q + R)$
    $+ \cos(R - P) \cos(R + P) = 2(\cos^2 P + \cos^2 Q + \cos^2 R)$ (A: 4.1)

7 A trapezium has parallel sides of lengths 6 cm and 14 cm and the remaining sides have lengths 5 cm and 7 cm. By regarding the trapezium as a parallelogram and a triangle, find the area of the triangle and deduce the area of the trapezium

Show that the smallest angle of the trapezium is $\cos^{-1}(11/14)$ and deduce the size of the largest angle   (A: 4.5, 4.6)

8 Express $\tan(45° + \theta)$ in terms of $\tan\theta$. Hence or otherwise:

(a)   express $\tan 75°$ in the form $a + b\sqrt{3}$ where a and b are integers
(b)   express $\tan(45° + \theta) + \cot(45° + \theta)$ in terms of $\cos 2\theta$
                                                                 (J: 4.2, 4.3, 1.6)

9 Show that, in triangle ABC, the length of the perpendicular from C to AB is given by $\dfrac{ab}{c}\sin\hat{C}$

The triangle ABC, in which $\hat{ACB} = 150°$ lies in a horizontal plane and D is the point at a distance 2 units vertically above C. Given that $\hat{DBC} = 30°$ and $\hat{DAC} = 45°$ find the lengths of the sides of the triangle ABC and the length of the perpendicular from C to AB   J(4.5, 4.6)

10 The smallest angle of a right angled triangle is θ and the shortest side is of length x. Given that the sum of the other two sides is kx, show that

$$k\sin\theta - \cos\theta = 1$$

Hence find the value of $\tan\frac{1}{2}\theta$ in terms of k, and deduce that $k \geqslant \cot\pi/8$
                                                                 (A: 4.5, 4.8)

11 The triangle ABC is horizontal with $AB = 25a$, $AC = 26a$ and $BC = 17a$. The point P lies on BC such that angle APB is 90°. Calculate:

(a)   the area of triangle ABC
(b)   the length of AP

The point Y lies **vertically** above A, where $YA = 18a$. Calculate, to 0.1°:

(c)   the acute angle between YB and the horizontal
(d)   the acute angle between plane YBC and the horizontal.
                                                                 (A: 4.6, 4.7)

*12 In the triangle ABC the sides AB and AC are of lengths $2k$ and $3k$ respectively. The point P on the side AB at a distance x from A is joined to the point Q on the side AC at a distance y from A. If x and y vary in such a way that the area of triangle APQ is always equal to half the area of triangle ABC, show that $y = 3k^2/x$. State the range of possible values of x. Find the minimum value of $x + y$ in terms of k (J: 3.6, 4.6)

*13

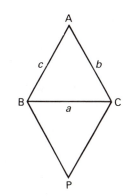

ABC is a triangle with sides of lengths a, b, c opposite $\hat{A}, \hat{B}, \hat{C}$ respectively. The point P is on the opposite side of BC to A, as shown in the diagram, and the triangle BCP is equilateral. Write down an expression for AP^2 in terms of a, c and the angle ABC. If the area of the triangle ABC is S, deduce that $AP^2 = \lambda(a^2 + b^2 + c^2) + \mu S$ where λ and μ are numerical constants and find λ and μ (J: 4.1, 4.6)

*14 The points A, B and C have coordinates $(0, 6)(1, -1)$ and $(9, 3)$ respectively. Find the coordinates of the point D for which each of the angles BAD and BCD is 90°. Hence, or otherwise, find the coordinates of the centre and the radius of the circle through A, B and C. Calculate the angle ABC to 2 d.p. Find also, coorect to 3 s.f., the area of the smaller sector cut off from the circle through A, B and C by the radii through A and C

(J: 1.2, 2.1, 2.2, 4.5)

5.1 Function of a Function (or Composite Functions)

Example 1   If $f(x) = x^2$, $g(x) = x + 1$ and $h(x) = \sin x$ find $fg(x)$, $gf(x)$, $gh(x)$, $hf(x)$, $fh(x)$, $f^2(x)$ and $fgh(x)$.

$$fg(x) = f(g(x)) = f(x+1) = (x+1)^2$$
$$gf(x) = g(f(x)) = g(x^2) = x^2 + 1$$
$$gh(x) = g(h(x)) = g(\sin x) = 1 + \sin x$$
$$hf(x) = h(f(x)) = h(x^2) = \sin(x^2)$$
$$fh(x) = f(h(x)) = f(\sin x) = (\sin x)^2 \text{ or } \sin^2 x$$
$$f^2(x) = f(f(x)) = f(x^2) = (x^2)^2 = x^4$$
$$fgh(x) = f(g(h(x))) = f(g(\sin x)) = f(1 + \sin x) = (1 + \sin x)^2$$

Example 2   Express each of the following functions as a composite of simpler functions:

(a) $y = (x + 1)^4$ (b) $y = \cos(x + 2)$ (c) $y = (x - 1)^{1/2}$
(d) $y = \cos(x - 1)^{1/2}$

(a) $y = (x + 1)^4$ is the same as $y = u^4$ where $u = x + 1$
(b) $y = \cos(x + 2)$ is the same as $y = \cos u$ where $u = x + 2$
(c) $y = (x - 1)^{1/2}$ is the same as $y = u^{1/2}$ where $u = x - 1$
(d) $y = \cos(x - 1)^{1/2}$ is the same as $y = \cos u$ where $u = v^{1/2}$ and where $v = x - 1$

Differentiation of Composite Functions

If $y = f(u)$ and $u = g(x)$
[that is, if y is a function of u and u is a function of x]
then

$$\frac{dy}{dx} = \frac{dy}{du} \times \frac{du}{dx} \quad - \text{ The Chain Rule}$$

The chain rule may be extended to fit chains of three or more functions.

e.g.
$$\frac{dy}{dx} = \frac{dy}{du} \times \frac{du}{dv} \times \frac{dv}{dw} \times \cdots \times \frac{d\sim}{dx}$$

Example 3   Find $\dfrac{dy}{dx}$ if $y = (2x + 1)^5$

Let
$$y = u^5 \quad \text{where} \quad u = 2x + 1$$

$$\frac{dy}{du} = 5u^4 \quad \text{and} \quad \frac{du}{dx} = 2$$

$$\therefore \frac{dy}{dx} = 5u^4 \times 2 = \underline{10(2x + 1)^4}$$

Example 4   Differentiate $\sqrt{\left(1 - \dfrac{1}{x^2}\right)^3}$

Let
$$y = \sqrt{(u^3)} = u^{3/2} \quad \text{where} \quad u = 1 - \frac{1}{x^2} = 1 - x^{-2}$$

$$\frac{dy}{du} = \frac{3}{2} \cdot u^{1/2} \quad \text{and} \quad \frac{du}{dx} = \frac{2}{x^3}$$

$$\therefore \frac{dy}{dx} = \frac{3}{2} \cdot u^{1/2} \times \frac{2}{x^3} = \underline{\frac{3}{x^3} \left(1 - \frac{1}{x^2}\right)^{1/2}}$$

Example 5   Given $\left. \begin{matrix} y = 5t^2 - t \\ x = 1 - 3t^3 \end{matrix} \right\}$ find $\dfrac{dy}{dx}$ when $t = 2$.

$$y = 5t^2 - t \qquad x = 1 - 3t^3$$

$$\frac{dy}{dt} = 10t - 1 \qquad \frac{dx}{dt} = -9t^2$$

$$\frac{dy}{dx} = \frac{dy}{dt} \times \frac{dt}{dx} = \frac{10t - 1}{-9t^2}$$

When $t = 2$, $\dfrac{dy}{dx} = \dfrac{20 - 1}{-9 \cdot 4} = \underline{-\dfrac{19}{36}}$

- **Exercise 5.1**   See page 86.

5.2 Alternative Notation

$$[fg(x)]' = \frac{d}{dx}(f(g(x))) = f'(g(x)) \times g'(x).$$

Example 6   Differentiate $\dfrac{1}{1 + x^3}$ w.r.t. x.

If
$$y = \frac{1}{1 + x^3} = (1 + x^3)^{-1}$$

then
$$\frac{dy}{dx} = -1(1 + x^3)^{-2} \times 3x^2 = \underline{-\frac{3x^2}{(1 + x^3)^2}}$$

Example 7   Find (a) $\dfrac{d}{dx}(x^4 + 3x^5)^3$ (b) $\dfrac{d}{dx}(x^2 + x + 1)^4$

(a) $\dfrac{d}{dx}(x^4 + 3x^5)^3 = 3(x^4 + 3x^5)^2 \times (4x^3 + 15x^4)$

(b) $\dfrac{d}{dx}(x^2 + x + 1)^4 = 4(x^2 + x + 1)^3 \times (2x + 1)$

Example 8   Differentiate $\{1 + (x^2 - 1)^3\}^{1/3}$ w.r.t. x

$$\frac{d}{dx}[\{1 + (x^2 - 1)^3\}^{1/3}] = \frac{1}{3}\{1 + (x^2 - 1)^3\}^{-2/3} \times 3(x^2 - 1)^2 \times 2x$$

$$= \frac{2x(x^2 - 1)^2}{\{1 + (x^2 - 1)^3\}^{2/3}}$$

●   Exercise 5.2   See page 87.

5.3 Integration as the Inverse of Differentiation

If $\dfrac{dy}{dx} = x^n$ then $y = \dfrac{x^{n+1}}{n+1} + c$ $n \neq -1$.

This can also be written as:

$$\int x^n \, dx = \frac{x^{n+1}}{n+1} + c$$

$\int \ldots dx$ is read as 'the integral of ... w.r.t. x'

Example 9   Find the equation of the curve, with gradient $1 - 3x^2$, which passes through the point $(0, 1)$.

We know that $\dfrac{dy}{dx} = 1 - 3x^2$

$$\therefore \quad y = x - x^3 + c$$

Since the curve passes through $(0, 1)$, $y = 1$ when $x = 0$ and so $c = 1$.
∴ the equation of the curve is $y = x - x^3 + 1$.

Example 10   The speed of a body moving in a straight line is given by $v = 12t - 6t^2$. Find s, the distance of the body from a fixed point after 2s.

We know that $v = \dfrac{ds}{dt} = 12t - 6t^2$

$$\therefore \quad s = 6t^2 - 2t^3 + c$$

when $r = 0$, $s = 0$

$$\therefore c = 0$$
$$\therefore s = 6t^2 - 2t^3$$

when $t = 2$

$$s = 8$$

Example 11   Integrate the following expressions with respect to x:

(a) $5x + 3$ (b) $x^4 + \dfrac{1}{x^2} - \sqrt{x}$ (c) $\left(x + \dfrac{1}{x}\right)^2$

(a) $\displaystyle\int (5x + 3)\,dx = \tfrac{5}{2}x^2 + 3x + c$

(b) $\displaystyle\int \left(x^4 + \dfrac{1}{x^2} - \sqrt{x}\right)dx = \int (x^4 + x^{-2} - x^{1/2})\,dx$

$$= \frac{x^5}{5} + \frac{x^{-2}}{-2} - \frac{x^{3/2}}{\frac{3}{2}} + c$$

$$= \frac{x^5}{5} - \frac{1}{2x^2} - \frac{2x^{3/2}}{3} + c$$

> Always write each term in the form x^n before integrating

(c) $\displaystyle\int \left(x + \dfrac{1}{x}\right)^2 dx$

$$= \int (x^2 + 2 + x^{-2})\,dx$$

$$= \frac{x^3}{3} + 2x - \frac{1}{x} + c$$

Integrating $(ax + b)^n$

$$\frac{d}{dx}(ax + b)^{n+1} = (n + 1)(ax + b)^n \times a$$

$$\therefore\quad \boxed{\int (ax + b)^n\,dx = \frac{1}{a(n + 1)}(ax + b)^{n+1} + c}$$

Example 12   Evaluate the following integrals.

(a) $\displaystyle\int (2x - 3)^5\,dx = \frac{1}{(6)(2)}(2x - 3)^6 + c = \tfrac{1}{12}(2x - 3)^6 + c$

(b) $\displaystyle\int (2 - 5x)^7\,dx = \frac{1}{(-5)(8)}(2 - 5x)^8 + c = -\tfrac{1}{40}(2 - 5x)^8 + c$

(c) $\displaystyle\int (\tfrac{1}{4} - \tfrac{2}{5}x)^3\,dx = \frac{1}{(4)(-\frac{2}{5})}(\tfrac{1}{4} - \tfrac{2}{5}x)^4 + c = -\tfrac{5}{8}(\tfrac{1}{4} - \tfrac{2}{5}x)^4 + c$

- **Exercise 5.3**   See page 87.

5.4 Definite Integrals: Area under a Curve

Definite Integrals

$$\int_a^b f(x)\,dx = \left[\text{value of }\int f(x)\,dx \text{ when } x = b\right]$$
$$-\left[\text{value of }\int f(x)\,dx \text{ when } x = a\right]$$

Example 13   Evaluate $\int_1^2 (4x - 1)\,dx$.

$$\int_1^2 (4x - 1)\,dx = [2x^2 - x + c]_1^2$$
$$= [\text{value of }(2x^2 - x + c) \text{ when } x = 2]$$
$$- [\text{value of }(2x^2 - x + c) \text{ when } x = 1$$
$$= [8 - 2 + c] - [2 - 1 + c] = \underline{5}$$

Note: the constant of integration always cancels out so in future we will omit it.

Example 14   Evaluate $\int_2^3 \left(x + \dfrac{1}{x^2}\right)dx$

$$\int_2^3 \left(x + \frac{1}{x^2}\right)dx = \int_2^3 (x + x^{-2})\,dx = \left[\frac{x^2}{2} + \frac{x^{-1}}{-1}\right]_2^3 = \left[\frac{x^2}{2} - \frac{1}{x}\right]_2^3$$
$$= [\tfrac{9}{2} - \tfrac{1}{3}] - [2 - \tfrac{1}{2}] = \underline{\underline{2\tfrac{2}{3}}}$$

Area under a Curve

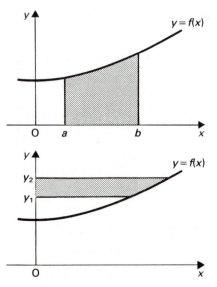

Area between the curve $y = f(x)$ and the x-axis and between $x = a$ and $x = b$ is given by:

$$\text{Area} = \int_a^b y\,dx = \int_a^b f(x)\,dx$$

Area between the curve $y = f(x)$ and the y-axis and between $y = y_1$ and $y = y_2$ is given by:

$$\text{Area} = \int_{y_1}^{y_2} x \cdot dy$$

Example 15 Find the area contained between the curve $y = 3x^2$ and
(a)   the x-axis and the ordinates $x = 1$
and $x = 3$ (b)   the y-axis and the abscissae $y = 1$ and $y = 4$.

(a) $A_1 = \displaystyle\int_a^b y\,dx$

$= \displaystyle\int_1^3 3x^2\,dx = [x^3]_1^3$

$= 27 - 1 = 26$ unit^2

(b) $A_2 = \displaystyle\int_{y_1}^{y_2} x\cdot dy$

$= \displaystyle\int_1^4 \sqrt{\frac{y}{3}}\cdot dy = \int_1^4 \frac{1}{\sqrt{3}}\cdot y^{1/2}\,dy = \frac{1}{\sqrt{3}}\left[\frac{y^{3/2}}{3/2}\right]_1^4$

$= \dfrac{1}{\sqrt{3}}[\frac{2}{3}\cdot 8 - \frac{2}{3}\cdot 1]$

$= \dfrac{14}{3\sqrt{3}} = \dfrac{14\sqrt{3}}{9}$ units^2

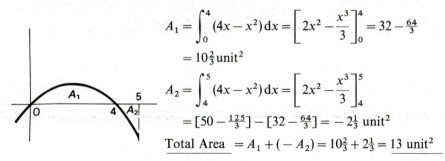

The Sign of the Area

Example 16   Find the area enclosed between the curve $y = x(4 - x)$, the
x-axis and the ordinates $x = 0$ and $x = 5$.

$A_1 = \displaystyle\int_0^4 (4x - x^2)\,dx = \left[2x^2 - \frac{x^3}{3}\right]_0^4 = 32 - \frac{64}{3}$

$= 10\frac{2}{3}$ unit^2

$A_2 = \displaystyle\int_4^5 (4x - x^2)\,dx = \left[2x^2 - \frac{x^3}{3}\right]_4^5$

$= [50 - \frac{125}{3}] - [32 - \frac{64}{3}] = -2\frac{1}{3}$ unit^2

Total Area $= A_1 + (-A_2) = 10\frac{2}{3} + 2\frac{1}{3} = 13$ unit^2

● **Exercise 5.4**   See page 89.

5.5 Volumes of Revolution

The volume obtained by rotating the portion of the curve between $x = a$ and $x = b$ about the x-axis is given by

$$V = \pi \int_a^b y^2 \, dx$$

The volume obtained by rotating the portion of the curve between $y = y_1$ and $y = y_2$ about the y-axis is given by

$$V = \pi \int_{y_1}^{y_2} x^2 \, dy$$

Example 17   The arc of the curve $y^2 = 4x$ between $(0,0)$ and $(1,2)$ is rotated about (a)   the x-axis and (b)   the y-axis. Find the volumes generated in each case.

(a)   Volume $= \pi \int_0^1 y^2 \, dx = \pi \int_0^1 4x \, dx = \pi[2x^2]_0^1 = 2\pi \text{ units}^3$

(b)   Volume $= \pi \int_0^2 x^2 \, dy = \pi \int_0^2 \frac{y^4}{16} \, dy = \pi\left[\frac{y^5}{80}\right]_0^2 = \frac{2\pi}{5} \text{ units}^3$

Example 18   The area contained between the curve $y^3 = x^2$ and the ordinates $x = 1$ and $x = 8$ is rotated about the x-axis. Find the volume of the solid formed.

$$\text{Volume} = \pi \int_1^8 y^2 \, dx = \pi \int_1^8 (x^{2/3})^2 \, dx = \pi \int_1^8 x^{4/3} \, dx$$

$$= \pi\left[\frac{3x^{7/3}}{7}\right]_1^8 = \frac{3\pi}{7}[8^{7/3} - 1^{7/3}] = \frac{3\pi}{7}[2^7 - 1]$$

$$= \frac{3\pi}{7} \times 127 = \frac{381\pi}{7} \text{ units}^3$$

- Exercise 5.5   See page 90.

5.6 Areas Between Curves and Lines: Revision of Unit

Example 19   Find the area of the segment between the line $y = \dfrac{x}{2}$ and the curve $y = (4 - x)x$.

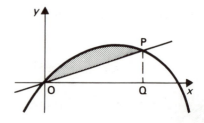

Shaded area = (area under curve between O and Q) − (area △ OPQ). First we need the coordinates of P.

At P
$$4x - x^2 = \frac{x}{2} \Rightarrow x^2 - \frac{7x}{2} = 0$$

$$\Rightarrow x(x - \tfrac{7}{2}) = 0$$

$$\Rightarrow x = 0 \quad \text{or} \quad x = \tfrac{7}{2}.$$

∴ P has x-coordinate $\tfrac{7}{2}$ and y-coordinate $\tfrac{7}{4}$.

Area under curve is
$$\int_0^{7/2} (4x - x^2)\,dx = \left[2x^2 - \frac{x^3}{3} \right]_0^{7/2}$$

$$= \tfrac{49}{2} - \tfrac{343}{24} = \tfrac{245}{24}$$

Area of △ OPQ is
$$\tfrac{1}{2} \cdot \tfrac{7}{2} \cdot \tfrac{7}{4} = \tfrac{49}{16}.$$

∴ Area of shaded segment $= \tfrac{245}{24} - \tfrac{49}{16} = \tfrac{343}{48}$ units^2

Example 20   Find the area enclosed between the curve $y = x^4$ and $y = x^3$

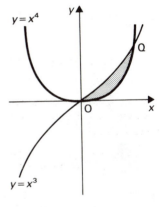

First we need the coordinates of Q.

At Q $x^4 = x^3 \Rightarrow x^3(x - 1) = 0$

$$\Rightarrow x = 0 \text{ (at O) and } x = 1 \text{ (at Q)}.$$

∴ Q is $(1, 1)$.

$$\text{Shaded area} = \int_0^1 x^3\,dx - \int_0^1 x^4\,dx$$

$$= \int_0^1 (x^3 - x^4)\,dx = \left[\frac{x^4}{4} - \frac{x^5}{5} \right]_0^1$$

$$= \tfrac{1}{4} - \tfrac{1}{5} \qquad\qquad = \tfrac{1}{20}$$

- Miscellaneous Exercise 5A   See page 91.

5.7 + A Level Questions

- Miscellaneous Exercise 5B   See page 93.

Unit 5 EXERCISES

Exercise 5.1

1 If $f(x) = x^2 - 1$ and $g(x) = \cos x$, find:

(a) $fg(x)$ (b) $gf(x)$ (c) $g^2(x)$

2 If $f(x) = 2x$, $g(x) = x^2$ and $h(x) = x + 1$, find:

(a) $fg(x)$ (b) $gf(x)$ (c) $hg(x)$ (d) $fgh(x)$

(e) $g^2(x)$ (f) $hgf(x)$ (g) $fhf(x)$

In questions 3–11 express each function as a composite of simpler functions:

3 $y = (x - 3)^2$ **4** $y = \sin(x^2)$

5 $y = \sin^2 x$ **6** $y = \cos(x - 1)$

7 $y = \cos(\sin x^2)$ **8** $y = e^{(x+3)}$

9 $y = \log(x - 3)^2$ **10** $y = \log(\cos(\sin(x + 1)))$

11 $y = (2x + 1)^2$

12 Express y in terms of u, given that $u = 2x + 1$ and:

(a) $y = (2x + 1)^3$ (b) $y = \sqrt{(2x + 1)}$ (c) $y = \dfrac{1}{2x + 1}$

(d) $y = x + \frac{1}{2}$ (e) $y = (x + \frac{1}{2})^2$

13 Using the chain rule, differentiate each of the functions in question 12 with respect to x

14 Find dy/dx in the following cases (in terms of x):

(a) $\begin{aligned} y &= 5u^{11} \\ u &= 4x + 3 \end{aligned}$ (b) $\begin{aligned} y &= 3u^3 \\ u &= 1 - 4x^2 \end{aligned}$ (c) $\begin{aligned} y &= 4u^4 \\ u &= 2x^2 - 1 \end{aligned}$

15 Differentiate:

(a) $(x + 5)^3$ (b) $(2x - 1)^5$ (c) $(1 - 3x)^6$

(d) $(3x^2 - 6)^4$ (e) $\sqrt{(1 + 3x)}$ (f) $\sqrt{(1 + 5x^2)}$

(g) $\sqrt[3]{(3x - x^2)}$

(h) $(x^5 + 1)^{2/3}$

16 If $y = 2t^2 - t$ and $x = 3t + 1$, find dy/dt and dx/dt. Deduce the value of dy/dx in terms of t.

17 Find the values of dy/dx when $t = -1$ in the following cases:

(a) $\begin{aligned} y &= 2t \\ x &= t^2 \end{aligned}$ (b) $\begin{aligned} y &= 4 - 3t^2 \\ x &= t^2 - t + 1 \end{aligned}$ (c) $\begin{aligned} y &= 1 - 1/t \\ x &= 1/t^2 \end{aligned}$

Exercise 5.2

Differentiate w.r.t. x:

1 $(x-1)^5$ **2** $(2x-1)^4$ **3** $(2-3x)^5$ **4** $(3x^2+x)^2$

5 $(x^2+1)^3$ **6** $(x+1)^{-1}$ **7** $\dfrac{1}{2x+1}$ **8** $(x^2-x)^{-2}$

9 $\dfrac{4}{(x^2-1)^3}$ **10** $\sqrt{(2x+1)}$ **11** $\sqrt[3]{(x^2-3x)}$ **12** $\dfrac{1}{\sqrt{x-1}}$

13 $\dfrac{1}{(4-x^2)^{1/3}}$ **14** $(1+\sqrt{x})^3$ **15** $\dfrac{1}{(1+2\sqrt{x})^{1/4}}$

16 $\dfrac{1}{(3+5\sqrt{x})^3}$ ***17** $\sqrt{\{(2x-3)^4-1\}}$ ***18** $\{(x^2+1)^5-2x\}^2$

*19 $\dfrac{1}{4+\sqrt{(4x^2+1)}}$ ***20** $\{(x+1)^3-2(3x-1)^2\}^{-2}$

21 Find the gradient of the curve $y=(2x^2-1)^3$ at the point where $x=-1$. Hence find the equation of the tangent to the curve at this point

22 Find the equation of the normal to the curve $y=\sqrt{(x^2+1)}$ at the point where $x=2$

*23 A particle moves along a straight line so that after t seconds its displacement s metres, from a fixed point on the line is given by $s^3=3t-2$. Find the velocity and acceleration of the particle when $t=1$

*24 An object moves along a straight line such that after t seconds its velocity is v m/s, where

$$v=t+\frac{4}{t+1}$$

Find an expression for its acceleration after t seconds. Find also the minimum speed of the object during its motion.

Exercise 5.3

1 Find y if:
   (a) $dy/dx=5x^4$ (b) $dy/dx=x^4$ (c) $dy/dx=3x^4$

2 If $dy/dx=5x^3$ find y, given that $y=1$ when $x=1$

3 If $ds/dt=3-t^2$, find s in terms of t, given that $s=5$ when $t=0$

In questions 4–15 integrate the following expressions with respect to x:

4 $x^5,\ x^{11},\ x^{-4},\ \dfrac{1}{x^3},\ x^{3/2},\ x^{-3/2},\ \sqrt[5]{x},\ \dfrac{2}{x^2},\ \dfrac{3}{x^7},\ \dfrac{1}{5x^4},\ \dfrac{5}{x^4},\ \sqrt{x},\ \dfrac{1}{\sqrt{x}},\ \dfrac{1}{2\sqrt{x}}$

5 $4x^3 - x^2$

6 $3 - x^{-3}$

7 $3x^2 - \dfrac{1}{x^2} + 3x$

8 $5x^3 + x^{-5} + 5$

9 $\dfrac{3}{x^4} - \dfrac{1}{x^2} - x$

10 $\sqrt{x} - (\sqrt{x})^3$

11 $2x^{3/2} - x^{-5/2}$

12 $4x^{-3} + x^{-4} + 2$

13 $(2x - 1)(1 - x)$

14 $\dfrac{x^3 - 1}{x^2}$

15 $\left(x + \dfrac{1}{x}\right)^2$

In questions 16–33 evaluate the following integrals:

16 $\displaystyle\int 3\,dx$

17 $\displaystyle\int 3x^4\,dx$

18 $\displaystyle\int (3x)^4\,dx$

19 $\displaystyle\int 3(x + 1)^4\,dx$

20 $\displaystyle\int (3x + 1)^4\,dx$

21 $\displaystyle\int x^2\,dx$

22 $\displaystyle\int (x + 1)^2\,dx$

23 $\displaystyle\int (x + 3)^2\,dx$

24 $\displaystyle\int (2x + 1)^2\,dx$

25 $\displaystyle\int (5x + 4)^2\,dx$

26 $\displaystyle\int (2 + 3x)^4\,dx$

27 $\displaystyle\int (3 - x)^3\,dx$

28 $\displaystyle\int 4(2 - x)^{1/2}\,dx$

29 $\displaystyle\int 5(3 + x)^{1/2}\,dx$

30 $\displaystyle\int \sqrt[3]{(1 + 2x)}\,dx$

31 $\displaystyle\int \dfrac{1}{(2x + 3)^2}\,dx$

32 $\displaystyle\int \dfrac{1}{\sqrt{(1 - 3x)}}\,dx$

33 $\displaystyle\int (qx + r)^p\,dx$

34 Find the equation of the curve with gradient $x^2 - x + 1$ which passes through the point $(6, 10)$

*35 Find $\displaystyle\int (4\sqrt{x} + \sqrt{4x + 1} - 4(1 - 2x)^4)\,dx$

*36 Find $\displaystyle\int \sqrt{(2 - x)} + \dfrac{1}{\sqrt{(2 - x)}} - \dfrac{1}{(2 - x)^2}\cdot dx$

Exercise 5.4

Evaluate the following definite integrals:

1 $\displaystyle\int_0^1 x^3\,dx$ **2** $\displaystyle\int_1^2 4x^5\,dx$ **3** $\displaystyle\int_1^9 \sqrt{x}\,dx$

4 $\displaystyle\int_{-2}^{-1} \frac{1}{x^2}\,dx$ **5** $\displaystyle\int_{-1}^1 (4x^2-1)\,dx$ **6** $\displaystyle\int_0^2 (2x^3-x)\,dx$

7 $\displaystyle\int_1^3 \left(x^2-\frac{1}{x^2}\right)dx$ **8** $\displaystyle\int_{-2}^2 (2x-1)(x+1)\,dx$

9 $\displaystyle\int_1^2 x(x-1)(x-2)\,dx$ **10** $\displaystyle\int_1^4 \frac{x+1}{\sqrt{x}}\,dx$

11 Sketch the curve $y=2x^2$. Find the area between the curve and
   (a) the x-axis and the ordinates $x=0$, $x=3$
   (b) the y-axis and the abscissae $y=1$, $y=4$

12 Find the areas bounded by the following curves, the x-axis and the ordinates $x=1$ and $x=3$. (None of these curves crosses the x-axis within this interval)
   (a) $y=4x^3$ (b) $y=2/x^2$ (c) $y=(1-x)^2$ (d) $y=4\sqrt{x}$

13

$y=x(2-x)$ (a) What are the coordinates of P?
   (b) Find the area under the curve above the x-axis

14 Sketch the curve $y=(1-x)(2+x)$ and find the area between the curve and the x-axis which lies above the axis

15 For what values of x is y negative if $y=(3x-1)(x-2)$? Find the area cut off by the curve below the x-axis

In questions 16–19 evaluate:

16 $\displaystyle\int_1^3 (s^2-s)\,ds$ **17** $\displaystyle\int_1^4 \left(t+\frac{1}{t}\right)^2 dt$

18 $\displaystyle\int_2^4 \frac{p^2-1}{p^2}\,dp$ **19** $\displaystyle\int_1^2 \frac{(q-1)^2}{q^4}\,dq$

20 Find the total area enclosed between the curve and the x-axis for $y=(x-1)(x-3)$

21   Sketch the curve $y = 3x - x^2$. Find the area contained between the curve, the x-axis and the ordinates $x = 0$ and $x = 5$

Evaluate $\int_0^5 (3x - x^2)\,dx$ and explain the result

Exercise 5.5

Find the volumes of the solids generated when the following areas are rotated about the given axis: (Leave answers as multiples of π).

1   The area between $y = 3x$ and the ordinates $x = 0$ and $x = 2$ is rotated about the x-axis

2   The area between $y = 2\sqrt{x}$ and the ordinates $x = 2$, $x = 3$ is rotated about the x-axis

3   The area between $y = 2/x$ and the ordinates $x = 1$ and $x = 3$ is rotated about the x-axis

4   The area between the line $y = 2x + 1$ and the ordinates $x = -1$ and $x = 1$, is rotated about the x-axis.

5   The area between $x^2 + y^2 = 16$ and the abscissae $y = 0$ and $y = 4$ is rotated about the y-axis

6   The area contained between the curve $y = 1 + \sqrt{x}$ and the ordinates $x = 0$, $x = 4$ is rotated about the x-axis

7   The area between the curve $y^2 = 1 + x^2$ and the abscissae $y = -2$ and $y = -1$ is rotated about the y-axis

8   The area between $y\sqrt{x^3} = 2$ and the ordinates $x = 1$, $x = 3$ is rotated about the x-axis

9   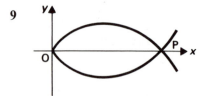   The diagram shows part of the curve $y^2 = x(x - 3)^2$. Find the coordinates of P and the volume of the solid formed when the loop is rotated about the x-axis

*10   Find the volume generated by revolving the curve $y = (x - 2)(x + 2)$ about the x-axis

*11   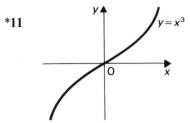   The curve $y = x^3$ between $x = -2$ to $x = +2$ is rotated through 360° around the y-axis, to form an hour glass shaped solid. Find its volume

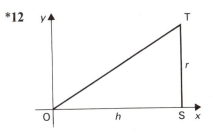

*12 $\triangle$ OST is right angled with OS $= h$, ST $= r$. If the $\triangle$ is rotated about the x-axis what solid is generated? What is the gradient of OT? What is the equation of OT? Deduce the volume of a right circular cone, base radius r and height h

Miscellaneous Exercise 5A

1 Find the area enclosed between the curves $y = x^2$ and $y = x^4$

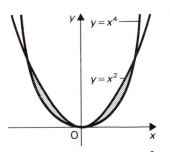

2 Find the area of the segment cut off on the curve $y = 6x - x^2$ by the line $y = 8$

3 Find the area of the segment cut off on the curve $y = x(4 - x)$ by the line $y = 2x$

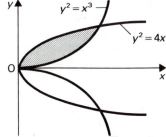

4 Find the shaded area between the curves $y^2 = 4x$ and $y^2 = x^3$

5 Differentiate:

(a) $(4x + 9)^7$ (b) $(x^3 + 1)^3$ (c) $\sqrt[3]{(3 - 2x^2)}$

(d) $\dfrac{2}{1 + \sqrt{x}}$ (e) $\sqrt{\left(\dfrac{x + 3}{x}\right)}$ (f) $\dfrac{4}{(1 - 2x)^2}$

6 If $u = 3x - 7$, state in terms of u

(a) $y = (3x - 7)^2$ (b) $y = \dfrac{3}{3x - 7}$ (c) $y = x - \frac{7}{3}$

(d) $y = 3x$

7 Integrate with respect to:

(a) $3x^2 + 2x - 1$ (b) $\sqrt{x} + \dfrac{4}{\sqrt{x}}$ (c) $\dfrac{1}{3x^2} + x + 3x^2$

8 Evaluate:

(a) $\displaystyle\int \frac{3x^3 + 4x^2 + 1}{x^2}\,dx$
(b) $\displaystyle\int_1^4 \sqrt{x}\,dx$
(c) $\displaystyle\int_1^3 (x^2 - x)\,dx$

(d) $\displaystyle\int_{-2}^{-1} \left(x - \frac{1}{x^2}\right)dx$
(e) $\displaystyle\int_a^b (ax^2 + bx)\,dx$

9 Calculate the area enclosed by $y = x^2/4 - x$, the ordinates $x = 0$ and $x = 6$ and the x-axis

**10 Calculate the area enclosed by the curve $xy^2 = c^3$, the y-axis and the lines $y = c$ and $y = 2c$. Calculate also the area above the x-axis enclosed by the curve, the x-axis and the lines $x = b$, $x = 2b$. If these two areas are equal calculate the ration $b:c$

*11

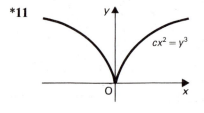

The area bounded by the curve $cx^2 = y^3$ and the line $y = k$ is rotated about the y-axis to give a volume of revolution equal to $\pi c^3/64$

Find the value of k in terms of c

**12 Calculate the area bounded by the curve $y = x^2 + 1$, the tangent to the curve at $(2, 5)$ and the x- and y-axis

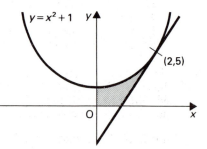

*13 The curve $y = (x + 2)(x + 1)(x - 3)$ cuts the x-axis at A, B and C in that order. Calculate the magnitude of the total area enclosed by the curve from A to C and the x-axis

***14 **A challenge** Find the area (shaded) between the curve and the line

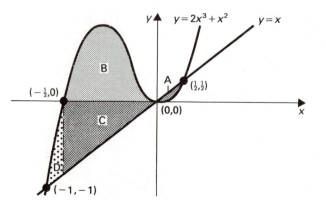

**15

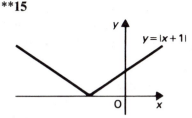

Find $\displaystyle\int_{-3}^{3} |x+1| \, dx$

Miscellaneous Exercise 5B

1 Show that the area of the finite region enclosed by the line $y = 4x$ and the curve $y^2 = 16x$ is $\frac{2}{3}$ (L: 5.4)

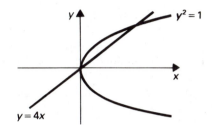

2 Find the equation of the chord which joins the points A$(-2, 3)$ and B$(0, 15)$ on the curve $y = 15 - 3x^2$

   (a) Show that the finite area enclosed by the curve and the chord AB is 4 square units

   (b) Find the volume generated when this area is rotated through $360°$ about the x-axis, leaving your answer in terms of π (A: 5.4, 5.5)

3 A closed hollow right-circular cone has internal height a and the internal radius of its base is also a. A solid circular cylinder of height h just fits inside the cone with the axis of the cylinder lying along the axis of the cone. Show that the volume of the cylinder is $V = \pi h(a - h)^2$

   If a is fixed, but h may vary, find h in terms of a when V is a maximum
    (J: 3.6)

4 A cylindrical vessel, closed at both ends, is made of thin material and contains a volume $16\pi \, m^3$. Given that the total exterior surface area of the vessel is a minimum, find its height and base radius   (L: 3.6)

5 A circular hollow cone of height h and semi-vertical angle α stands on a horizontal table. A circular cylinder placed in an upright position just fits inside the space between the cone and the table, as shown in the diagram. Prove that the maximum possible volume of the cylinder is

$$\frac{4\pi h^3 \tan^2 \alpha}{27}$$

(J: 3.6)

6   Show that the graph of the function $y = x^5 - 16(x - 1)^5$ has two stationary points only, one at $x = \frac{2}{3}$ and the other at $x = 2$. At each of these points, determine whether y has a maximum or minimum value. Sketch the graph of the function. Find the area of the region bounded by the graph, the x-axis and the ordinates $x = 0$ and $x = 1$ (J: 3.4, 5.4)

7   From a large thin plane sheet of metal it is required to make an open rectangular box with a square base so that the box would contain a given volume V. Express the area of the sheet used in terms of V and x, where x is the length of a side of the square base.

Hence find the ratio of the height of the box to x in order that the box consists of a minimum area of the metal sheet   (L: 3.6)

*8   With O as origin, the fixed point F is (a, b) where $a > 0$ and $b > 0$. The variable point P is $(p, 0)$, where $p > a$. The line PF when produced meets the y-axis at the point Q

(a)   Prove that the area of the triangle OPQ is $bp^2/2(p - a)$
(b)   As p varies, find the least area of the triangle OPQ in terms of a and b
                                                                                  (A: 2.2, 2.3, 3.6)
[Hint: If A is the area, it is easier to find the maximum value of $1/A$]

*9   The volume of a right circular cone is $18\pi\,\text{cm}^3$. Given that the vertical height is $h\,\text{cm}$, show that the slant height is $(h^2 + 54/h)^{1/2}\,\text{cm}$. As h varies, find the minimum slant height, verifying that it is a minimum. (Volume of a cone $= \frac{1}{3}\pi r^2 h$)   (A: 3.6)

6.1 Arithmetic Series (Arithmetic Progressions)

If the first term of an arithmetic progression is a and the common difference is d, the progression becomes

$$a, a + d, a + 2d, \ldots, \underbrace{a + (n - 1)d}_{n\text{th term}}$$

Example 1 Find (a) the first three terms
(b) the seventh term
(c) the nth term
of the arithmetic progression whose first term is -5 and whose common difference is 3.

(a) First three terms are $-5, -5 + 3, -5 + 6$, i.e. $\underline{-5, -2, 1}$

(b) The seventh term is $-5 + 6 \cdot (3) = \overline{13}$

(c) The nth term is $-5 + (n - 1)3 = \underline{3n - 8}$

The Sum of an A.P.

If an A.P. has first term a, last term l and common difference d, then the sum of n terms is

$$S_n = a + (a + d) + (a + 2d) + \cdots + (l - d) + l$$

Also $\qquad S_n = l + (l - d) + (l - 2d) + \cdots + (a + d) + a$

Adding $\qquad 2S_n = n(a + l)$

$$\therefore \boxed{S_n = \tfrac{1}{2}n(a + l)}$$

But if we only know the first term and the common difference then $l = n$th term $= a + (n - 1)d$.

$$\therefore S_n = \tfrac{1}{2}n[a + (a + (n - 1)d)]$$

$$\therefore \boxed{S_n = \tfrac{1}{2}n[2a + (n - 1)d]}$$

Note These results may be quoted, unless you are asked to derive them.

Example 2   How many terms are there in the A.P. $3, 9, 15, \ldots, 69$.

$$a = 3, \quad d = 6 \quad \therefore \text{nth term} = 3 + (n-1)6 = 69$$
$$\Rightarrow \quad (n-1)6 = 66$$
$$n - 1 \quad = 11$$
$$\therefore n \quad = 12$$

$\therefore$ there are 12 terms in the A.P.

Example 3   If the seventh term of an A.P. is 15 and the twelfth term is 35, find the first term, the common difference and the sum to 15 terms.

$$\text{Seventh term} = a + 6d = 15$$
$$\text{twelfth term} = a + 11d = 35$$
$$\text{subtracting} \qquad \overline{5d = 20}$$
$$d = 4$$

Hence $a = 15 - 6d = 15 - 6 \cdot 4 = -9$.
$\therefore$ first term is -9 and the common difference is 4

Sum to 15 terms is

$$S_{15} = \tfrac{15}{2}[2(-9) + 14.4]$$
$$S_{15} = \tfrac{15}{2}[-18 + 56] = 285$$
$$\therefore S_{15} = 285$$

Example 4   The nth term of an A.P. is given by $5 + 2n$. Find the first term and the common difference.

When $n = 1$ 1st term $= 5 + 2 \cdot 1 = 7$
When $n = 2$ 2nd term $= 5 + 2 \cdot 2 = 9$
$\therefore$ 1st term is 7 and common difference is 2

● Exercise 6.1   See page 108.

6.2 Geometric Series (Geometric Progressions)

In a G.P. each term is found by multiplying the preceding term by a constant, called the common ratio. If a is the first term and r is the common ratio, the series is

$$\boxed{a, ar, ar^2, \ldots, \underbrace{ar^{n-1}}_{\text{nth term}}}$$

The Sum of a G.P.

Let $S_n = a + ar + ar^2 + \cdots + ar^{n-1}$
and $rS_n = \qquad ar + ar^2 + \cdots + ar^{n-1} + ar^n$
Subtracting: $(1 - r)S_n = a - ar^n = a(1 - r^n)$

$$\therefore \boxed{S_n = \frac{a(1 - r^n)}{(1 - r)} \quad r < 1} \quad \text{and} \quad \boxed{S_n = \frac{a(r^n - 1)}{(r - 1)} \quad r > 1}$$

Example 5   The fifth term of a G.P. is 48, the third term is 12 and the fourth term is positive. Find the first term, the common ratio and the sum of the first ten terms.

$$5\text{th term} = ar^4 = 48 \tag{1}$$
$$3\text{rd term} = ar^2 = 12. \tag{2}$$

Dividing (1) by (2) gives

$$\frac{ar^4}{ar^2} = \frac{48}{12}$$

$$\Rightarrow r^2 = 4$$

$$\therefore r = \pm 2.$$

but, the 3rd, 4th and 5th terms are all positive $\therefore r$ is positive.
$\therefore$ common ratio $= 2$

$(2) \Rightarrow a = $ 1st term $= 3$.

$$S_{10} = \frac{3(2^{10} - 1)}{2 - 1} = 3069$$

$\therefore$ sum of first ten terms $= 3069$

The Sum of an Infinite G.P.

$$\boxed{S_\infty = \frac{a}{1 - r} \quad \text{provided } |r| < 1}$$

Example 6   Find the sum to infinity of the series $50 + 25 + 12\frac{1}{2} + \cdots$

$$r = \tfrac{1}{2} \quad \therefore |r| < 1.$$

$\therefore$ the series converges and the sum to infinity exists.

$$S_\infty = \frac{50}{1 - \frac{1}{2}} = 100$$

$\therefore$ the sum to infinity $= 100$

Example 7   For the series $x + 1 + \dfrac{1}{x} + \dfrac{1}{x^2} + \cdots$

(a) find the sum of the first n terms
(b) find the range of values of x for which the sum to infinity exists
(c) find the sum to infinity.

(a) $x + 1 + \dfrac{1}{x} + \dfrac{1}{x^2} + \cdots$ is a G.P. with first term x and common ratio $\dfrac{1}{x}$.

$$\therefore S_n = \frac{x\left(\left(\dfrac{1}{x}\right)^n - 1\right)}{\left(\dfrac{1}{x} - 1\right)} = \frac{x\left(\dfrac{1 - x^n}{x^n}\right)}{\dfrac{1 - x}{x}} = \frac{x(1 - x^n)}{x^n} \times \frac{x}{1 - x}$$

$$\therefore S_n = \frac{(1 - x^n)}{x^{n-2}(1 - x)}$$

(b) S_∞ exists provided $\left|\dfrac{1}{x}\right| < 1$ i.e. $\underline{|x| > 1}$

(c) $\underline{S_\infty} = \dfrac{x}{1 - \dfrac{1}{x}} = \dfrac{x}{\dfrac{x - 1}{x}} = \underline{\dfrac{x^2}{x - 1}}$

- **Exercise 6.2**   See page 109.

6.3 The $\sum$ notation

Example 8   What expressions are represented by:

(a) $\displaystyle\sum_{r=1}^{r=4} r^2$ (b) $\displaystyle\sum_{r=1}^{r=2n} (r - 1)$ (c) $\displaystyle\sum_{1}^{20} (r + 1)(r + 2)$

(a) $\displaystyle\sum_{r=1}^{r=4} r^2 = 1^2 + 2^2 + 3^2 + 4^2$

(b) $\displaystyle\sum_{r=1}^{r=2n} (r - 1) = 0 + 1 + 2 + \cdots + (2n - 1)$

(c) $\displaystyle\sum_{1}^{20} (r + 1)(r + 2) = 2\cdot3 + 3\cdot4 + 4\cdot5 + \cdots + 21\cdot22$

Example 9   Use the $\sum$ notation to express the sums of the following series:

(a) $4 + 5 + 6 + \cdots + n$ $= \displaystyle\sum_{4}^{n} r$

(b) $1 + \frac{1}{2} + \frac{1}{3} + \frac{1}{4} + \cdots$ $= \displaystyle\sum_{1}^{\infty} \frac{1}{r}$

(c) $1 - 2 + 3 - 4 + 5 + \cdots + (-50) = \displaystyle\sum_{1}^{50} (-1)^{r-1} r$

(d) $2 + 5 + 8 + 11 + \cdots + 38$ $= \displaystyle\sum_{1}^{13} (3r - 1)$

Example 10   What is the 6th term of the series $\sum_{1}^{n}(-1)^r r^2$?

$r = 6$ $\therefore$ the sixth term is $(-1)^6 \, 6^2 = 36$

Example 11   Write down the first four terms of the series given by $\sum_{1}^{n}(3r + 2)$.
Evaluate $\sum_{1}^{n}(3r + 2)$.

$$r = 1 \Rightarrow \text{1st term} = \ \ 3 + 2 = \ \ 5$$
$$r = 2 \Rightarrow \text{2nd term} = \ \ 6 + 2 = \ \ 8$$
$$r = 3 \Rightarrow \text{3rd term} = \ \ 9 + 2 = 11$$
$$r = 4 \Rightarrow \text{4th term} = 12 + 2 = 14$$

$\therefore$ first four terms are $5, 8, 11, 14$

These terms form an A.P. with first term 5 and common difference 3. Hence $\sum_{1}^{n}(3r + 2)$ is the sum of n terms of this A.P.

$$\therefore \ \sum_{1}^{n}(3r + 2) = \frac{n}{2}[10 + (n - 1)3] = \frac{n}{2}[3n + 7]$$

Example 12   Evaluate $\sum_{1}^{12}3(0.5)^r$ to 3.d.p.

$$\sum_{1}^{12}3(0.5)^r = 3(0.5) + 3(0.5)^2 + 3(0.5)^3 + \cdots + 3(0.5)^{12}$$

which is a G.P. with first term 1.5, common ratio 0.5 and has 12 terms.

$$\therefore \sum_{1}^{12}3(0.5)^r = \frac{1.5(1 - 0.5)^{12})}{1 - 0.5} \simeq 2.999$$

- Exercise 6.3   See page 111.

6.4 Proof by Induction

Example 13   Prove, by induction, that $\sum_{1}^{n}r^2 = \frac{1}{6}n(n + 1)(2n + 1)$.

Proof

Stage 1   Let $n = 1$.

$$\left. \begin{array}{l} \text{LHS} = 1^2 = 1 \\ \text{RHS} = \frac{1}{6} \cdot 1 \cdot 2 \cdot 3 = 1 = \text{LHS} \end{array} \right\} \Rightarrow \text{true for } n = 1$$

Stage 2   Assume true for $n = k$
then

$$\sum_1^k r^2 = 1^2 + 2^2 + 3^2 + \cdots + k^2 \qquad = \tfrac{1}{6}k(k+1)(2k+1)$$

$$\sum_1^{k+1} r^2 = 1^2 + 2^2 + 3^2 + \cdots + k^2 + (k+1)^2 = \tfrac{1}{6}k(k+1)(2k+1) + (k+1)^2$$

$$= \tfrac{1}{6}(k+1)[k(2k+1) + 6(k+1)]$$

Take out any common factors and any fractions

$$= \tfrac{1}{6}(k+1)(2k^2 + 7k + 6)$$
$$= \tfrac{1}{6}(k+1)(k+2)(2k+3)$$
$$= \tfrac{1}{6}(k+1)[(k+1)+1][2(k+1)+1]$$

Thus, if it is true for $n = k$, it is also true for $n = k+1$ and, since it is true for $n = 1$, it is also true for $n = 2$, and hence for $n = 3$, $n = 4$…. i.e. all n.

Example 14   Prove that $\dfrac{1}{1 \cdot 2} + \dfrac{1}{2 \cdot 3} + \dfrac{1}{3 \cdot 4} + \cdots + \dfrac{1}{n(n+1)} = \dfrac{n}{n+1}$.

Proof

Stage 1   Let $n = 1$

$$\left.\begin{array}{l} \text{LHS} = \tfrac{1}{1 \cdot 2} = \tfrac{1}{2} \\ \text{RHS} = \tfrac{1}{1+1} = \tfrac{1}{2} = \text{LHS} \end{array}\right\} \Rightarrow \text{true for } n = 1$$

Stage 2   Assume true for $n = k$
then

$$\frac{1}{1 \cdot 2} + \frac{1}{2 \cdot 3} + \frac{1}{3 \cdot 4} + \cdots + \frac{1}{k(k+1)} \qquad = \frac{k}{k+1}$$

and

$$\frac{1}{1 \cdot 2} + \frac{1}{2 \cdot 3} + \frac{1}{3 \cdot 4} + \cdots + \frac{1}{k(k+1)} + \frac{1}{(k+1)(k+2)} = \frac{k}{k+1} + \frac{1}{(k+1)(k+2)}$$

$$= \frac{k(k+2) + 1}{(k+1)(k+2)}$$

$$= \frac{k^2 + 2k + 1}{(k+1)(k+2)}$$

$$= \frac{(k+1)^2}{(k+1)(k+2)}$$

$$= \frac{k+1}{k+2}$$

Thus, if it is true for $n = k$, it is also true for $n = k+1$ and, since it is true for $n = 1$, it is also true for $n = 2$, and hence for $n = 3$, $n = 4$…. i.e. all n.

●   **Exercise 6.4**   See page 113.

6.5 Sums of Powers of Natural Numbers

$$\sum_{1}^{n} r = \tfrac{1}{2}n(n+1)$$

$$\sum_{1}^{n} r^2 = \tfrac{1}{6}n(n+1)(2n+1)$$

$$\sum_{1}^{n} r^3 = \tfrac{1}{4}n^2(n+1)^2 = \left(\sum_{1}^{n} r\right)^2$$

Example 15 Find the sum of the squares of the integers 10 to 20 inclusive.

Sum is

$$10^2 + 11^2 + 12^2 + \cdots + 20^2$$

> Fewer errors occur if you start by writing out the terms to be summed.

$$= \sum_{1}^{20} r^2 - \sum_{1}^{9} r^2$$

$$= \tfrac{1}{6} \cdot 20 \cdot 21 \cdot 41 - \tfrac{1}{6} \cdot 9 \cdot 10 \cdot 19$$

$$= \underline{2585}$$

Example 16 Find the sum of the series $2 + 4 + 6 + \cdots + 90$.

$$2 + 4 + 6 + \cdots + 90 = 2(1 + 2 + 3 + \cdots + 45)$$

$$= 2 \sum_{1}^{45} r$$

$$= 2 \times \tfrac{1}{2} \times 45 \times 46$$

$$= \underline{2070}$$

Sums of Series with rth terms of the form $ar^3 + br^2 + cr + d$

Sum is

$$a \cdot 1^3 + b \cdot 1^2 + c \cdot 1 + d$$
$$+ a \cdot 2^3 + b \cdot 2^2 + c \cdot 2 + d$$
$$+ a \cdot 3^3 + b \cdot 3^2 + c \cdot 3 + d$$
$$+ \cdots$$
$$\vdots$$
$$+ a \cdot n^3 + b \cdot n^2 + c \cdot n + d$$

$$\Rightarrow \text{Sum} = a \sum_{1}^{n} r^3 + b \sum_{1}^{n} r^2 + c \sum_{1}^{n} r + nd$$

Example 17 Evaluate $\sum_{1}^{n} (3r^2 + r - 2)$.

$$\sum_{1}^{n} (3r^2 + r - 2) = 3 \sum_{1}^{n} r^2 + \sum_{1}^{n} r - 2n$$

$$= 3 \cdot \tfrac{1}{6} \cdot n(n+1)(2n+1) + \tfrac{1}{2}n(n+1) - 2n$$
$$= \tfrac{1}{2}n[(n+1)(2n+1) + (n+1) - 4]$$
$$= \tfrac{1}{2}n[2n^2 + 4n - 2]$$
$$= n(n^2 + 2n - 1)$$

Example 18   Find (a)   the rth term and (b)   the sum to n terms of the series $1^2 + 3^2 + 5^2 + \cdots$.

(a) rth term $= (2r - 1)^2 = 4r^2 - 4r + 1$

(b) $S_n = \displaystyle\sum_{1}^{n}(4r^2 - 4r + 1)$

$$= 4\sum_{1}^{n} r^2 - 4\sum_{1}^{n} r + n$$

$$= 4 \cdot \tfrac{1}{6} \cdot n(n+1)(2n+1) - 4 \cdot \tfrac{1}{2} \cdot n(n+1) + n$$

$$= \tfrac{1}{3}n[2(n+1)(2n+1) - 6(n+1) + 3]$$

$$= \tfrac{1}{3}n(4n^2 - 1)$$

- **Exercise 6.5**   See page 113.

6.6 Revision of Positive and Negative Rational Indices

Rules of Indices

1   $a^m \times a^n = a^{m+n}$

2   $a^m \div a^n = a^{m-n}$

3   $a^{-n} = \dfrac{1}{a^n}$

4   $a^0 = 1$

5   $(a^m)^n = a^{mn} = (a^n)^m$

6   $a^{1/n} = \sqrt[n]{a}$ or the nth root of a

7   $a^{m/n} = \sqrt[n]{a^m}$ or $(\sqrt[n]{a})^m$ ⟵ | easiest form to use |

Example 19   Simplify $\left(\dfrac{125}{8}\right)^{-2/3}$.

$$\left(\frac{125}{8}\right)^{-2/3} = \left(\frac{8}{125}\right)^{2/3} = \left(\sqrt[3]{\frac{8}{125}}\right)^2 = \left(\frac{2}{5}\right)^2 = \frac{4}{25}$$

Example 20   Simplify $\dfrac{2x^{-1/2}(x-1)^{1/2} + x^{1/2}(x-1)^{-1/2}}{x^{3/2}}$.

The technique is to remove all the powers with negative indices. So we multiply the numerator and the denominator by powers with positive indices. That is

$$\frac{2x^{-1/2}(x-1)^{1/2}+x^{1/2}(x-1)^{-1/2}}{x^{3/2}} = \left[\frac{2x^{-1/2}(x-1)^{1/2}+x^{1/2}(x-1)^{-1/2}}{x^{3/2}}\right]$$

$$\times\left(\frac{x^{1/2}(x-1)^{1/2}}{x^{1/2}(x-1)^{1/2}}\right)$$

$$=\frac{2(x-1)+x}{x^2(x-1)^{1/2}}$$

$$=\frac{3x-2}{x^2(x-1)^{1/2}}$$

Example 21   Simplify $\dfrac{x^{1/2}(x+1)^{-1/2}+x^{3/2}(x+1)^{-3/2}}{x^{-1/2}}$

> Here we multiply numerator and denominator by $x^{1/2}(x+1)^{3/2}$ as the term $(x+1)^{3/2}$ will simplify $(x+1)^{-1/2}$ and $(x+1)^{-3/2}$.

$$\therefore \frac{x^{1/2}(x+1)^{-1/2}+x^{3/2}(x+1)^{-3/2}}{x^{-1/2}} = \left[\frac{x^{1/2}(x+1)^{-1/2}+x^{3/2}(x+1)^{-3/2}}{x^{-1/2}}\right]$$

$$\times\left(\frac{x^{1/2}(x+1)^{3/2}}{x^{1/2}(x+1)^{3/2}}\right)$$

$$=\frac{x(x+1)+x^2}{(x+1)^{3/2}}$$

$$=\frac{2x^2+x}{(x+1)^{3/2}}$$

- **Exercise 6.6** See page 114.

6.7 Binomial Expansions Using Pascal's △

Pascal's △

$(1+x)^0 = 1$					1			
$(1+x)^1 = 1+x$				1		1		
$(1+x)^2 = 1+2x+x^2$			1		2		1	
$(1+x)^3 = 1+3x+3x^2+x^3$		1		3		3		1
$(1+x)^4 = 1+4x+6x^2+4x^3+x^4$	1		4		6		4		1

This triangle can be extended. Each term in the △ is the sum of the two terms above it.

Example 22   Expand $(1 + 3y)^3$.

From Pascal's $\triangle$ $(1 + x)^3 = 1 + 3x + 3x^2 + x^3$
Replacing x by $3y \Rightarrow (1 + 3y)^3 = 1 + 3(3y) + 3(3y)^2 + (3y)^3$
$$\therefore (1 + 3y)^3 = 1 + 9y + 27y^2 + 27y^3$$

Example 23   Expand $(a + b)^4$.

$$(1 + x)^4 = 1 + 4x + 6x^2 + 4x^3 + x^4$$

$$(a + b)^4 = a^4\left(1 + \frac{b}{a}\right)^4 = a^4\left(1 + 4\left(\frac{b}{a}\right) + 6\left(\frac{b}{a}\right)^2 + 4\left(\frac{b}{a}\right)^3 + \left(\frac{b}{a}\right)^4\right)$$

$$\therefore (a + b)^4 = a^4 + 4a^3b + 6a^2b^2 + 4ab^3 + b^4$$

Note:
Compare $(1 + x)^4 = 1 + 4x + 6x^2 + 4x^3 + x^4$

and $(a + b)^4 = a^4 + 4a^3b + 6a^2b^2 + 4ab^3 + b^4$

(a)   The numerical coefficients of both expansions are the same.
(b)   The sum of the powers of a and b is always 4 in each term
(c)   As the powers of a decrease, the powers of b increase.

Example 24   Expand $(a + b)^6$.

By continuing Pascal's $\triangle$ we find that:

$$(1 + x)^6 = 1 + 6x + 15x^2 + 20x^3 + 15x^4 + 6x^5 + x^6$$
$$\therefore (a + b)^6 = a^6 + 6a^5b + 15a^4b^2 + 20a^3b^3 + 15a^2b^4 + 6ab^5 + b^6$$

Example 25   Expand $(3p - 4q)^3$.

$$(a + b)^3 = a^3 + 3a^2b + 3ab^2 + b^3$$

Replacing a by $3p$ and b by $(-4q)$ we get

$$(3p - 4q)^3 = (3p)^3 + 3(3p)^2(-4q) + 3(3p)(-4q)^2 + (-4q)^3$$
$$\therefore (3p - 4q)^3 = 27p^3 - 108p^2q + 144pq^2 - 64q^3$$

• Exercise 6.7   See page 115.

6.8 The Binomial Theorem I

$$(1 + x)^n = 1 + nx + \frac{n(n - 1)}{2!}x^2 + \frac{n(n - 1)(n - 2)}{3!}x^3 + \cdots + \binom{n}{r}x^r + \cdots$$

where $\dbinom{n}{r} = \dfrac{n(n - 1)(n - 2)\cdots(n - r + 1)}{r!} = \dfrac{n!}{(n - r)!r!}$ and $\dbinom{n}{0} = 1$

If n is a positive integer, the series terminates and is convergent for all x.
If n is not a positive integer, the series is infinite and converges for $|x| < 1$.

Example 26   Evaluate (a) $\dfrac{7!}{4!}$ (b) $\dfrac{10!}{7!\,3!}$.

(a) $\dfrac{7!}{4!} = \dfrac{7 \cdot 6 \cdot 5 \cdot \cancel{4} \cdot \cancel{3} \cdot \cancel{2} \cdot \cancel{1}}{\cancel{4} \cdot \cancel{3} \cdot \cancel{2} \cdot \cancel{1}} = \underline{210}$ (b) $\dfrac{10!}{7!\,3!} = \dfrac{10 \cdot 9 \cdot 8}{1 \cdot 2 \cdot 3} = \underline{120}$

Example 27   Evaluate (a) $\begin{pmatrix} 7 \\ 4 \end{pmatrix}$ (b) $\begin{pmatrix} -1 \\ 3 \end{pmatrix}$ (c) $\begin{pmatrix} -\frac{1}{2} \\ 2 \end{pmatrix}$

(a) $\begin{pmatrix} 7 \\ 4 \end{pmatrix} = \dfrac{7 \cdot \cancel{6} \cdot 5 \cdot \cancel{4}}{1 \cdot \cancel{2} \cdot \cancel{3} \cdot \cancel{4}} = \underline{35}$

(b) $\begin{pmatrix} -1 \\ 3 \end{pmatrix} = \dfrac{(-1)(-2)(-3)}{1 \cdot 2 \cdot 3} = \underline{-1}$

(c) $\begin{pmatrix} -\frac{1}{2} \\ 2 \end{pmatrix} = \dfrac{(-\frac{1}{2}) \cdot (-\frac{3}{2})}{1 \cdot 2} = \underline{\dfrac{3}{8}}$

Example 28   Write down the first terms in the expansions in ascending powers of x of (a) $(1 - x/3)^9$ (b) $(1 - 2x)^{-1/2}$. In each case state the values of x for which the expansion is valid.

(a) $\left(1 - \dfrac{x}{3}\right)^9 = 1 + 9\left(-\dfrac{x}{3}\right) + \dfrac{9 \cdot 8}{1 \cdot 2}\left(-\dfrac{x}{3}\right)^2 + \dfrac{9 \cdot 8 \cdot 7}{1 \cdot 2 \cdot 3}\left(-\dfrac{x}{3}\right)^3$

$\qquad\qquad + \dfrac{9 \cdot 8 \cdot 7 \cdot 6}{1 \cdot 2 \cdot 3 \cdot 4}\left(-\dfrac{x}{3}\right)^4 + \cdots$

$\therefore \left(1 - \dfrac{x}{3}\right)^9 = \underline{1 - 3x + 4x^2 - \tfrac{28}{9}x^3 + \tfrac{14}{9}x^4 + \cdots}$ Valid for all x

(b) $(1 + 2x)^{-1/2} = 1 + (-\tfrac{1}{2})(2x) + \dfrac{(-\frac{1}{2})(-\frac{3}{2})}{1 \cdot 2}(2x)^2$

$\qquad\qquad + \dfrac{(-\frac{1}{2})(-\frac{3}{2})(-\frac{5}{2})(2x)^3}{1 \cdot 2 \cdot 3} + \dfrac{(-\frac{1}{2})(-\frac{3}{2})(-\frac{5}{2})(-\frac{7}{2})(2x)^4}{1 \cdot 2 \cdot 3 \cdot 4} + \cdots$

$\therefore (1 + 2x)^{-1/2} = \underline{1 - x + \tfrac{3}{2}x^2 - \tfrac{5}{2}x^3 + \tfrac{35}{8}x^4 + \cdots}$ Valid for $|2x| < 1$

$\qquad\qquad\qquad\qquad\qquad\qquad\qquad\qquad\qquad$ i.e. $\underline{|x| < \tfrac{1}{2}}$

● Exercise 6.8   See page 116.

6.9 The Binomial Theorem II

Example 29   Write down the first three terms in the expansions of
(a) $(2 + 3x)^7$ (b) $(3 - 2x)^{-3}$. For what values of x is (b) valid?

(a) $(2+3x)^7 = 2^7(1+\tfrac{3}{2}x)^7 = 2^7\left[1 + 7\cdot(\tfrac{3}{2}x) + \dfrac{7\cdot\overset{3}{\cancel{6}}}{1\cdot\cancel{2}}\left(\dfrac{3x}{2}\right)^2 + \cdots\right]$

$\qquad\qquad = 2^7\left[1 + \dfrac{21x}{2} + \dfrac{189}{2^2}x^2 + \cdots\right]$

$\therefore (2+3x)^7 = 2^7 + 21\cdot2^6 x + 189\cdot2^5 x^2 + \cdots$

$\therefore (2+3x)^7 = 128 + 1344x + 6048x^2 + \cdots$

(b) $(3-2x)^{-3} = 3^{-3}(1-\tfrac{2}{3}x)^{-3}$

$\qquad\qquad = 3^{-3}\left[1 + (-3)(-\tfrac{2}{3}x) + \dfrac{(-3)(-4)}{1\cdot2}(-\tfrac{2}{3}x)^2 + \cdots\right]$

$\qquad\qquad = \tfrac{1}{27}[1 + 2x + \tfrac{8}{3}x^2 + \cdots]$ Valid for $\left|\dfrac{2x}{3}\right| < 1$

$\therefore (3-2x)^{-3} = \tfrac{1}{27} + \tfrac{2}{27}x + \dfrac{8x^2}{81} + \cdots$ for $|x| < \tfrac{3}{2}$.

Example 30   Find the coefficient of x^5 in the following expansions
(a) $(1+3x)^{15}$ (b) $(2-5x)^{-2}$.

(a)   Term in $x^5 = \dfrac{\overset{3}{\cancel{15}}\cdot\overset{7}{\cancel{14}}\cdot13\cdot\overset{2}{\cancel{12}}\cdot11}{1\cdot\cancel{2}\cdot\cancel{3}\cdot\cancel{4}\cdot\cancel{5}}(3x)^5 = 729729x^5$

$\therefore$ coeff of $x^5 = 729729$

(b) $(2-5x)^{-2} = 2^{-2}[1-\tfrac{5}{2}x]^{-2}$

$\therefore$ Term in $x^5 = \dfrac{\overset{(-1)}{\cancel{1}}\overset{(-1)}{(\cancel{2})}\overset{(-1)}{(\cancel{3})}\overset{(-1)}{(\cancel{4})}\overset{}{(\cancel{5})}(-6)]}{4\cdot1\cdot\cancel{2}\cdot\cancel{3}\cdot\cancel{4}\cdot\cancel{5}}\left(-\dfrac{5x}{2}\right)^5 = \tfrac{9375}{64}x^5$

$\therefore$ coeff of $x^5 = \tfrac{9375}{64}$

Example 31   Use the Binomial Theorem to find the value of $1/\sqrt{(0.97)}$ to 4.d.p.

$\dfrac{1}{\sqrt{0.97}} = (1-0.03)^{-1/2} = 1 + (-\tfrac{1}{2})(-0.03) + \dfrac{(-\tfrac{1}{2})(-\tfrac{3}{2})}{1\cdot2}(-0.03)^2 + \cdots$

$\qquad\qquad\qquad = 1 + 0.015 + 0.0003375 + \cdots$

$\therefore \dfrac{1}{\sqrt{0.97}} = 1.0153$

● Exercise 6.9   See page 116.

6.10 Arithmetic and Geometric Means: Revision of Unit

Arithmetic Mean

If three numbers a, x, b are in arithmetic progression then x is called the **arithmetic mean** of a and b.

Example 32   Find the arithmetic mean of two numbers a and b.

Let x be the arithmetic mean.
Then a, x, b form an A.P.

$$\Rightarrow x - a = b - x \qquad \boxed{= \text{common difference}}$$
$$\Rightarrow 2x = a + b$$
$$\Rightarrow x = \frac{a + b}{2}$$

Geometric Mean

If three numbers a, x, b are in geometric progression then x is called the **geometric mean** of a and b.

Example 33   Find the geometric mean of two numbers a and b.

Let x be the geometric mean.
Then a, x, b form a G.P.

$$\Rightarrow \frac{x}{a} = \frac{b}{x} \qquad \boxed{= \text{common ratio}}$$
$$\Rightarrow x^2 = ab$$
$$\Rightarrow x = \sqrt{ab}$$

> The **arithmetic mean** of a and b is $\frac{1}{2}(a + b)$
> The **geometric mean** of a and b is $\sqrt{ab}$

- Miscellaneous Exercise 6A   See page 117.

6.11 + A Level Questions

- Miscellaneous Exercise 6B   See page 118.

Unit 6 EXERCISES

Exercise 6.1

1 In each of the following cases, state whether the series is an Arithmetic Progression:

 (a) $5, 9, 13, 17, \ldots$
 (b) $6, 4, 2, 0, -2, \ldots$
 (c) $1, -1, 1, -1, 1, \ldots$
 (d) $\frac{1}{4}, \frac{5}{4}, \frac{9}{4}, \frac{13}{4}, \ldots$
 (e) $1, 4, 9, 16, \ldots$
 (f) $21, 17, 13, 9, \ldots$
 (g) $1, \frac{1}{2}, \frac{1}{4}, \frac{1}{8}, \ldots$

2 Write down the seventh term and the nth term in each of the following A.P.s:

 (a) $1, 6, \ldots$
 (b) $3, 2\frac{1}{2}, \ldots$
 (c) first term 3, common difference -2
 (d) first term p, common difference $2q$

3 How many terms are there in each of the following A.P.s:

 (a) $11, 14, 17, \ldots, 89$
 (b) $20, 18, 16, \ldots, -30$
 (c) $151, 155, 159, \ldots, 291$

4 Find the sums of each of the following A.P.s:

 (a) $15, 20, 25, \ldots,$ to 15 terms
 (b) $51, 48, 45, \ldots,$ to 20 terms
 (c) $37, 30, 23, \ldots, -40$
 (d) $-5, -2, 1, \ldots, 34$

5 Find the sum of all the odd numbers between 30 and 120

6 Find the sum of all the even numbers between 1 and 77

7 Find the twentieth term and the sum to twenty-five terms of the series $3, 5, 7, 9, \ldots$

8 The fifth term of an A.P. is 28 and the tenth term is 58. Find the first term and the common difference

9 The sixth term of an A.P. is twice the third term and the first term is 3. Find the common difference and the ninth term

10 How many terms of the A.P. $13 + 16 + 19 + \cdots$ must be taken for the sum to equal 455

11 The first term of an A.P. is 2 and the sum to seventy-six terms is 4484. Find the last term

12 Find the least number of terms of the A.P. $5, 8, 11, 14, \ldots$ that are required so that their sum exceeds 5000

13 The nth term of an A.P. is given by $10 - 3n$. Find the first term, the second term and the common difference

14 If $S_n = 4n + 20$ and the series is $12, 16, 20, \ldots$ find n

15 For an A.P., $S_{10} = 120$ and the seventh term is three times the fourth term. Find the first term and S_{20}

*16 For a series, $S_n = n^2 - 3n$ find the first three terms of the series and show that it is an A.P.

*17 The sum of the first n terms of a series is $n^2 + 5n$. Find the fourth term and the nth term

*18 The first and last terms of an A.P. are -3 and 25 and the sum of all the terms is 1837. Find

    (a) the number of terms
    (b) the common difference
    (c) the middle term

*19 Find five numbers in arithmetic progression whose sum is 155 and whose last term is 47

Exercise 6.2

1 In each of the following cases state whether the series is a Geometric Progression:

    (a) $4, 12, 36, 108, \ldots$
    (b) $\frac{1}{2}, \frac{1}{4}, \frac{1}{8}, \frac{1}{16}, \ldots$
    (c) $1, 2, 4, 6, \ldots$
    (d) $a, 4a, 16a^2, 64a^3, \ldots$
    (e) $36, 24, 16, 10\frac{2}{3}, \ldots$

2 Write down the common ratio, the tenth term and the nth term in each of the following G.P.s:

    (a) $2, 6, 18, \ldots$
    (b) $27, 18, 12, 8, \ldots$
    (c) $3x, 3x^2, 3x^3, \ldots$
    (d) $\frac{1}{2}, \frac{1}{6}, \frac{1}{18}, \frac{1}{54}, \ldots$

3 Fill in the missing terms in the following G.P.s:

    (a) $5, \ldots, 20$
    (b) $27, \ldots, 3, \ldots$
    (c) $12, \ldots, 3, \ldots, \ldots$
    (d) $3, -6, \ldots, \ldots$

4 How many terms are there in each of the following G.P.s:

    (a) $4, 8, 16, \ldots, 256$

(b) $\frac{1}{3},\frac{1}{6},\frac{1}{12},\ldots,\frac{1}{384}$

(c) $x,x^3,x^5,\ldots,x^{17}$

5 Find the sums of each of the following G.P.s:

(a) $3,6,12,\ldots$, to 6 terms

(b) $2,-4,8,-16,\ldots$, to 9 terms

(c) $1,5,25,125,\ldots$, to 6 terms

(d) $1,1.02,(1.02)^2,(1.02)^3,(1.02)^4,\ldots,(1.02)^8$

(e) $10,1,0.1,0.01,\ldots$, to 6 terms

(f) $x,x^3,x^5,\ldots,x^{15}$

6 The seventh term of a G.P. is 32 and the third term is 2. Find the first term and the common ratio

7 The first term of a G.P. is $\frac{3}{11}$ and the fifth term is $4\frac{4}{11}$. Find the common ratio and the sum to ten terms, given that the common ratio is positive

8 The first term of a G.P. is $\frac{2}{5}$ and the fourth term is $9\frac{3}{8}$. Find the fifth term

9 The second term of a G.P. is -2 and the third term is $6\frac{3}{4}$. Find the first term and the common ratio

10 Sum to infinity the following series:

(a) $1-\frac{1}{3}+\frac{1}{9}-\frac{1}{27}+\cdots$

(b) $4+1+\frac{1}{4}+\frac{1}{16}+\cdots$

(c) $1+0.1+0.01+0.001+\cdots$

(d) $0.3+0.03+0.003+\cdots$

(e) $1+x+x^2+x^3+\cdots$ where $|x|<1$

(f) $1+\sin^2\theta+\sin^4\theta+\sin^6\theta+\cdots$ where $\theta\neq\dfrac{n\pi}{2}$

11 Find the range of values of x for which the following series converge:

(a) $1+x+x^2+\cdots$

(b) $x+\dfrac{1}{x}+\dfrac{1}{x^3}+\cdots$

(c) $1+3x+9x^2+27x^3+\cdots$

12 The sum to infinity of a G.P. is three times the first term. Find the common ratio

13 Find the sum to infinity of each of the following:

(a) $x+x^2+x^3+\cdots$ $|x|<1$

(b) $1-y+y^2-\cdots$ $|y|<1$

14 Find the range of values of z for which the sum to infinity exists of

$$z+\frac{z^2}{3}+\frac{z^3}{9}+\frac{z^4}{27}+\cdots$$

and find the sum to infinity

*15 What is the value of the first term of the series $5, 20, 80, \ldots$ which exceeds 5000?

*16 The second and third terms of a G.P. are 24 and $12(b + 1)$ respectively. What is the first term? If the sum of these three terms is 76 find the possible values of b

Exercise 6.3

What expressions are represented by:

1 $\displaystyle\sum_{r=1}^{r=4} r$ **2** $\displaystyle\sum_{r=1}^{r=3} 3r^2$ **3** $\displaystyle\sum_{r=1}^{r=5} r(r-1)$

4 $\displaystyle\sum_{r=3}^{r=9} r^3$ **5** $\displaystyle\sum_{r=2}^{r=8} \frac{1}{r}$ **6** $\displaystyle\sum_{r=10}^{r=13} \frac{1}{(r+1)(r+3)}$

7 $\displaystyle\sum_{r=1}^{r=n} r^4$ **8** $\displaystyle\sum_{r=3}^{r=n} (r+1)(r+4)$ **9** $\displaystyle\sum_{r=1}^{r=n} x^r$

10 What is the 8th term in the series $\displaystyle\sum_{1}^{n} \frac{1}{r(r+2)}$?

11 What is the 11th term in the series $\displaystyle\sum_{1}^{n} \frac{r+2}{r-1}$?

12 How many terms are there in the series $\displaystyle\sum_{1}^{3n} (r^2 - r + 2)$? Write down the $(n+1)$th term

Use the $\sum$ notation to express the sums of the following series:

13 $2 + 4 + 6 + 8 + \cdots$ 20 terms

14 $1 + \frac{1}{2} + \frac{1}{3} + \frac{1}{4} + \cdots$ 30 terms

15 $1 + 4 + 7 + 10 + \cdots$ 15 terms

16 $1^3 + 3^3 + 5^3 + 7^3 + \cdots + 27^3$

17 $3 \cdot 4 + 4 \cdot 5 + 5 \cdot 6 + 6 \cdot 7 + \cdots + 21 \cdot 22$

18 $\dfrac{1}{1^2} + \dfrac{1}{2^2} + \dfrac{1}{3^2} + \dfrac{1}{4^2} + \cdots$

19 $\dfrac{1}{4 \cdot 5} + \dfrac{1}{5 \cdot 6} + \dfrac{1}{6 \cdot 7} + \dfrac{1}{7 \cdot 8} + \cdots \dfrac{1}{99 \cdot 100}$

20 $\dfrac{2}{3 \cdot 4} + \dfrac{3}{4 \cdot 5} + \dfrac{4}{5 \cdot 6} + \quad + \cdots 2n$ terms

21 $2 - 4 + 6 - 8 + 10 - \cdots 11$ terms

22 $-1 + \frac{1}{3} - \frac{1}{5} + \frac{1}{7} - \cdots$

23   What is the 5th term of the series $\sum_{1}^{n}(-1)^{r+1}\dfrac{1}{r(r-1)}$?

24   What is the $(n-1)$th term of the series $\sum_{1}^{2n}(r^2-1)$?

What are the values of each of the following series:

25 $\sum\limits_{2}^{4}\dfrac{1}{r}$ **26** $\sum\limits_{1}^{3}r^2$ **27** $\sum\limits_{5}^{7}3r$ **28** $\sum\limits_{10}^{12}(r^2+1)$

29   Write down the first five terms of the series given by $\sum_{1}^{n}(2r-1)$. Show that they form an A.P.

30   Find the sum of the A.P. given by $\sum_{1}^{n}3(r-1)$

31   Evaluate $\sum\limits_{1}^{8}(1.3)^r$

32   Evaluate $\sum\limits_{1}^{10}5\left(\dfrac{2}{3}\right)^r$

33   Evaluate $\sum\limits_{1}^{\infty}\left(\dfrac{3}{4}\right)^r$

*For the following series
(a)   write down the term asked for, and
(b)   give the number of terms in the series:

34 $\sum\limits_{r=1}^{7}3^r$, 4th term   **35** $\sum\limits_{r=-2}^{9}(3r-1)$, 3rd term

36 $\sum\limits_{-5}^{1}\dfrac{1}{r(2r-1)}$, 4th term   **37** $\sum\limits_{0}^{\infty}\left(\dfrac{1}{4}\right)^r$, nth term

*38   Find the first four terms of the series $\sum\limits_{r=0}^{\infty}(-1)^{r+1}3^r x^{-r}$

*39   If $|x| < \frac{1}{2}$ find the sum of the geometric series given by $\sum\limits_{r=1}^{\infty}2x^r$

*40   Express the series $-x + 2x^2 - 3x^3 + 4x^4 + \cdots$ using the $\sum$ notation

Exercise 6.4

Prove, by induction:

1 $\sum_{1}^{n} r = \frac{1}{2}n(n+1)$

2 $1 + 3 + 5 + \cdots + (2n-1) = n^2$

3 $1 + 4 + 7 + \cdots + (3n-2) = \frac{1}{2}n(3n-1)$

4 $\sum_{1}^{n} r^3 = \frac{1}{4}n^2(n+1)^2$

5 $1 + 5 + 25 + \cdots + 5^{n-1} = \frac{1}{4}(5^n - 1)$

6 $1 \cdot 2 + 2 \cdot 3 + 3 \cdot 4 + \cdots + n(n+1) = \frac{1}{3}n(n+1)(n+2)$

7 $\dfrac{1}{1 \cdot 4} + \dfrac{1}{4 \cdot 7} + \dfrac{1}{7 \cdot 10} + \cdots + \dfrac{1}{(3n-2)(3n+1)} = \dfrac{n}{3n+1}$

8 $\sum_{1}^{n} \dfrac{1}{r(r+2)} = \dfrac{3}{4} - \dfrac{2n+3}{2(n+1)(n+2)}$

9 $\sum_{1}^{n} (3r^2 + r) = n(n+1)^2$

10 $\sum_{1}^{n} (r+1)2^{r-1} = n \cdot 2^n$

11 $1 \cdot 4 + 2 \cdot 9 + 3 \cdot 16 + \cdots + n(n+1)^2 = \frac{1}{12}n(n+1)(n+2)(3n+5)$

Exercise 6.5

In questions 1–4 find the sums of the following series:

1 $1^2 + 2^2 + 3^2 + \cdots + 24^2$

2 $2 + 3 + 4 + \cdots + 28$

3 $15 + 16 + 17 + \cdots + 40$

4 $6^3 + 7^3 + \cdots + 15^3$

5 Find the sum of the even numbers between 11 and 121

6 Find the sum of the cubes of the natural numbers from 10 to 20 inclusive

7 Evaluate $101 + 103 + 105 + \cdots + 199$

8 Find the sums of:
 (a) $2 + 4 + 6 + \cdots + 50$

(b) $2^2 + 4^2 + 6^2 + \cdots + 50^2$
(c) $2^3 + 4^3 + 6^3 + \cdots + 50^3$

9   Show that the sum of the series $1 - 2 + 3 - 4 + 5 - \cdots - 100$ is -50

10   What is the sum of the squares of the first $2n$ natural numbers?

11   Show that the sum of the series

$$n^2 + (n+1)^2 + (n+2)^2 + \cdots + (2n)^2$$

is $\frac{1}{6}n(n+1)(14n+1)$

In questions 12–14 evaluate:

12 $\sum_1^n (4r^2 + 1)$ **13** $\sum_1^n r(r+1)$ **14** $\sum_1^n (r-2)^2$

In questions 15–20 in each of the following series, find the rth term and the sum to n terms.

15 $4 + 8 + 12 + \cdots$

16 $1^2 + 3^2 + 5^2 + \cdots$

17 $2 \cdot 3 + 3 \cdot 4 + 4 \cdot 5 + \cdots$

18 $1^2 + 4^2 + 7^2 + \cdots$

19 $3^3 + 6^3 + 9^3 + \cdots$

20 $3 \cdot 5 + 7 \cdot 9 + 11 \cdot 13 + \cdots$

*21   The sum to n terms of a series is $2n^2 + n$. By putting $n = 1$, find the first term. Find the second and third terms. Deduce the rth term

Exercise 6.6

Evaluate:

1 $\left(\dfrac{1}{3}\right)^{-2}$ **2** $\dfrac{1}{4^{-1}}$ **3** $27^{1/3}$ **4** $27^{-1/3}$

5 $\left(\dfrac{25}{81}\right)^{1/2}$ **6** $625^{1/4}$ **7** $64^{-1/3}$ **8** $144^{3/2}$

9 $\left(\dfrac{1}{9}\right)^{5/2}$ **10** $\left(\dfrac{1}{4}\right)^{-3/2}$ **11** $\left(\dfrac{100}{81}\right)^{-3/2}$ **12** $\left(\dfrac{8}{27}\right)^{2/3}$

13 $\left(-\dfrac{1}{5}\right)^{-3}$ **14** $\left(-\dfrac{2}{3}\right)^{-2}$ **15** $0.25^{1/2}$ **16** $(0.09)^{-2}$

17 $(1.96)^{-1/2}$ **18** $\left(2\dfrac{14}{25}\right)^{1/2}$ **19** $\left(6\dfrac{1}{4}\right)^{-1/2}$ **20** $\left(\dfrac{256}{25}\right)^{0}$

21 $4^{-1}\cdot 3^{2}\cdot 2^{0}$ **22** $18^{1/2}\cdot 2^{1/2}$ **23** $9^{1/3}\cdot 3^{1/3}$ **24** $32^{1/5}\cdot 2^{-1}$

25 $81^{1/4}\cdot 64^{-1/3}$ **26** $\dfrac{25^{1/2}\cdot 8^{1/2}}{2^{1/2}}$ **27** $\dfrac{5^{2/3}\cdot 5^{0}\cdot 25^{2/3}}{125^{2/3}}$ **28** $\dfrac{8^{1/3}\cdot 16^{1/4}}{32^{-1/5}}$

29 $\dfrac{100^{-1/2}\cdot 25^{1/2}}{64^{1/3}}$ **30** $27^{-1/3}\cdot 125^{1/3}$

Simplify:

31 $\dfrac{x^{1/6}\cdot x^{1/2}}{x^{-1/3}}$ **32** $\dfrac{y^{1/2}\cdot y^{-1/4}}{y^{-3/4}}$ **33** $\dfrac{\sqrt{p}\sqrt{p^{5}}}{p^{-2}}$

34 $\dfrac{(\sqrt{q})^{3}\cdot q^{3}}{\sqrt{(q)^{7}}}$ **35** $\dfrac{(x-1)^{-1/2}+(x-1)^{1/2}}{(x-1)^{1/2}}$

36 $\dfrac{(x+3)^{-1/2}+3(x+3)^{-3/2}}{x+3}$ **37** $\dfrac{x(x-1)^{1/2}-(x-1)^{-1/2}}{(x-1)^{1/2}}$

38 $\dfrac{(p^{2}+2)^{1/2}-p^{2}(p^{2}+2)^{-1/2}}{(p^{2}+2)^{-1/2}}$ **39** $\dfrac{(y+1)^{1/3}-\frac{1}{2}y(y+1)^{-2/3}}{(y+1)^{2/3}}$

40 $\dfrac{3(p^{2}-q^{2})^{1/2}-p^{2}(p^{2}-q^{2})^{-1/2}}{(p+q)^{1/2}}$

Exercise 6.7

$$\begin{array}{ccc} & 1 & \\ 1 & 1 & 1 \\ 1 & 2 & 1 \end{array}$$

Copy and continue Pascal's triangle until it has eight lines.

In questions 2–13 use Pascal's triangle to expand:

2 $(1+x)^{7}$ **3** $(1+2x)^{4}$ **4** $(1-3x)^{3}$

5 $\left(1-\dfrac{x}{2}\right)^{3}$ **6** $(2+x)^{4}$ **7** $(2+3x)^{5}$

8 $(2x-5)^{4}$ **9** $\left(t+\dfrac{1}{t}\right)^{4}$ **10** $(a-3b)^{6}$

11 $(x^{2}-y^{3})^{7}$ **12** $\left(x^{2}-\dfrac{1}{x^{2}}\right)^{3}$ **13** $\left(x+\dfrac{2}{x}\right)^{5}-\left(x-\dfrac{2}{x}\right)^{5}$

In questions 14–16 use Pascal's triangle to simplify:

14 $(1+\sqrt{3})^{3}$ **15** $(\sqrt{5}-\sqrt{3})^{4}$ **16** $(1+\sqrt{2})^{3}+(1-\sqrt{2})^{3}$

17   Find a and b if $(2 + \sqrt{3})^5 \equiv a + b\sqrt{3}$

18   Use Pascal's triangle to calculate the value of

$$(2\sqrt{3} - 1)^5 + 12(\sqrt{3} - 2)^4 - 6(3 - \sqrt{3})^3$$

Exercise 6.8

1   Evaluate:

(a) $\dfrac{6!}{4!}$ (b) $\dfrac{5!}{2!}$ (c) $\dfrac{12!}{10!}$ (d) $\dfrac{12!}{10!2!}$ (e) $\dfrac{9!}{5!4!}$ (f) $\dfrac{6!}{4!2!}$

2   Evaluate:

(a) $\dbinom{4}{2}$ (b) $\dbinom{7}{3}$ (c) $\dbinom{13}{6}$ (d) $\dbinom{-1}{2}$

(e) $\dbinom{\frac{1}{2}}{4}$ (f) $\dbinom{-\frac{3}{4}}{2}$

Use the Binomial Theorem to expand the following:

3 $(1 + x)^4$ **4** $(1 - y)^5$ **5** $(1 + 2x)^6$ **6** $\left(1 - \dfrac{x}{3}\right)^3$

Write down the first four terms of each of the following expansions:

7 $(1 + x)^{1/2}$ **8** $(1 + x)^{-2}$ **9** $\dfrac{1}{\sqrt{(1 + x)}}$ **10** $(1 - x)^{-3/2}$

Write down the first four terms of each of the following expansions. In each case state the values of x for which the expansion is valid.

11 $\dfrac{1}{(1 + 4x)^4}$ **12** $\sqrt[3]{(1 - 3x)}$ **13** $\left(1 + \dfrac{x}{3}\right)^{2/3}$ **14** $\dfrac{1}{2x + 1}$

15   Find the coefficient of x^5 in the expansion of $(1 + 3x)^{15}$

Write down the first five terms of each of the following expansions:

16 $(1 + x)^{-1}$ **17** $(1 - x)^{-1}$ **18** $(1 - x)^{-2}$

19 $(1 + x^2)^{-1}$ **20** $(1 - 2x)^{-1}$

**21   If x is so small that x^4 and higher powers of x can be neglected, show that

$$(1 + x)\sqrt{(1 - x)} \approx 1 + \tfrac{1}{2}x - \tfrac{5}{8}x^2 - \tfrac{3}{16}x^3$$

Exercise 6.9

Write down the first four terms in each of the following expansions. In each case state the values of x for which the expansion is valid.

1 $(2+x)^5$ **2** $(4-x)^{1/4}$ **3** $\dfrac{1}{(3+2x)^{1/2}}$ **4** $\dfrac{1}{(3x-1)^2}$

Find the coefficient of x^4 in each of the following expansions:

5 $(1+x)^{12}$ **6** $(1-3x)^9$ **7** $\left(3+\dfrac{x}{5}\right)^{10}$ **8** $(1+2x^2)^6$

Use the Binomial Theorem to find the values of the following expressions, correct to 4 d.p.

9 $(1.003)^4$ **10** $\sqrt{1.03}$ **11** $(0.998)^5$ **12** $1/(2.002)^2$

13 If x is so small that x^4 and higher powers can be neglected, show that

(a) $(2+3x)^{-1} \approx \dfrac{1}{2} - \dfrac{3x}{4} + \dfrac{9x^2}{8} - \dfrac{27x^3}{16}$

(b) $(3-4x)^{-2} \approx \dfrac{1}{9} + \dfrac{8x}{27} + \dfrac{16x^2}{27} + \dfrac{256x^3}{243}$

Find the coefficients of x^3 in the following expansions:

14 $(1+2x)^{1/4}$ **15** $\dfrac{1}{(1-3x)^{2/3}}$ **16** $(3-x)^{-2}$ **17** $\sqrt{(p+x)}$

Find the first four terms in the expansions of:

*18 $\dfrac{2}{1+x} + \dfrac{3}{1-2x}$ ***19** $\dfrac{1+x}{1-x}$ ***20** $\dfrac{2+x}{(1+x)^2}$

*Find, and simplify, the middle terms of the expansions of:

21 $(1+2x)^6$ **22** $(3-x)^{10}$ **23** $\left(x-\dfrac{1}{x}\right)^6$ **24** $\left(3a+\dfrac{1}{a}\right)^8$

*25 If x is so small that x^3 and higher powers of x are negligible, show that
$$(2x+3)(1-2x)^{10} \approx 3 - 58x + 500x^2$$

*26 Expand $(a-x)^5$
If $(\sqrt{5}-\sqrt{3})^5 = A\sqrt{5} - B\sqrt{3}$ find the values of A and B

Miscellaneous Exercise 6A

1 Find the arithmetic and geometric means of
(a) -6 and -12 (b) x and $4x$

2 The arithmetic mean of two numbers, x and y, is 10 and their geometric mean is 8. Find the values of x and y

3 Evaluate: (a) $\left(\dfrac{4}{9}\right)^{-1/2}$ (b) $\left(\dfrac{16}{81}\right)^{1/4}$ (c) $\left(\dfrac{18}{32}\right)^{-1/2}$

(d) $2^{-3} \times 4^2 \div 8^{-2}$

4 Simplify: (a) $\dfrac{p^{1/2}(p+q)^{-1/2} + p^{-1/2}(p+q)^{1/2}}{(p+q)^{3/2}}$ (b) $\dfrac{5p^{-4}q^{2/3}}{p^{3/2}q^{-2/3}}$

5 Find the tenth term in the series $2, 7, 12, 17, \ldots$

6 Find the sum of the first eight terms of the series $-2,\ 1,\ -\frac{1}{2}, \frac{1}{4}, \ldots$

7 The sum of the first eleven terms of an A.P. is 22 and the common difference is $\frac{3}{5}$. Find the nth term and the sum to 46 terms

8 Find (a) $\displaystyle\sum_{1}^{20}(2r-1)$ (b) $\displaystyle\sum_{1}^{10}(2.3)^r$

9 The first and last terms of an A.P. are 7 and 127 and the sum of all the terms is 1675. Find the number of terms and the middle term of the progression

10 The twelfth term of an A.P. is seven times the second term, and the sixth term is 17. Find the first term and the common difference

11 A pendulum is set swinging. Its first oscillation is $36°$ and each succeeding oscillation is $\frac{2}{3}$ of the one before it. What is the total angle described before it stops?

*12 How many terms of the A.P. $10, 16, 22, \ldots$ must be taken so that the sum exceeds 150?

*13 How many terms of the G.P. $5, 2\frac{1}{2}, 1\frac{1}{4}, \frac{5}{8}, \ldots$ must be taken in order that the sum may differ from the sum to infinity by less than 0.01?

**14 Find the sum of all the integers from 1000 to 2000 inclusive which are not multiples of 7

**15 A ball rebounds to $\frac{2}{5}$ of the height it was dropped from. If it was dropped from $10\,\text{m}$ find

(a) the height of the rebound after the first bounce
(b) the height of the rebound after the sixth bounce
(c) the total distance covered before the ball comes to rest

Miscellaneous Exercise 6B

1 Evaluate $\displaystyle\sum_{r=1}^{100}(3r+2)$ (L: 6.5)

2 (a) The 4th term of an arithmetic series is 7 and the 7th term is 4. Find the sum of the first 29 terms of this series

(b) The 4th term of a geometric series is 7 and the 7th term is 4. Find the sum to infinity of this series, giving your answer to the nearest integer (A: 6.1, 6.2)

3 An infinite geometric series with first term 2 converges to the sum 3. Find the fourth term in the series (L: 6.2)

4 Evaluate, giving your answer correct to 2 s.f. $\sum_{1}^{20} (1.1)^r$ (J: 6.2, 6.3)

5 Expand (a) $\sqrt[3]{(1 + px)}$ $|px| < 1$

      (b) $\dfrac{1 + 2qx}{1 + qx}$ $|qx| < 1$,

where p and q are constants, in terms of ascending powers of x up to and including the terms in x^2

Given that these terms of the expansions are the same, show that $p = 3q$ (L: 6.8, 6.9)

6 Prove that $\sum_{r=1}^{n} r(r + 1) = \frac{1}{3}n(n + 1)(n + 2)$

Evaluate $\sum_{r=1}^{20} r(r - 1)$ (L: 6.5)

7 Find the positive constants a and b such that 0.25, a, 9 are in geometric progression and 0.25, a, $9 - b$ are in arithmetic progression (L: 6.1, 6.2, 6.10)

8 Give the first four terms in the binomial expansion of $(1 + ax)^n$ in ascending powers of x. Show that the ratio of the coefficient of x^{r+1} to that of x^r

is $\dfrac{a(n - r)}{(r + 1)}$

Given that the ratio of the coefficient of x^6 to that of x^5 is 30 and that the ratio of the coefficient of x^9 to that of x^8 is 15, find integer values for a and n (A: 6.8)

9 The first and third terms of an arithmetic series are a and b respectively. The sum of the first n terms of this series is denoted by S_n. Find S_4 in terms of a and b

Given that S_4, S_5 and S_7 are consecutive terms of a geometric series, show that $7a^2 = 13b^2$. (L: 6.1, 6.10)

10 Show that the first three terms in the expansion in ascending powers of x of $(1 + 8x)^{1/4}$ are the same as the first three terms in the expansion of $\left(\dfrac{1 + 5x}{1 + 3x}\right)$

Use the corresponding approximation

$$(1 + 8x)^{1/4} \approx \frac{(1 + 5x)}{(1 + 3x)}$$

to obtain an approximation to $(1.16)^{1/4}$ as a rational fraction in its lowest terms (J: 6.9)

11   The sum of the first twenty terms of an A.P. is 45 and the sum of the first forty terms is 290. Find the first term and the common difference. Find the number of terms in the progression which are less than 100 (J:6.1)

12   Given that $|x| < 1$, find in ascending powers of x, up to and including the terms in x^3, series expansions in which the coefficients are simplified for

(a) $(1 + x)^{1/2}$,   (b) $\left(1 + \dfrac{x}{4}\right)^{-1}$

Prove that if x were sufficiently small for terms in x^4 and higher powers of x to be neglected, then

$$(1 + x)^{1/2} - \frac{4 + 3x}{4 + x} = \tfrac{1}{32}x^3 \qquad\qquad \text{(A: 6.8, 6.9)}$$

13   The first three terms in the expansion of $(1 - kx)^p$ in ascending powers of x are 1, $-x$ and $-\tfrac{1}{2}x^2$. Find the values of the constants k and p and the coefficient of x^4 in the expansion. State the set of values of x for which the expansion is valid (L: 6.8)

14   The first and last terms of an A.P. are a and l respectively. If the progression has n terms, prove from first principles that its sum is $\tfrac{1}{2}n(a + l)$. An array consists of 100 rows of numbers, in which the rth row contains $(r + 1)$ numbers. In each row, the numbers form an A.P. whose first and last terms are 1 and 99, respectively. Calculate the sum of the numbers in the rth row and hence determine the sum of the numbers in the array (J: 6.1)

15   Starting from first principles, prove that the sum of the first n terms of a G.P., whose first term is a and whose common ratio is r $(r \neq 1)$ is

$$S_n = \frac{a(1 - r^n)}{1 - r}$$

Show that $\dfrac{S_{3n} - S_{2n}}{S_n} = r^{2n}$

Given that $r = \tfrac{1}{2}$, find $\displaystyle\sum_{1}^{\infty}\left(\frac{S_{3n} - S_{2n}}{S_n}\right)$ (J: 6.2, 6.3)

16   Write down the expansion of $(1 + y)^n$ in ascending powers of y, giving the first four terms. Given that x is so small that its cube and higher powers are negligible, compared with unity, find the constants a, b and

c in the approximate formula

$$\left(\frac{1+3x}{8-3x}\right)^{1/3} \approx a + bx + cx^2 \qquad \text{(J:6.9)}$$

*17 The first, second and third terms of an arithmetic series are p, q, p^2 respectively, where $p < 0$. The first, second and third terms of a geometric series are p, p^2, q respectively.

(a) Show that $p = -\frac{1}{2}$ and find the value of q
(b) Find the sum to infinity of the geometric series
(c) Find the seventeenth term of the arithmetic series $\qquad$ (A: 6.1, 6.2)

*18 Find without simplification the coefficient of x in the binomial expansion

of $\left(\dfrac{1}{2x} - \dfrac{x^2}{3}\right)^{11}$ $\qquad$ (L: 6.9)

*19 The midpoints of the sides of an equilateral triangle $\triangle_1$ are joined to form another triangle $\triangle_2$; the midpoints of the sides of the triangle $\triangle_2$ are then joined to form another triangle $\triangle_3$ and so on. Denoting the

areas of $\triangle_1, \triangle_2, \ldots$ by $A_1, A_2, \ldots$ respectively and given that $\displaystyle\sum_{n=1}^{\infty} A_n = a^2$

show that $A_1 = \frac{3}{4}a^2$

Find, in terms of a, the length of the perimeter of $\triangle_n$ $\qquad$ (J: 6.2, 6.3)

*20 The three real, distinct and nonzero numbers a, b, c are such that

a, b, c are in arithmetic progression
a, c, b are in geometric progression.

Find the numerical value of the common ratio of the G.P.
Hence find an expression, in terms of a, for the sum to infinity of the
G.P. whose first terms are a, c, b $\qquad$ (J: 6.1, 6.2, 6.10)

*21 The series $\displaystyle\sum_{r=1}^{n+1} u_r$ is a geometric series with common ratio k, where $k^2 \neq 1$.

Show that the series $\displaystyle\sum_{r=1}^{n} (u_r u_{r+1})$ is a geometric series and that its sum is

equal to

$$\frac{u_1^2 k(1 - k^{2n})}{(1 - k^2)} \qquad \text{(L: 6.2, 6.3)}$$

*22 Obtain the first three non-zero terms in the expansion of $(1 + y)^{1/2}$, where $|y| < 1$, as a series of ascending powers of y.

Given that

$$f(x) \equiv (1 + x^{1/2})^{1/2}$$

write down the coefficients a_0, a_1, a_2 in the expansion

$$a_0 + a_1 x^{1/2} + a_2 x + \cdots$$

of $f(x)$ as a series of ascending powers of $x^{1/2}$

Using this expansion, or otherwise, estimate the value of

$$\int_0^{0.01} f(x)\,dx$$

giving your answer to 4 decimal places.   (L: 6.8, 5.3)

*23   In the $\triangle ABC$, $BC = a$, $CA = b$, $AB = c$ and the angle C is obtuse. Prove that $c^2 = a^2 + b^2 - 2ab\cos C$

The perimeter of the triangle is $3l$ and the largest angle is $120°$. Given that the lengths of the sides of the triangle are in arithmetic progression, find, in terms of l, the length of each side   (J: 4.5, 6.1)

*24   The diagram shows two straight lines OX and OY such that the angle XOY is θ. $(0° < \theta < 90°)$

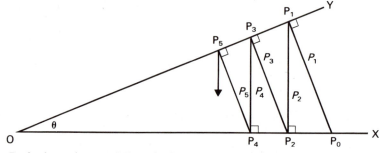

P_0 is the point on OX such that $OP_0 = l$
P_1 is the foot of the perpendicular from P_0 to OY
P_2 is the foot of the perpendicular from P_1 to OX
P_3 is the foot of the perpendicular from P_2 to OY; and so on.
The length of the line joining P_{r-1} to P_r is denoted by P_r $(r = 1, 2, 3, \ldots)$ and the area of the triangle $OP_{r-1}P_r$ is denoted by A_r $(r = 1, 2, 3, \ldots)$
Show that the lengths $P_1, P_2, P_3, \ldots$ form a G.P. with common ratio $\cos\theta$ and prove that

$$\sum_{r=1}^{\infty} P_r = l\cot\theta/2$$

Show that the areas $A_1, A_2, A_3, \ldots$ also form a G.P. and prove that

$$\sum_1^{\infty} A_r = \tfrac{1}{2}l^2 \cot\theta \qquad\qquad (J: 4.6, 6.2, 6.3)$$

*25   Given that $x > 2$, use the binomial expansion to express $\left(\dfrac{x+2}{x}\right)^{-1/2}$ in

the form $a + \dfrac{b}{x} + \dfrac{c}{x^2} + \dfrac{d}{x^3} + \ldots$, evaluating the constants a, b, c and d

Taking $x = 100$, use your series to find an approximation for $\left(\dfrac{450}{51}\right)^{1/2}$

giving your answer to 4 decimal places.   (A: 6.9)

*26 Write down expressions for the roots of the quadratic equation $x^2 - x + a = 0$, where a is a constant. If a is so small that a^4, and higher powers of a, may be neglected, show that the roots are approximately

$$1 - a - a^2 - 2a^3 \quad \text{and} \quad a + a^2 + 2a^3$$

Use these approximations to estimate to 3 d.p. the roots of the equation

$$100x^2 - 100x + 3 = 0 \qquad \qquad \text{(J: 1.8, 6.6)}$$

7.1 Factor Formulae

$$\sin S + \sin T = 2\sin\left(\frac{S+T}{2}\right)\cos\left(\frac{S-T}{2}\right)$$

$$\sin S - \sin T = 2\cos\left(\frac{S+T}{2}\right)\sin\left(\frac{S-T}{2}\right)$$

$$\cos S + \cos T = 2\cos\left(\frac{S+T}{2}\right)\cos\left(\frac{S-T}{2}\right)$$

$$\cos S - \cos T = -2\sin\left(\frac{S+T}{2}\right)\sin\left(\frac{S-T}{2}\right)$$

Note   These are easier remembered in words:

e.g. $\sin + \sin = 2\sin\cos$
$\sin - \sin = 2\cos\sin$
$\cos + \cos = 2\cos\cos$
$\cos - \cos = -2\sin\sin$

Example 1   Factorise:

(a) $\sin 5A + \sin A = 2\sin\left(\frac{5A+A}{2}\right)\cos\left(\frac{5A-A}{2}\right) = \underline{2\sin 3A\cos 2A}$

(b) $\cos 5A - \cos 2A = -2\sin\left(\frac{5A+2A}{2}\right)\sin\left(\frac{5A-2A}{2}\right) = \underline{-2\sin\frac{7A}{2}\sin\frac{3A}{2}}$

(c) $\sin 2A - \sin 4A = 2\cos\left(\frac{2A+4A}{2}\right)\sin\left(\frac{2A-4A}{2}\right) = 2\cos 3A\cdot\sin(-A)$

$$= \underline{-2\cos 3A\cdot\sin A}$$

Example 2   Factorise, and hence evaluate without tables or calculator, the expression $\sin 75° - \sin 15°$,

$$\sin 75° - \sin 15° = 2\cos\left(\frac{75°+15°}{2}\right)\sin\left(\frac{75°-15°}{2}\right)$$

$$= 2\cos 45° \sin 30°$$

$$= 2\cdot\frac{1}{\sqrt 2}\cdot\frac{1}{2} = \underline{\frac{1}{\sqrt 2}}$$

Example 3   Express $2 \sin 2\theta \cos \theta$ as the sum of two sines.

Let
$$2 \sin 2\theta \cos \theta = 2 \sin \left(\frac{P+Q}{2} \right) \cdot \cos \left(\frac{P-Q}{2} \right)$$

$$\therefore \left. \begin{array}{l} P+Q = 4\theta \\ \text{and } P-Q = 2\theta \end{array} \right\} \Rightarrow P = 3\theta \quad \text{and} \quad Q = \theta$$

$$\therefore 2 \sin 2\theta \cos \theta = \sin 3\theta + \sin \theta$$

Example 4   Solve the equation $\cos 5\theta - \cos \theta = \sin 3\theta$ for $0° \leqslant \theta \leqslant 180°$.

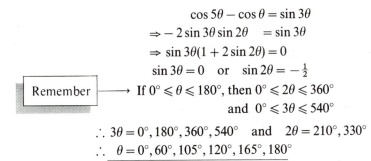

$$\cos 5\theta - \cos \theta = \sin 3\theta$$
$$\Rightarrow -2 \sin 3\theta \sin 2\theta = \sin 3\theta$$
$$\Rightarrow \sin 3\theta (1 + 2 \sin 2\theta) = 0$$
$$\sin 3\theta = 0 \quad \text{or} \quad \sin 2\theta = -\tfrac{1}{2}$$

Remember $\longrightarrow$ If $0° \leqslant \theta \leqslant 180°$, then $0° \leqslant 2\theta \leqslant 360°$
and $0° \leqslant 3\theta \leqslant 540°$

$$\therefore 3\theta = 0°, 180°, 360°, 540° \quad \text{and} \quad 2\theta = 210°, 330°$$
$$\therefore \quad \theta = 0°, 60°, 105°, 120°, 165°, 180°$$

Example 5   Factorise $\cos \theta - \cos 3\theta - \cos 5\theta + \cos 7\theta$.

$$\cos \theta - \cos 3\theta - \cos 5\theta + \cos 7\theta$$
$$= (\cos 7\theta + \cos \theta) - (\cos 5\theta + \cos 3\theta)$$
$$= 2 \cos 4\theta \cdot \cos 3\theta - 2 \cos 4\theta \cdot \cos \theta$$
$$= 2 \cos 4\theta [\cos 3\theta - \cos \theta]$$
$$= 2 \cos 4\theta [-2 \sin 2\theta \cdot \sin \theta]$$
$$= -4 \cos 4\theta \cdot \sin 2\theta \cdot \sin \theta$$

Example 6   Prove that $\dfrac{\sin 3\theta + \sin \theta}{\cos 3\theta + \cos \theta} \equiv \tan 2\theta$.

$$\frac{\sin 3\theta + \sin \theta}{\cos 3\theta + \cos \theta} \equiv \frac{2 \sin 2\theta \cos \theta}{2 \cos 2\theta \cos \theta} \equiv \tan 2\theta \qquad\qquad \text{Q.E.D}$$

● **Exercise 7.1**   See page 133.

7.2 The Expression $a \cos \theta + b \sin \theta$

Let
$$a \cos \theta + b \sin \theta \equiv R \cos (\theta - \alpha)$$
$$= R[\cos \theta \cdot \cos \alpha + \sin \theta \cdot \sin \alpha]$$
$$= (R \cos \alpha) \cos \theta + (R \sin \alpha) \sin \theta$$
$$\Rightarrow a = R \cos \alpha \quad \text{and} \quad b = R \sin \alpha$$

Hence

$$a \cos \theta + b \sin \theta = R \cos (\theta - \alpha)$$

$$\text{where } R = \sqrt{a^2 + b^2} \text{ and } \cos \alpha = \frac{a}{R}, \sin \alpha = \frac{b}{R}$$

Similar results can be found using $R \cos (x + \alpha)$, $R \sin (x + \alpha)$, $R \sin (x - \alpha)$.

Example 7   Find the maximum and minimum values of the fraction $f(\theta) = 2 \cos \theta - 3 \sin \theta$.

Let
$$2 \cos \theta - 3 \sin \theta \equiv R \cos (\theta + \alpha)$$
$$= R \cos \theta \cos \alpha - R \sin \theta \sin \alpha$$
$$\Rightarrow R \cos \alpha = 2 \quad \text{and} \quad R \sin \alpha = 3$$
$$\Rightarrow R = \sqrt{4 + 9} = \sqrt{13} \quad \text{and} \quad \tan \alpha = \tfrac{3}{2} \Rightarrow \alpha = 56.3°$$
$$\therefore f(\theta) = 2 \cos \theta - 3 \sin \theta = \sqrt{13} \cos (\theta + 56.3°)$$

The maximum value of $\cos (\theta + 56.3°)$ is 1 $\therefore$ Max $f(\theta) = \sqrt{13}$

The minimum value of $\cos (\theta + 56.3°)$ is -1 $\therefore$ Min $f(\theta) = -\sqrt{13}$

Example 8   Solve $\cos x + 2 \sin x = 1$ for $0° \leqslant x \leqslant 360°$.

Let
$$\cos x + 2 \sin x \equiv R \cos (x - \alpha)$$
$$= R \cos x \cos \alpha + R \sin x \sin \alpha$$
$$\Rightarrow R \cos \alpha = 1 \quad \text{and} \quad R \sin \alpha = 2$$
$$\Rightarrow R = \sqrt{5} \quad \text{and} \quad \tan \alpha = 2 \Rightarrow \alpha = 63.4°$$

$\therefore$ equation becomes

$$\sqrt{5} \cos (x - 63.4°) = 1$$
$$\Rightarrow \cos (x - 63.4°) \quad = 1/\sqrt{5}$$

[Note: If $0° \leqslant x \leqslant 360°$, then $-63.4° \leqslant x - 63.4° \leqslant 296.6°$]

Hence
$$x - 63.4° = -63.4°, 63.4°, 296.6°$$
$$\therefore x = 0°, 126.8°, 360°$$

Example 9   Sketch the curve $y = 3 \sin x - 4 \cos x$ for $-90° \leqslant x \leqslant 360°$.

Let
$$3 \sin x - 4 \cos x \equiv R \sin (x - \alpha)$$
$$= R \sin x \cdot \cos \alpha - R \cos x \cdot \sin \alpha$$

$$\Rightarrow R\cos\alpha = 3 \quad \text{and} \quad R\sin\alpha = 4$$
$$\Rightarrow R = 5 \quad \text{and} \quad \tan\alpha = \tfrac{4}{3} \Rightarrow \alpha = 53.1°$$

∴ the equation of the curve can be written as

$$y = 5\sin(x - 53.1°)$$

The maximum value of y is 5 and this occurs when $\sin(x - 53.1°) = 1$.
i.e. when $x - 53.1° = 90°, 450°, \ldots$
i.e. when $x = 143.1°, 503.1°, \ldots$ (every 360°).

The minimum value of y is -5 and this occurs when $\sin(x - 53.1°) = -1$.
i.e. when $x - 53.1° = -90°, 270°, \ldots$
i.e. when $x = -36.9, 323.1°, \ldots$ (every 360°).

$y = 0$ when $\sin(x - 53.1°) = 0$
i.e. when $x - 53.1° = 0°, 180°, 360°$
i.e. when $x = 53.1°, 233.1°, 413.1°, \ldots$ (every 180°).

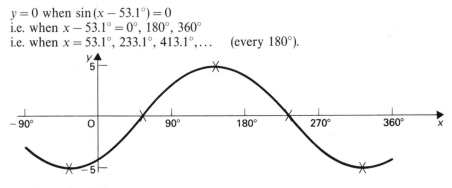

● Exercise 7.2   See page 134.

7.3 General Solutions of Equations

Angles with the same sine:

$$\boxed{\theta = n\pi + (-1)^n\alpha}$$

Principal value: $\alpha \in [-\pi/2, \pi/2]$

Angles with the same cosine:

$$\boxed{\theta = 2n\pi \pm \alpha}$$

Principal value: $\alpha \in [0, \pi]$

Angles with the same tangent:

$$\boxed{\theta = n\pi + \alpha}$$

Principal value: $\alpha \in [-\pi/2, \pi/2]$

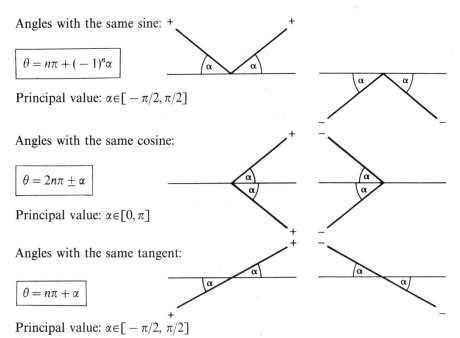

Example 10   Write down the general solutions of the equations (a) $\sin \theta = \frac{1}{2}$

(b) $\cos \theta = -\dfrac{\sqrt{3}}{2}$ (c) $\cot \theta = -1$

(a) $\sin \theta = \frac{1}{2}$ The value of θ between $-\pi/2$ and $\pi/2$ is $\pi/6$.
   i.e. $\alpha = \pi/6$.
   $\therefore$ general solution is $\theta = n\pi + (-1)^n \pi/6$

(b) $\cos \theta = -\dfrac{\sqrt{3}}{2}$ The value of θ between 0 and π is $5\pi/6$.

   i.e. $\alpha = 5\pi/6$.
   $\therefore$ general solution is $\theta = 2n\pi \pm 5\pi/6$

(c) $\cot \theta = -1 \Rightarrow \tan \theta = -1 \Rightarrow \alpha = -\pi/4$
   $\therefore$ general solution is $\theta = n\pi - \pi/4$

Example 11   Find the general solution of $2\cos\dfrac{\theta}{3} = 1$.

$$\cos\frac{\theta}{3} = \frac{1}{2} \Rightarrow \alpha = \frac{\pi}{3}$$

$\therefore$ general solution is $\dfrac{\theta}{3} = 2n\pi \pm \dfrac{\pi}{3}$

or $\theta = 6n\pi \pm \pi$

Example 12   Find the general solution of $\tan 2\theta = 1$ expressed in degrees.
Hence find all solutions in the range $-180° \leqslant \theta \leqslant 180°$.

$$\tan 2\theta = 1 \Rightarrow \alpha = 45°$$

$\therefore$ general solution is $2\theta = n180° + 45°$
$\Rightarrow$ $\theta = n90° + 22\frac{1}{2}°$

$$
\begin{aligned}
n &= 0 &&\Rightarrow \theta = 22\tfrac{1}{2}° \\
n &= 1 &&\Rightarrow \theta = 112\tfrac{1}{2}° \\
n &= 2 &&\Rightarrow \theta = 202\tfrac{1}{2}° \text{ (out of range)} \\
n &= -1 &&\Rightarrow \theta = -67\tfrac{1}{2}° \\
n &= -2 &&\Rightarrow \theta = -157\tfrac{1}{2}° \\
n &= -3 &&\Rightarrow \theta = -247\tfrac{1}{2}° \text{ (out of range)}
\end{aligned}
$$

$\therefore$ solutions are $-157\frac{1}{2}°,\ -67\frac{1}{2}°,\ 22\frac{1}{2}°,\ 112\frac{1}{2}°$

Example 13   Find the general solution of $\sin 2\theta \cdot \cos 3\theta = 0$.

$$\sin 2\theta \cos 3\theta = 0$$
$$\Rightarrow \sin 2\theta = 0 \quad \text{and} \quad \cos 3\theta = 0$$

$$\Rightarrow \quad 2\theta = n\pi \quad \text{and} \quad 3\theta = 2n\pi \pm \frac{\pi}{2}$$

$$\Rightarrow \quad \theta = \frac{n\pi}{2} \quad \text{and} \quad \theta = 2n\frac{\pi}{3} \pm \frac{\pi}{6}$$

$\therefore$ general solution is $\theta = \dfrac{n\pi}{2}$ and $\theta = 2n\dfrac{\pi}{3} \pm \dfrac{\pi}{6}$

• Exercise 7.3 See page 135.

7.4 Applications of General Solutions

Example 14 Find the general solution of the equation $\tan 4x = \tan x$ and hence deduce the solutions between 0^c and π^c.

As $4x$ and x have the same tangent

$$4x = n\pi + x$$

$$\Rightarrow 3x = n\pi$$

$$\therefore \quad x = \frac{n\pi}{3} \text{ is the general solution}$$

$$n = 0 \Rightarrow x = 0$$
$$n = 1 \Rightarrow x = \pi/3$$
$$n = 2 \Rightarrow x = 2\pi/3$$
$$n = 3 \Rightarrow x = \pi$$

$\therefore$ solutions are $0, \pi/3, 2\pi/3, \pi$

Example 15 Find all solutions of $\sin 2\theta = \cos \theta$.

Method 1 This can be written as

$$\sin 2\theta = \sin (\pi/2 - \theta)$$
$$\Rightarrow 2\theta = n\pi + (-1)^n [\pi/2 - \theta]$$
$$\Rightarrow 2\theta + (-1)^n \theta = n\pi + (-1)^n \pi/2$$

$\Rightarrow$ general solution is $\theta = \dfrac{n\pi + (-1)^n \pi/2}{2 + (-1)^n}$

Method 2 Alternatively

$$\cos \theta = \cos (\pi/2 - 2\theta)$$
$$\Rightarrow \quad \theta = 2n\pi \pm (\pi/2 - 2\theta)$$
$$\Rightarrow \quad \theta = 2n\pi + \pi/2 - 2\theta \quad \text{and} \quad \theta = 2n\pi - \pi/2 + 2\theta$$
$$\Rightarrow \quad 3\theta = 2n\pi + \pi/2 \quad \text{and} \quad \theta = \pi/2 - 2n\pi$$

$\Rightarrow$ general solution is $\theta = 2n\pi/3 + \pi/6$ and $\theta = \pi/2 - 2n\pi$

> **Note** It is quite possible to have two (or more) equivalent general solutions. But, putting in the various values of *n* will still produce the same set of solutions.

Example 16 Find values in the range $0° \leqslant \theta \leqslant 360°$ which satisfy $\cos 3\theta = \cos \theta$.

General solution is $3\theta = 360°n \pm \theta$
Hence $4\theta = 360°n$ and $2\theta = 360°n \Rightarrow \theta = n90°$ and $n180°$
$\therefore$ for $0° \leqslant \theta \leqslant 360°$ we get $\underline{\theta = 0°, 90°, 180°, 270°, 360°}$

● **Exercise 7.4** See page 136.

7.5 Miscellaneous Trigonometric Equations: Revision of Unit

1 *Equations of a quadratic form*

Use
$$\sin^2 \theta + \cos^2 \theta = 1$$
$$\tan^2 \theta + 1 = \sec^2 \theta$$
$$1 + \cot^2 \theta = \csc^2 \theta$$
and
$$\cos 2\theta = 2\cos^2 \theta - 1 = 1 - 2\sin^2 \theta$$

See also 1.6 and 4.4

Example 17 Solve $6\sin^2 x + 5\cos x = 7$ giving values of x between $-180°$ and $180°$.

$$6\sin^2 x + 5\cos x = 7$$
$$\therefore \ 6(1 - \cos^2 x) + 5\cos x = 7$$
$$\therefore \ 6\cos^2 x - 5\cos x + 1 = 0$$
$$(3\cos x - 1)(2\cos x - 1) = 0$$
$$\cos x = \tfrac{1}{3} \quad \text{and} \quad \cos x = \tfrac{1}{2}$$
$$\therefore \ \underline{x = \pm 70.5° \text{ and } \pm 60°}$$

Example 18 Find the general solution of the equation $\cos 2\theta - 3\cos \theta + 2 = 0$.

$$\cos 2\theta - 3\cos \theta + 2 = 0$$
$$(2\cos^2 \theta - 1) - 3\cos \theta + 2 = 0$$
$$2\cos^2 \theta - 3\cos \theta + 1 = 0$$
$$(2\cos \theta - 1)(\cos \theta - 1) = 0$$
$$\cos \theta = \tfrac{1}{2} \quad \text{and} \quad \cos \theta = 1$$
$$\underline{\theta = 2n\pi \pm \pi/3 \quad \text{and} \quad \theta = 2n\pi}$$

2 Using the factor formulae

See also 7.1

Example 19 Solve $\cos x - \cos 4x = \cos 2x - \cos 3x$ given $-\pi \leqslant x \leqslant \pi$.

The equation can be written as

$$\cos x + \cos 3x = \cos 2x + \cos 4x$$
$$\Rightarrow \qquad 2\cos 2x \cdot \cos x = 2\cos 3x \cdot \cos x$$
$$\Rightarrow \cos x(\cos 3x - \cos 2x) = 0$$
$$\Rightarrow \cos x \cdot \sin\frac{5x}{2} \cdot \sin\frac{x}{2} = 0$$
$$\Rightarrow \cos x = 0 \quad \text{or} \quad \sin\frac{5x}{2} = 0 \quad \text{or} \quad \sin\frac{x}{2} = 0$$

$$\cos x = 0 \Rightarrow x = 2n\pi \pm \pi/2$$
$$\sin\frac{5x}{2} = 0 \Rightarrow \frac{5x}{2} = n\pi \Rightarrow x = \frac{2n\pi}{5}$$
$$\sin\frac{x}{2} = 0 \Rightarrow \frac{x}{2} = n\pi \Rightarrow x = 2n\pi$$

Hence roots between $-\pi$ and π are 0, $\pm\dfrac{2\pi}{5}$, $\pm\dfrac{4\pi}{5}$

3 Equations of the form $a\cos\theta + b\sin\theta = c$

Example 20 (Method 1)
Solve $6\cos\theta + 8\sin\theta = 9$ for $0° \leqslant \theta \leqslant 360°$.

See also 7.2

Let
$$6\cos\theta + 8\sin\theta \equiv R\cos(\theta - \alpha)$$
$$= R\cos\theta \cdot \cos\alpha + R\sin\theta \cdot \sin\alpha$$
$$\Rightarrow R\cos\alpha = 6 \quad \text{and} \quad R\sin\alpha = 8$$
$$\Rightarrow R = 10 \quad \text{and} \quad \tan\alpha = \tfrac{4}{3} \Rightarrow \alpha = 53.1°$$

$\therefore$ Equation becomes
$$10\cos(\theta - 53.1°) = 9$$
$$\therefore \quad \cos(\theta - 53.1°) = 0.9$$

and if $0° \leqslant \theta \leqslant 360°$ then $-53.1° \leqslant \theta - 53.1° \leqslant 306.9°$
$$\therefore \theta - 53.1° = \pm 25.8°,$$
Hence
$$\theta = 53.1° \pm 25.8°$$

$\therefore$ Solutions are $\theta = 27.3°$ and $78.9°$

Example 21   (Method 2) Solve $20 \cos x - 25 \sin x = 1$ for $-180° \leqslant x \leqslant 180°$.

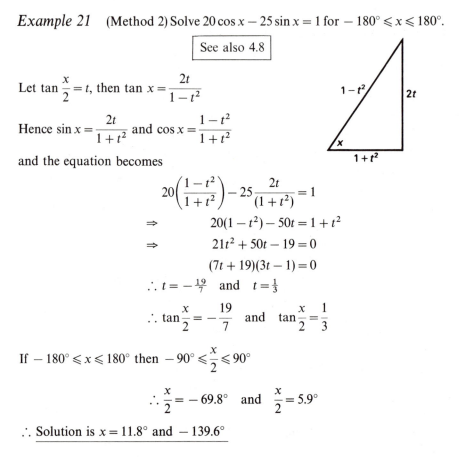

See also 4.8

Let $\tan \dfrac{x}{2} = t$, then $\tan x = \dfrac{2t}{1-t^2}$

Hence $\sin x = \dfrac{2t}{1+t^2}$ and $\cos x = \dfrac{1-t^2}{1+t^2}$

and the equation becomes

$$20\left(\frac{1-t^2}{1+t^2}\right) - 25\frac{2t}{(1+t^2)} = 1$$

$$\Rightarrow \qquad 20(1-t^2) - 50t = 1+t^2$$

$$\Rightarrow \qquad 21t^2 + 50t - 19 = 0$$

$$(7t+19)(3t-1) = 0$$

$$\therefore t = -\tfrac{19}{7} \quad \text{and} \quad t = \tfrac{1}{3}$$

$$\therefore \tan\frac{x}{2} = -\frac{19}{7} \quad \text{and} \quad \tan\frac{x}{2} = \frac{1}{3}$$

If $-180° \leqslant x \leqslant 180°$ then $-90° \leqslant \dfrac{x}{2} \leqslant 90°$

$$\therefore \frac{x}{2} = -69.8° \quad \text{and} \quad \frac{x}{2} = 5.9°$$

$\therefore$ Solution is $x = 11.8°$ and $-139.6°$

Note   When some of the constants are not integral, method 1 is usually simpler.

● **Miscellaneous Exercise 7A**   See page 137.

7.6 + Small Angles: A Level Questions

For small angles

| $\sin \theta \approx \theta$ |
| $\cos \theta \approx 1 - \theta^2/2$ |
| $\tan \theta \approx \theta$ |

θ in radians

Example 22   If θ is so small that θ^3 and higher powers can be ignored, find approximations in terms of θ for:

(a) $\sin \dfrac{\theta}{2}$ (b) $\sin 2\theta$ (c) $\sin\left(\dfrac{\pi}{6} + \theta\right).$

(a) If θ is small, then $\dfrac{\theta}{2}$ is even smaller $\therefore$ $\sin\dfrac{\theta}{2}\approx\dfrac{\theta}{2}$

(b) $\sin 2\theta = 2\sin\theta\cdot\cos\theta$

$$\approx 2\cdot\theta\cdot\left(1-\dfrac{\theta^2}{2}\right)$$

$$\approx 2\theta - \theta^3$$

but θ^3 is negligible $\therefore$ $\sin 2\theta \approx 2\theta$

(c) $(\pi/6 + \theta)$ is not a small angle so $\sin(\pi/6+\theta) \not\approx (\pi/6+\theta)$

$$\sin\left(\dfrac{\pi}{6}+\theta\right) = \sin\dfrac{\pi}{6}\cdot\cos\theta + \cos\dfrac{\pi}{6}\cdot\sin\theta$$

$$\approx \dfrac{1}{2}\left(1-\dfrac{\theta^2}{2}\right) + \dfrac{\sqrt3}{2}\cdot\theta$$

$$= \dfrac{1}{2} - \dfrac{\theta^2}{4} + \dfrac{\sqrt{3}\theta}{2}$$

$$\therefore \sin\left(\dfrac{\pi}{6}+\theta\right) \approx \dfrac{1}{2} + \dfrac{\sqrt{3}\theta}{2} - \dfrac{\theta^2}{4}$$

Example 23 If θ is so small that θ^2 and higher powers can be ignored, find an approximation in terms of θ for $\cos(\theta + \pi/3)$.

Here $\sin\theta \approx \theta$ and $\cos\theta \approx 1$

$$\cos(\theta + \pi/3) = \cos\theta\cdot\cos\pi/3 - \sin\theta\cdot\sin\pi/3$$

$$\approx 1\cdot\dfrac{1}{2} - \theta\cdot\dfrac{\sqrt3}{2}$$

$$\therefore \cos(\theta + \pi/3) \approx \tfrac{1}{2}(1 - \sqrt3\theta)$$

- Miscellaneous Exercise 7B See page 137.

Unit 7 EXERCISES

Exercise 7.1

Factorise:

1 $\sin 3A + \sin A$ **2** $\cos 5A + \cos A$ **3** $\cos 3A - \cos A$

4 $\cos 7A + \cos A$ **5** $\sin 3A - \sin 2A$ **6** $\cos 2A - \cos 4A$

7 $\cos 2A + \cos 2B$ **8** $\sin 5A - \sin 3B$ **9** $\sin 30° + \sin 60°$

10 $\cos 65° - \cos 55°$

Express as a sum or difference:

11 $2\cos 3\theta \cdot \cos 2\theta$ **12** $2\cos \theta \cdot \sin 4\theta$ **13** $-2\sin 4\theta \cdot \sin 2\theta$

14 $2\sin 3\theta \cdot \sin \theta$ **15** $\cos 5\theta \cdot \cos \theta$ **16** $\cos 60° \cdot \sin 30°$

17   Factorise and hence evaluate:
(a) $\cos 75° + \cos 15°$ (b) $\cos 75° - \cos 15°$

Solve the following equations, giving roots between $0°$ and $180°$ inclusive:

18 $\cos 3x + \cos x = 0$ **19** $\sin 5x - \sin x = 0$

20 $\cos 7x - \cos x = 0$ **21** $\cos 5x + \cos x = \cos 3x$

22 $\sin 9x + \sin 3x = \sin 6x$

Simplify the following fractions:

23 $\dfrac{\sin 2x + \sin x}{\cos 2x + \cos x}$ **24** $\dfrac{\sin 2x - \sin x}{\cos 2x - \cos x}$

25 $\dfrac{\sin 3\theta + \sin \theta}{\cos 3\theta - \cos \theta}$ **26** $\dfrac{\cos 7\theta + \cos 3\theta}{\sin 7\theta + \sin 3\theta}$

27 $\dfrac{\sin 3\theta/2 + \sin \theta/2}{\cos 5\theta/2 + \cos 3\theta/2}$ **28** $\dfrac{\cos X - \cos Y}{\sin X + \sin Y}$

Prove the following identities:

29 $\dfrac{\sin 3x + \sin x}{\cos 3x + \cos x} \equiv \tan 2x$ **30** $\dfrac{\sin 3A - \sin A}{\cos 3A - \cos A} \equiv -\cot 2A$

31 $\dfrac{\sin 7\theta - \sin 5\theta}{\cos 7\theta + \cos 5\theta} \equiv \tan \theta$ **32** $\dfrac{\sin A + \sin 2A}{\cos A - \cos 2A} \equiv \cot \dfrac{A}{2}$

Solve, giving values between $0°$ and $180°$ inclusive:

*33 $\cos \theta + \cos 3\theta = \sin \theta + \sin 3\theta$ ***34** $\sin 3\theta + \sin 6\theta + \sin 9\theta = 0$

*35 $\sin 3x - \sin x = \cos 2x$ ***36** $\cos 5y = \cos 3y - \cos y$

**37 $\cos 2\phi = \cos (30° - \phi)$

Exercise 7.2

1   Express $\cos x + \sin x$ in the form $R\cos (x - \alpha)$

2   Express $2\cos x - \sqrt{2}\sin x$ in the form $R\cos (x + \alpha)$

3   Express $2\sin x - \cos x$ in the form $R\sin (x - \alpha)$

4 Find the maximum and minimum values of $4 \cos x + 3 \sin x$

5 Find the value(s) of x between $0°$ and $540°$ for which $3 \cos x + \sqrt{3} \sin x$ is a maximum

6 Find the values of x between $-180°$ and $180°$ for which $3 \sin x + 4 \cos x$ is zero

7 Solve $3 \cos x + 4 \sin x = 2$ for $0° \leqslant x \leqslant 360°$

8 Solve $2 \cos x - \sin x = 1$ for $0° \leqslant x \leqslant 360°$

9 Sketch the graph of $y = \sqrt{3} \cos x - \sin x$ for $-180° \leqslant x \leqslant 180°$

10 Sketch the graph of $y = \cos x + 3 \sin x$ for $-180° \leqslant x \leqslant 180°$

*11 Express $2 \sin 2x + \cos 2x$ in the form $R \sin(2x + \alpha)$. Hence solve $2 \sin 2x + \cos 2x = 1$ for $0° \leqslant x \leqslant 180°$

*12 Solve $5 \cos 3x - 12 \sin 3x = 7.5$ for $0° \leqslant x \leqslant 180°$

**13 Express $7 \cos \theta + 24 \sin \theta$ in the form $R \cos(\theta - \alpha)$. Hence find the maximum and minimum values of the following functions, stating in each case, the values of θ between $0°$ and $360°$ at which the stationary values occur:

(a) $f(\theta) = 7 \cos \theta + 24 \sin \theta - 3$ (b) $g(\theta) = (7 \cos \theta + 24 \sin \theta)^2$

(c) $h(\theta) = \dfrac{1}{(7 \cos \theta + 24 \sin \theta)^2}$

Exercise 7.3

Find the general solutions of each of the following equations:

1 $\sin \theta = \dfrac{1}{\sqrt{2}}$ **2** $\cos \theta = \dfrac{1}{2}$ **3** $\tan \theta = 1$

4 $\sin \theta = -\dfrac{\sqrt{3}}{2}$ **5** $\cos \theta = -\dfrac{1}{\sqrt{2}}$ **6** $\tan \theta = \dfrac{1}{\sqrt{3}}$

7 $\operatorname{cosec} \theta = 2$ **8** $\cot \theta = -\dfrac{1}{\sqrt{3}}$ **9** $\sec \theta = \sqrt{2}$

10 $\sin \theta = 0.2474$ **11** $\cos \theta = 0.9492$ **12** $\tan \theta = 1.5574$

13 $\sin 2\theta = \dfrac{1}{2}$ **14** $\cos 3\theta = \dfrac{\sqrt{3}}{2}$ **15** $\tan 4\theta = -1$

16 $\sin \dfrac{\theta}{3} = 0.5$ **17** $\tan \dfrac{5\theta}{2} = 0$ **18** $\sin^2 \theta = 1$

19 $\cos^2 \theta = \dfrac{1}{2}$ **20** $3 \tan^2 \theta = 1$ **21** $3 \cot^2 \theta - 1 = 0$

22 $(2 \sin \theta - 1)(\sin \theta - 1) = 0$ **23** $\sin 3\theta (\sin 3\theta + 1) = 0$

24 Find all the solutions of $\tan^2 \theta - 2 \tan \theta + 1 = 0$

25 Find the general solution of $\sin 2\theta = \sqrt{3}/2$. Hence find all solutions of the equation which lie between 0^c and π^c

*26 Find the general solution of $\cot^2 (\theta/2) = 1$ and hence find all solutions lying between $-360°$ and $360°$

*27 Find the general solution of $\tan^2 2\theta - 3 \tan 2\theta + 2 = 0$

Exercise 7.4

Find the general solutions of each of the following equations:

1 $\sin 2\theta = \sin \theta$ **2** $\cos 3\theta = \cos \theta$

3 $\tan 3\theta = \tan \theta$ **4** $\sin 5\theta = \sin 3\theta$

5 $\tan (\pi/4 - \theta) = \tan \theta$

6 Use the fact that $\cot \alpha = \tan (\pi/2 - \alpha)$ to solve $\tan 2\theta = \cot \theta$

7 Find the general solution of $\sin 3\theta = \cos \theta$

In questions 8–11 find the general solution and hence find all the solutions from 0^c to π^c inclusive:

8 $\sin 3\theta = \sin \dfrac{\pi}{5}$ **9** $\cos \dfrac{3\theta}{2} = \sin \dfrac{\pi}{6}$

10 $\cos 7\theta = \cos 3\theta$ **11** $\tan 3\theta = \cot \theta$

In questions 12–14 find the general solutions and hence find all the solutions from $0°$ to $360°$ inclusive:

12 $\sin 2\theta = \sin 60°$ **13** $\tan 3\theta = \tan (\theta - 60°)$

*14 $\cos (2\theta + 30°) = \cos \theta$

If $0^c \leqslant \theta \leqslant \pi^c$ solve the following:

15 $\sin 5\theta = \sin (\theta - \pi/6)$ ****16 $\sin 3\theta = \cos 2\theta$

If $0° \leqslant \theta \leqslant 360°$ solve the following:

*17 $\tan (\theta + 30°) = \tan (2\theta - 40°)$

*18 $\sin 2\theta = \cos (\theta + 45°)$

Miscellaneous Exercise 7A

[Give angles correct to 0.1° where applicable]

1 Use $\sin^2\theta + \cos^2\theta = 1$, $\tan^2\theta + 1 = \sec^2\theta$, $1 + \cot^2\theta = \operatorname{cosec}^2\theta$ to solve the following equations, given that $0° \leqslant \theta \leqslant 360°$:

(a) $3\cos^2\theta + 5\sin\theta = 1$ (b) $8\sin^2\theta + 2\cos\theta = 5$
(c) $\tan^2\theta = \sec\theta + 5$ (d) $7\sin^2\theta + \cos^2\theta = 5\sin\theta$
(e) $2\sec^2\theta + 3\tan\theta = 4$ (f) $4\cot^2\theta - 3\operatorname{cosec}^2\theta = 2\cot\theta$

2 Use the double angle formulae to solve the following equations for $0° \leqslant x \leqslant 360°$:

(a) $\cos 2x = 5\cos x + 2$ (b) $3\cos 2x + 1 = 2\sin x$
(c) $\sin 2x = \cos x$ (d) $\cos 2x = \cos x$
(e) $\tan 2x + \tan x = 0$ (f) $1 - 2\sin x = 4\cos 2x$

Solve the following equations, given that $-180° \leqslant x \leqslant 180°$

3 $3\sin 2x = 2\cos x$ **4** $\tan x + 2\cot x = 3$

5 $\sin x + \sin 3x + \sin 2x = 0$ **6** $\sin 5x - \sin x = \cos 3x$

7 $2\cos x + 5\sin x = 4$ **8** $3\cos x + 7\sin x + 3 = 0$

9 $3\sec^2 x - 5\tan x = 5$ **10** $15\sin^2 x + 2\cos x = 14$

11 $\tan 2x = 3\tan x$ ***12** $4\sin 2x - 3\cos 2x = 3$

*13 $\cos 2x + \sin x = 0.8$ ***14** $\cos 3x - \sin 3x = \cos 2x - \sin 2x$

**13 $2\cos 2x = \cos x - \sin x$

Express in radians the general solutions of the following equations:

16 $\sin 2\theta = \sin\theta$ **17** $\sin 2\theta = \cos\theta$

18 $\tan 3\theta = \cot\theta$ ***19** $\cos 3\theta + 2\cos 5\theta + \cos 7\theta = 0$

*20 $\sin\theta + \sqrt{3}\cos\theta = 1$ ****21** $6\cos\theta - 3\sin\theta + 4 = 0$

Miscellaneous Exercise 7B

1 If θ is so small that θ^3 and higher powers can be ignored, find approximations in terms of θ for:

(a) $\cos\dfrac{\theta}{2}$ (b) $\cos 3\theta$ (c) $\cos\left(\dfrac{\pi}{2} - \theta\right)$ (d) $\sin\left(\theta - \dfrac{\pi}{4}\right)$

2 Express $\sin x - \cos x$ in the form $R\sin(x - \alpha)$, where R is positive and α is an acute angle. Hence, or otherwise, find, in radians, the general solution of the equation $\sin x - \cos x = 1$ (L: 7.2)

3 Find all the values of θ in the range $0 \leqslant \theta \leqslant 2\pi$ for which
$\sin\theta + \sin 3\theta = \cos\theta + \cos 3\theta$ (J: 7.1, 7.5)

4   Find all the values of x for which $0 \leqslant x \leqslant 2\pi$ and $\cos 2x = 1 + \sin x$

(L: 4.4, 7.4)

5   Write down the expressions for $\sin \theta$ and $\cos \theta$ in terms of t, where $t = \tan \frac{1}{2}\theta$. Hence, or otherwise, obtain the solutions of the equation $2 \sin \theta° - 3 \cos \theta° = 1$ in the range $0 < \theta < 360$. Give your solutions to the nearest integer

(L: 4.8, 7.5)

6   Express $\sqrt{3} \sin \theta - \cos \theta$ in the form $R \sin (\theta - \alpha)$ where R is positive. Find all values of θ in the range $0° \leqslant \theta \leqslant 360°$ which satisfy the equation $4 \sin \theta \cdot \cos \theta = \sqrt{3} \sin \theta - \cos \theta$

(J: 7.2)

7   By means of the substitution $\tan \theta = t$, or otherwise, find the value of θ in the range $0 \leqslant \theta \leqslant \pi/2$ such that

$$(2 - \tan \theta)(1 + \sin 2\theta) - 2 = 0$$

Show that, when θ is small enough for powers above the first to be negligible $(2 - \tan \theta)(1 + \sin 2\theta) - 2 \approx 3\theta$

(J: 7.6)

8   Prove that $\sin 5A - \sin A + \sin 2A = \sin 2A(2 \cos 3A + 1)$

Hence solve the equation $\sin 5\theta + \sin 2\theta = \sin \theta$ for values of θ in the interval $0° < \theta < 180°$

(A: 7.1)

9   Given that $3 \sin x - \cos x \equiv R \sin (x - \alpha)$, where $R > 0$ and $0° < \alpha < 90°$, find the values of R and α correct to one decimal place.

Hence find one value of x between $0°$ and $360°$ for which the curve $y = 3 \sin x - \cos x$ has a turning point

(L: 7.2)

10 (a)   Find the general solution, in radians, of the equation $\sin 3x = -\frac{1}{2}$
    (b)   By putting $\tan (\theta/2) = t$, or otherwise, find the general solution of the equation $2 \cos \theta - \sin \theta = 1$, giving your answers to the nearest tenth of a degree

(L: 7.3, 7.5)

11 Find, to 0.1 of a degree, the acute angle α for which

$$4 \cos \theta - 3 \sin \theta \equiv 5 \cos (\theta + \alpha)$$

Calculate the values of θ, in the interval $-180° \leqslant \theta \leqslant 180°$, for which the function

$$f(\theta) = 4 \cos \theta - 3 \sin \theta - 4$$

attains its greatest value, its least value and the value zero

(J: 7.2)

12 Find the general solutions of
    (a) $\tan 7\theta = \tan \theta$
    (b) $\sin 5\theta - \sin \theta = \cos 3\theta$

(J: 7.1, 7.4)

13 Express $f(\theta) = 2 \cos \frac{1}{2}\theta \cdot \cos \frac{1}{2}(\theta + \pi/3)$ as the sum of two cosines. Deduce the maximum and minimum values of $f(\theta)$ as θ varies over the range $-\pi \leqslant \theta \leqslant \pi$, and the values of θ, in this range, at which they occur.

If θ is small, find the values of the constants a, b and c in the approximate formula $f(\theta) \approx a + b\theta + c\theta^2$ (J: 7.1, 7.6)

14 Find all the solutions, in terms of π, in the interval $0 \leqslant \theta \leqslant 2\pi$ of *each* of the following equations:

(a) $\cos 2\theta + \cos \theta = 0$
(b) $\tan 2\theta - 3 \tan \theta = 0$
(c) $\sqrt{3} \cos \theta + \sin \theta + 1 = 0$ (J: 7.5)

15 Given that $y = 3 \sin \theta + 3 \cos \theta$, express y in the form $R \sin(\theta + \alpha)$ where $R > 0$ and $0° < \alpha < 90°$

Hence find (a) the greatest and least values of y^2, and (b) the values of θ in the interval $0°$ to $90°$ for which $y = 3\sqrt{6}/2$ (A: 7.2)

16 Use the trigonometrical formula for $\sin(A + B)$ and the numerical data $\sin 1 \approx 0.84147$ and $\cos 1 \approx 0.54030$ to estimate, *without* the use of tables or a calculator, $\sin(1.0002)$ to 4 decimal places   (L: 4.1, 7.6)

17 In the $\triangle ABC$, $BC = a$, $CA = b$, $AB = c$ and angle $B\hat{A}C = \theta^c$. Given that θ is small enough for its cube and higher powers to be negligible, use the cosine rule to show that $\theta^2 \approx \dfrac{a^2 - (b - c)^2}{bc}$

In the case when $a = 3$, $b = 50$, $c = 49$ show without using any calculating aids that (a) $\theta \approx \frac{2c}{35}$ (b) $\sin B \approx \frac{20}{21}$ (J: 4.5, 7.6)

18 Write down expressions for $\tan \theta$ and $\sec \theta$ in terms of t where $t = \tan \theta/2$ and show that
$$\sec \theta + \tan \theta = \tan(45° + \theta/2)$$
Find a solution in the interval $0° < \theta < 90°$ of the equation $\sec \theta + \tan \theta = \cot 2\theta$

19 Given that $x = \sin(\theta + 105°) + \sin(\theta - 15°) + \sin(\theta + 45°)$
        and $y = \cos(\theta + 105°) + \cos(\theta - 15°) + \cos(\theta + 45°)$

show that $\dfrac{x}{y} = \tan(\theta + 45°)$

Express $\dfrac{x}{y} - \dfrac{y}{x}$ in terms of $\tan 2\theta$ (J: 7.1, 4.1)

20

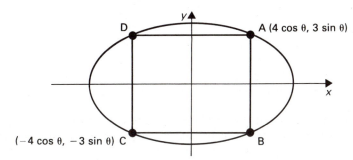

A rectangle has vertices on the ellipse $\dfrac{x^2}{16} + \dfrac{y^2}{9} = 1$. The sides of the rectangle are parallel to the coordinate axes and two vertices are $(4\cos\theta, 3\sin\theta)$ and $(-4\cos\theta, -3\sin\theta)$. Express the perimeter of the rectangle in the form $R\sin(\theta + \alpha)$. Determine the value of θ, to the nearest $0.1°$, for which the perimeter of the rectangle is a maximum and state the value of the maximum perimeter   (J: 7.2)

21 Find the general solution of the equation

$$\cos 2x - \sin 2x = \cos x - \sin x - 1 \qquad \text{(J: 7.3, 7.5)}$$

*22 Given that $y = \dfrac{\sin\theta - 2\sin 2\theta + \sin 3\theta}{\sin\theta + 2\sin 2\theta + \sin 3\theta}$, prove that $y = -\tan^2\dfrac{\theta}{2}$.

Find

(a) the exact value of $\tan^2 15°$ in the form $p + q\sqrt{r}$, where p, q and r are integers

(b) the values of θ between $0°$ and $360°$ for which $2y + \sec^2\dfrac{\theta}{2} = 0$

(A: 7.1, 4.3, 1.6)

*23 Given that $\cos\theta + \sin\theta = \sin 2\theta$, show, by squaring both sides of the equation, or otherwise, that $\sin 2\theta = \frac{1}{2}(1 - \sqrt{5})$

Find the general solution of the second equation and hence find the smallest positive solution of the first equation, giving your answers correct to the nearest degree   (J: 4.3, 1.7)

*24 Find the range of values of a for which the equation

$$\cos(x + 90°) + \cos x = a$$

has real solutions. For the case when $a = 0$ find all the solutions in the interval $0° \leqslant x \leqslant 360°$. Sketch the graph of $y = \cos(x + 90°) + \cos x$ for $0° \leqslant x \leqslant 360°$ (J: 7.2)

*25

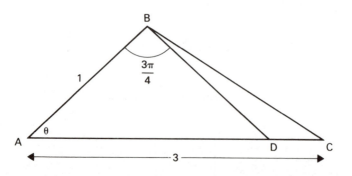

In the $\triangle$ ABC, AB $= 1$ unit, AC $= 3$ units and the angle BAC is θ radians. D is the point on AC such that the angle ABD is $3\pi/4$. Prove that

(a) $BC = (10 - 6\cos\theta)^{1/2}$
(b) $AD = 1/(\cos\theta - \sin\theta)$

Given that θ is small enough for the approximations

$$\sin\theta \approx \theta \quad \text{and} \quad \cos\theta \approx 1 - \tfrac{1}{2}\theta^2$$

to apply, show that

(d) $BC \approx p + q\theta^2$
(e) $AD \approx 1 + r\theta + s\theta^2$

where p, q, r and s are constants to be determined   (J: 4.5, 6.8, 7.6)

*26   Express $\cos\theta - \sin\theta$ in the form $R\cos(\theta + \alpha)$ where R is positive and $0 < \alpha < \pi/2$.

(a)   Find both solutions in the interval $0 \leqslant x \leqslant 2\pi$ of the equation

$$\cos 2x - \sin 2x = \sqrt{2}$$

(b)   Calculate, in the interval $0 \leqslant x \leqslant 2\pi$, the coordinates of the maximum and minimum points of the curves given by the following equations,

(i) $y = 1 - \cos x + \sin x$

(ii) $y = \dfrac{1}{\cos x - \sin x}$

stating, in each case, the x-coordinates in terms of π and the y-coordinates in surd form.
Sketch the curve $y = 1 - \cos x + \sin x$ in the interval $0 \leqslant x \leqslant 2\pi$
          (J: 7.2)

8 UNIT CALCULUS III

8.1 Differentiation and Integration of Trigonometric Functions

1 Let
$$y = \sin x$$
$$y + \delta y = \sin(x + \delta x)$$
$$\frac{\delta y}{\delta x} = \frac{\sin(x + \delta x) - \sin x}{\delta x}$$

Using the factor formulae

$$\frac{\delta y}{\delta x} = \frac{2\cos\left(x + \frac{\delta x}{2}\right)\sin\left(\frac{\delta x}{2}\right)}{\delta x}$$

$$\frac{\delta y}{\delta x} = \cos\left(x + \frac{\delta x}{2}\right)\left[\frac{\sin\left(\frac{\delta x}{2}\right)}{\left(\frac{\delta x}{2}\right)}\right]$$

As $\delta x \to 0$, $\frac{\delta x}{2} \to 0$ and $\frac{\sin\left(\frac{\delta x}{2}\right)}{\left(\frac{\delta x}{2}\right)} \to 1$

$$\therefore \frac{dy}{dx} = \lim_{\delta x \to 0} \frac{\delta y}{\delta x} = \cos x$$

2 Let $y = \cos x$
$$y + \delta y = \cos(x + \delta x)$$
$$\frac{\delta y}{\delta x} = \frac{\cos(x + \delta x) - \cos x}{\delta x}$$

$$\frac{\delta y}{\delta x} = \frac{-2\sin\left(x + \frac{\delta x}{2}\right) \cdot \sin\left(\frac{\delta x}{2}\right)}{\delta x}$$

$$\frac{\delta y}{\delta x} = -\sin\left(x + \frac{\delta x}{2}\right)\left[\frac{\sin\left(\frac{\delta x}{2}\right)}{\left(\frac{\delta x}{2}\right)}\right]$$

As $\delta x \to 0$, $\frac{dy}{dx} = \lim_{\delta x \to 0} \frac{\delta y}{\delta x} = -\sin x$

3 Let $y = \tan x$

$y + \delta y = \tan(x + \delta x)$

$$\frac{\delta y}{\delta x} = \frac{\tan(x + \delta x) - \tan x}{\delta x}$$

$$\frac{\delta y}{\delta x} = \frac{\dfrac{\sin(x + \delta x)}{\cos(x + \delta x)} - \dfrac{\sin x}{\cos x}}{\delta x}$$

$$\frac{\delta y}{\delta x} = \frac{\sin(x + \delta x)\cos x - \sin x \cos(x + \delta x)}{\delta x \cdot \cos(x + \delta x)\cos x}$$

$$\frac{\delta y}{\delta x} = \frac{\sin \delta x}{\delta x \cos(x + \delta x)\cos x}$$

$$\frac{\delta y}{\delta x} = \frac{1}{\cos(x + \delta x)\cos x}\left[\frac{\sin \delta x}{\delta x}\right]$$

As $\delta x \to 0$, $\dfrac{dy}{dx} = \lim\limits_{\delta x \to 0}\dfrac{\delta y}{\delta x} = \sec^2 x$

$\dfrac{d}{dx}(\sin x) = \cos x$	$\displaystyle\int \sin x\,dx = -\cos x + c$
$\dfrac{d}{dx}(\cos x) = -\sin x$	$\displaystyle\int \cos x\,dx = \sin x + c$
$\dfrac{d}{dx}(\tan x) = \sec^2 x$	$\displaystyle\int \sec^2 x\,dx = \tan x + c$

Note x must be in RADIANS.

Example 1   Differentiate $5 \sin x - 6 \cos x$.

$$\frac{d}{dx}(5 \sin x - 6 \cos x) = 5 \cos x - 6(-\sin x) = \underline{5 \cos x + 6 \sin x}$$

Example 2   Evaluate $\displaystyle\int_0^{\pi}[3 \sin x - \sec^2 x]\,dx$

$$\int_0^{\pi}[3 \sin x - \sec^2 x]\,dx = [-3 \cos x - \tan x]_0^{\pi}$$

$$= (-3(-1) - 0) - (-3(1) - 0) = \underline{6}$$

Example 3   If $s = 3 \sin t + 2 \cos t$ show that $\dfrac{d^2 s}{dt^2} + s = 0$

$$s = 3 \sin t + 2 \cos t$$

$$\frac{ds}{dt} = 3\cos t - 2\sin t$$

$$\frac{d^2s}{dt^2} = -3\sin t - 2\cos t = -(3\sin t + 2\cos t) = -s$$

$$\therefore \frac{d^2s}{dt^2} + s = 0$$

Example 4   Find the area contained between the x-axis and one loop of the curve $y = \sin x$, where x is in radians.

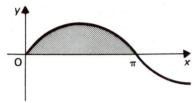

$$\text{Area of one loop} = \int_0^\pi \sin x \, dx$$

$$[-\cos x]_0^\pi$$

$$= -(-1) + (1) = \underline{2}$$

● Exercise 8.1   See page 155.

8.2 Trig. Functions and the Chain Rule

Example 5   Differentiate w.r.t x: (a) $\sin(2x + 3)$ (b) $\cos\dfrac{3x}{4}$

(c) $\tan(3x - 4)$

(a) $\dfrac{d}{dx}(\sin(2x + 3)) = 2\cos(2x + 3)$

derivative of bracket

(b) $\dfrac{d}{dx}\left(\cos\dfrac{3x}{4}\right) = \dfrac{3}{4}\left(-\sin\dfrac{3x}{4}\right) = -\dfrac{3}{4}\sin\dfrac{3x}{4}$

(c) $\dfrac{d}{dx}(\tan(3x - 4)) = 3\sec^2(3x - 4)$

Example 6   Integrate w.r.t. x: (a) $\sin(2x - 3)$ (b) $\sec^2(5x + 1)$

(a) $\displaystyle\int \sin(2x - 3)\,dx = -\tfrac{1}{2}\cos(2x - 3) + c$

(b) $\displaystyle\int \sec^2(5x + 1)\,dx = \tfrac{1}{5}\tan(5x + 1) + c$

Always check answers to integration by differentiating your answer.

Example 7   Differentiate w.r.t. x: (a) $\cos^2 x$ (b) $\tan^3(2x)$

(a) $\dfrac{d}{dx}(\cos^2 x) = \dfrac{d}{dx}(\cos x)^2 = 2\cos x(-\sin x) = \underline{-2\sin x \cos x}$

(b) $\dfrac{d}{dx}(\tan^3(2x)) = \dfrac{d}{dx}(\tan 2x)^3 = 3(\tan 2x)^2\cdot(\sec^2 2x) \times 2 = \underline{6\tan^2 2x\cdot\sec^2 2x}$

● Exercise 8.2   See page 156.

8.3 Differentiating Products and Quotients

Product

$$\dfrac{d}{dx}(uv) = v\dfrac{du}{dx} + u\dfrac{dv}{dx}$$

or

$$(uv)' = vu' + uv'$$

Quotient

$$\dfrac{d}{dx}\left(\dfrac{u}{v}\right) = \dfrac{v\dfrac{du}{dx} - u\dfrac{dv}{dx}}{v^2}$$

or

$$\left(\dfrac{u}{v}\right)' = \dfrac{vu' - uv'}{v^2}$$

Eample 8   Differentiate (a) $3x^2 \sin x$ (b) $\dfrac{4}{1-2x}$.

(a) $\left(\underset{u}{\underbrace{3x^2}}\cdot\underset{v}{\underbrace{\sin x}}\right)' = \overset{u'\ v}{6x\sin x} + \overset{u\ v'}{3x^2\cos x}$

(b) $\left(\underset{v}{\underbrace{\dfrac{4}{1-2x}}}\right)'\overset{u}{} = \dfrac{\overset{v}{(1-2x)}\cdot\overset{u'}{0} - \overset{u}{4}\cdot\overset{v'}{(-2)}}{(1-2x)^2} = \dfrac{8}{(1-2x)^2}$

Example 9   Differentiate $\dfrac{x^2 \sin 2x}{1-x}$.

$$\dfrac{d}{dx}\left(\dfrac{x^2\sin 2x}{1-x}\right) = \dfrac{(1-x)\dfrac{d}{dx}(x^2\sin 2x) - x^2\sin 2x\cdot(-1)}{(1-x)^2}$$

$$= \frac{(1-x)[x^2 \cdot 2\cos 2x + 2x \cdot \sin 2x] + x^2 \sin 2x}{(1-x)^2}$$

$$= \frac{2x}{1-x}(x\cos 2x + \sin 2x) + \frac{x^2 \sin 2x}{(1-x)^2}$$

Example 10   Differentiate $\sec \theta$.

$$\frac{d}{d\theta}(\sec \theta) = \frac{d}{d\theta}\left(\frac{1}{\cos \theta}\right)\frac{u}{v} = \frac{\cos \theta \cdot 0 - 1 \cdot (-\sin \theta)}{\cos^2 \theta}$$

$$= \frac{\sin \theta}{\cos^2 \theta} = \frac{\sin \theta}{\cos \theta} \times \frac{1}{\cos \theta}$$

$$\therefore \frac{d}{d\theta}(\sec \theta) = \sec \theta \cdot \tan \theta$$

Example 11   Find, and classify, any turning points on the curve $y = \dfrac{1}{x(x-2)}$.

$$y = \frac{1}{x^2 - 2x}$$

$$\frac{dy}{dx} = \frac{(x^2 - 2x) \cdot 0 - 1 \cdot (2x - 2)}{(x^2 - 2x)^2} = \frac{2 - 2x}{(x^2 - 2x)^2}$$

For turning points, $\dfrac{dy}{dx} = 0 \Rightarrow 2 - 2x = 0 \Rightarrow \left.\begin{array}{l} x = 1 \\ y = -1 \end{array}\right\}$

$\therefore$ there is a turning point at $(1, -1)$

$$\frac{d^2y}{dx^2} = \frac{(x^2 - 2x)^2(-2) - (2 - 2x)2(x^2 - 2x)(2x - 2)}{(x^2 - 2x)^4}$$

where $x = 1$ $\dfrac{d^2y}{dx^2} = \dfrac{(-1)^2(-2) - 0}{(-2)^4} < 0 \Rightarrow$ max t.p.

Note   In this kind of question there is no need to simplify $\dfrac{d^2y}{dx^2}$ before putting the value of x into it.

$\therefore$ there is a maximum t.p. at $(1, -1)$

● **Exercise 8.3**   See page 158.

8.4 Differentiation of Implicit Functions

If $y = x^2 + 2x - 3$ then we say that y is given **explicitly** in terms of x.

If $y^3 + 2xy + x^2 - x - 3 = 0$ then we say that y is given **implicitly** in terms of x.

In cases like this, it is often difficult, if not impossible, to express y explicitly in terms of x. When we have implicit functions, the method of differentiation in the following examples is used.

Example 12   Differentiate $y^2 + x^2 = 16$.

Differentiate each term w.r.t. x

i.e.
$$\frac{d}{dx}(y^2) + \frac{d}{dx}(x^2) = \frac{d}{dx}(16)$$

but
$$\frac{d}{dx}(y^2) = \frac{d}{dy}(y^2) \times \frac{dy}{dx} \qquad \text{(chain rule)}$$

$$= 2y \cdot \frac{dy}{dx}$$

$$\therefore\ 2y \cdot \frac{dy}{dx} + 2x = 0$$

If, however, you are asked to find $\dfrac{dy}{dx}$, then

$$\frac{dy}{dx} = -\frac{2x}{2y} = -\frac{x}{y}$$

Example 13   If $3x^2 - y = 2y^2$ find $\dfrac{dy}{dx}$.

Differentiating w.r.t. x

$$\Rightarrow 6x - \frac{dy}{dx} = 4y \cdot \frac{dy}{dx}$$

$$\frac{dy}{dx}(4y + 1) = 6x$$

$$\frac{dy}{dx} = \frac{6x}{1 + 4y}$$

Example 14   Find the gradient of the curve $xy + x^3 = 3y^2$ at the point $(2, 2)$.

Hence $\left[\ \underbrace{x\dfrac{dy}{dx} + y \cdot 1}\ \right] + 3x^2 = 6y \cdot \dfrac{dy}{dx}.$ | Note: xy is a product. |

derivative of xy

When $x = 2$, $y = 2$ this becomes

$$2\frac{dy}{dx} + 2 + 12 = 12\frac{dy}{dx}$$

$$\Rightarrow \qquad 10\frac{dy}{dx} = 14$$

$$\frac{dy}{dx} = 1.4$$

Example 15   If $x^2 - y^2 + 10x - 5y + 19 = 0$ (1)

find the points where the gradient is zero and find the value of $\dfrac{d^2y}{dx^2}$ at these points.

Differentiating implicitly $\Rightarrow 2x - 2y\dfrac{dy}{dx} + 10 - 5\dfrac{dy}{dx} = 0$ (2)

When $\dfrac{dy}{dx} = 0$ the equation
$$\begin{aligned} &\Rightarrow 2x - 2y(0) + 10 - 5(0) = 0 \\ &\Rightarrow 2x + 10 = 0 \\ &\Rightarrow x = -5 \end{aligned}$$

Putting $x = -5$ in (1)
$$\begin{aligned} &\Rightarrow y^2 + 5y + 6 = 0 \\ &(y + 2)(y + 3) = 0 \\ &\Rightarrow y = -2 \quad \text{and} \quad y = -3 \end{aligned}$$

$\therefore$ gradient is zero at $(-5, -2)$ and $(-5, -3)$

We need to find $\dfrac{d^2y}{dx^2}$

Method 1
Differentiating (2) implicitly $\Rightarrow$

$$2 - 2\left[y\frac{d^2y}{dx^2} + \left(\frac{dy}{dx}\right)^2\right] - 5\frac{d^2y}{dx^2} = 0$$

At both these points, $\dfrac{dy}{dx} = 0$

$$\Rightarrow \qquad 2 - 2y\frac{d^2y}{dx^2} - 5\frac{d^2y}{dx^2} = 0$$

$$\Rightarrow \qquad \frac{d^2y}{dx^2} = \frac{2}{5 + 2y}$$

Method 2

Rearranging (2) $\Rightarrow \dfrac{dy}{dx} = \dfrac{2x + 10}{2y + 5}$

Differentiating implicitly: $\dfrac{d^2 y}{dx^2} = \dfrac{(2y+5)2 - (2x+10)2\dfrac{dy}{dx}}{(2y+5)^2}$

but $\dfrac{dy}{dx} = 0 \Rightarrow \dfrac{d^2 y}{dx^2} = \dfrac{2}{2y+5}$

$\therefore$ at $(-5, -2)$ $\dfrac{d^2 y}{dx^2} = 2$ and at $(-5, -3)$ $\dfrac{d^2 y}{dx^2} = -2$

- Exercise 8.4 See page 159.

8.5 Parametric Coordinates

Suppose x and y can each be given as a function of a third variable t, that is, $x = f(t)$ and $y = g(t)$.
Then t is called a **parameter** and the equations are said to be the **parametric equations** of the curve.

Example 16 Find the distance between the points $\phi = 0$ and $\phi = 3\pi/4$ on the locus $(1 + \cos\phi, \sin\phi)$.

1st point: $x = 1 + \cos 0 = 2$, $y = \sin 0 = 0$ $\Rightarrow (2, 0)$

2nd point: $x = 1 + \left(-\dfrac{1}{\sqrt 2}\right)$, $y = \dfrac{1}{\sqrt 2}$ $\Rightarrow \left(1 - \dfrac{1}{\sqrt 2}, \dfrac{1}{\sqrt 2}\right)$

$d^2 = \left(\dfrac{1}{\sqrt 2} - 0\right)^2 + \left(1 - \dfrac{1}{\sqrt 2} - 2\right)^2 = \dfrac{1}{2} + \left(1 + \dfrac{1}{\sqrt 2}\right)^2 = 3.4142$

$\therefore$ distance $= 1.85$ units

Example 17 Find the coordinates of the points of intersection of the curve $x = 2t^2 - 1$, $y = 5(t + 1)$ and the line $5x - 2y = 5$.

Substituting for x and y in the equation of the line

$$\Rightarrow 5(2t^2 - 1) - 10(t + 1) = 5$$
$$\Rightarrow 2t^2 - 2t - 4 = 0$$
$$\Rightarrow t^2 - t - 2 = 0$$
$$\Rightarrow (t + 1)(t - 2) = 0$$
$$t = -1 \quad \text{and} \quad 2$$

Coordinates of points of intersection are found by putting $t = -1$ and $t = 2$ in the parametric equations.

Thus the points of intersection are $(1, 0)$ and $(7, 15)$

Example 18   Find the equation of the tangent to the curve $x = at^2$, $y = 2at$ at the point with parameter p.

$$\left. \begin{array}{l} x = at^2 \Rightarrow \dfrac{dx}{dt} = 2at \\[2mm] y = 2at \Rightarrow \dfrac{dy}{dt} = 2a \end{array} \right\} \quad \frac{dy}{dx} = \frac{dy}{dt} \times \frac{dt}{dx} = \frac{2a}{2at} = \frac{1}{t}$$

At the point with parameter p, $x = ap^2$, $y = 2ap$ and gradient $= 1/p$.

Equation of tangent: $\dfrac{y - 2ap}{x - ap^2} = \dfrac{1}{p} \Rightarrow \underline{py = x + ap^2}$

Example 19   Find the Cartesian equation of each of the following curves:

(a) $x = 2t$, $y = 3t^2$ (b) $x = 3\cos\theta$, $y = 4\sin\theta$ (c) $x = t - \dfrac{1}{t}, y = t + \dfrac{1}{t}$.

(a) $\left. \begin{array}{l} x = 2t \Rightarrow t = x/2 \\ y = 3t^2 \Rightarrow t^2 = y/3 \end{array} \right\} \Rightarrow \dfrac{y}{3} = \left(\dfrac{x}{2}\right)^2$

$\qquad\qquad\qquad\qquad \Rightarrow \dfrac{y}{3} = \dfrac{x^2}{4}$

$\qquad\qquad\qquad\qquad \Rightarrow \underline{4y = 3x^2}$

(b) $\left. \begin{array}{l} x = 3\cos\theta \Rightarrow \cos\theta = x/3 \\ y = 4\sin\theta \Rightarrow \sin\theta = y/4 \end{array} \right\} \therefore \sin^2\theta + \cos^2\theta = 1$

$\qquad\qquad\qquad\qquad \Rightarrow (y/4)^2 + (x/3)^2 = 1$

$\qquad\qquad\qquad\qquad \Rightarrow \underline{\dfrac{y^2}{16} + \dfrac{x^2}{9} = 1}$

(c) $\qquad\qquad x = t - \dfrac{1}{t}$

$\qquad\qquad\qquad y = t + \dfrac{1}{t}$

Adding $x + y = 2t$

$\qquad \Rightarrow \qquad\qquad t = \dfrac{x + y}{2}$

$\qquad$ Hence $x = \left(\dfrac{x + y}{2}\right) - \left(\dfrac{2}{x + y}\right)$

$\Rightarrow 2x(x + y) = (x + y)^2 - 4$

$\Rightarrow 2x^2 + 2xy = x^2 + y^2 + 2xy - 4$

$\Rightarrow \underline{y^2 - x^2 \quad = 4}$

Example 20   If $x = t^2$, $y = t^3$ find $\dfrac{dy}{dx}$ and $\dfrac{d^2y}{dx^2}$ in terms of t.

$$\frac{dx}{dt} = 2t, \quad \frac{dy}{dt} = 3t^2$$

$$\frac{dy}{dx} = \frac{dy}{dt} \times \frac{dt}{dx} = \frac{3t^2}{2t} = \frac{3t}{2}$$

$$\frac{d^2y}{dx^2} = \frac{d}{dx}\left(\frac{dy}{dx}\right) = \frac{d}{dx}\left(\frac{3t}{2}\right) = \frac{d}{dt}\left(\frac{3t}{2}\right) \times \frac{dt}{dx} \qquad \text{(chain rule)}$$

$$= \frac{3}{2} \times \frac{1}{2t} \qquad \therefore \frac{d^2y}{dx^2} = \frac{3}{4t}$$

● Exercise 8.5   See page 160.

8.6 Inverse Circular Functions and their Derivatives

Principal Values of Inverse Circular Functions

$$\left.\begin{array}{l} -\pi/2 \leqslant \arcsin x \leqslant \pi/2 \\ -\pi/2 \leqslant \arctan x \leqslant \pi/2 \\ 0 \leqslant \arccos x \leqslant \pi \end{array}\right\} \quad \text{or} \quad \left\{\begin{array}{l} -\pi/2 \leqslant \sin^{-1} x \leqslant \pi/2 \\ -\pi/2 \leqslant \tan^{-1} x \leqslant \pi/2 \\ 0 \leqslant \cos^{-1} x \leqslant \pi \end{array}\right.$$

Example 21   Find the principal value of $\cot^{-1} 1$.

We need the angle lying between $-\pi/2$ and $\pi/2$ whose cot is 1, that is, whose tan is 1.

$\therefore \cot^{-1} 1 = \pi/4$

Differentiation of Inverse Circular Functions

1   Let $y = \sin^{-1} x$ then $\sin y = x$

Differentiating w.r.t. $x \Rightarrow \cos y \cdot \dfrac{dy}{dx} = 1$

$$\therefore \frac{dy}{dx} = \frac{1}{\cos y} = \frac{1}{\sqrt{1 - \sin^2 y}} = \frac{1}{\sqrt{1 - x^2}}$$

2   Let $y = \cos^{-1} x$ then $\cos y = x$

Differentiating w.r.t. $x \Rightarrow -\sin y \dfrac{dy}{dx} = 1$

$$\therefore \frac{dy}{dx} = \frac{-1}{\sin y} = \frac{-1}{\sqrt{1 - \cos^2 y}} = \frac{-1}{\sqrt{1 - x^2}}$$

3   Let $y = \tan^{-1} x$ then $\tan y = x$

differentiating w.r.t. $x \Rightarrow \sec^2 y \dfrac{dy}{dx} = 1$

$$\therefore \frac{dy}{dx} = \frac{1}{\sec^2 y} = \frac{1}{1 + \tan^2 y} = \frac{1}{1 + x^2}$$

$$\frac{d}{dx}(\sin^{-1} x) = \frac{1}{\sqrt{1 - x^2}}$$

$$\frac{d}{dx}(\cos^{-1} x) = \frac{-1}{\sqrt{1 - x^2}}$$

$$\frac{d}{dx}(\tan^{-1} x) = \frac{1}{1 + x^2}$$

Example 22   Differentiate: (a) $\sin^{-1} 3x$ (b) $\cos^{-1} x/2$

(c) $\tan^{-1} \sqrt{x}$ (d) $\sec^{-1} x$ (e) $\sin^{-1}\left(\dfrac{x}{1+x}\right)$

(a) $\dfrac{d}{dx}(\sin^{-1} 3x) = \dfrac{1}{\sqrt{1 - (3x)^2}} \times 3$ using the chain rule

$\therefore \dfrac{d}{dx}(\sin^{-1} 3x) = \dfrac{3}{\sqrt{1 - 9x^2}}$

(b) $\dfrac{d}{dx}(\cos^{-1} x/2) = \dfrac{-1}{\sqrt{1 - \left(\dfrac{x}{2}\right)^2}} \times \dfrac{1}{2} = \dfrac{-1}{2\sqrt{1 - \dfrac{x^2}{4}}}$

$$= \frac{-1}{\sqrt{4\left(1 - \dfrac{x^2}{4}\right)}}$$

$\therefore \dfrac{d}{dx}(\cos^{-1} x/2) = \dfrac{-1}{\sqrt{4 - x^2}}$

(c) $\dfrac{d}{dx}(\tan^{-1} \sqrt{x}) = \dfrac{1}{1 + (\sqrt{x})^2} \times \tfrac{1}{2} x^{-1/2}$

$\therefore \dfrac{d}{dx}(\tan^{-1} \sqrt{x}) = \dfrac{1}{2\sqrt{x}(1 + x)}$

(d) $\dfrac{d}{dx}(\sec^{-1} x) = \left(\cos^{-1} \dfrac{1}{x}\right)' = \dfrac{-1}{\sqrt{1 - \left(\dfrac{1}{x}\right)^2}} \times \left(-\dfrac{1}{x^2}\right)$

$$= \frac{1}{x^2 \sqrt{1 - \dfrac{1}{x^2}}} = \frac{1}{x \sqrt{x^2 \left(1 - \dfrac{1}{x^2}\right)}}$$

$$\therefore \frac{d}{dx}(\sec^{-1} x) = \frac{1}{x\sqrt{x^2 - 1}}$$

(e) $\dfrac{d}{dx}\left(\sin^{-1}\dfrac{x}{1+x}\right) = \dfrac{1}{\sqrt{1 - \left(\dfrac{x}{1+x}\right)^2}} \times \dfrac{d}{dx}\left(\dfrac{x}{1+x}\right)$

$$= \frac{1 + x}{\sqrt{(1 + x)^2 - x^2}} \times \left[\frac{(1 + x) \cdot 1 - x \cdot 1}{(1 + x)^2}\right]$$

$$= \frac{1}{(1 + x)\sqrt{1 + 2x}}$$

- **Exercise 8.6**   See page 161.

8.7 Miscellaneous Calculus Techniques

$$\boxed{\begin{array}{l} \displaystyle\int \frac{dx}{\sqrt{a^2 - x^2}} = \sin^{-1}\frac{x}{a} + c \\[3mm] \displaystyle\int \frac{dx}{a^2 + x^2} = \frac{1}{a}\tan^{-1}\frac{x}{a} + c \end{array}}$$

Example 23

(a) $\displaystyle\int \frac{dx}{\sqrt{9 - x^2}} = \sin^{-1}\frac{x}{3} + c$

(b) $\displaystyle\int \frac{dx}{9 + x^2} = \tfrac{1}{3}\tan^{-1}\frac{x}{3} + c$

(c) $\displaystyle\int \frac{dx}{\sqrt{1 - 9x^2}} = \frac{1}{3}\int \frac{dx}{\sqrt{\frac{1}{9} - x^2}} = \tfrac{1}{3}\sin^{-1} 3x + c$

(d) $\displaystyle\int \frac{dx}{1 + 9x^2} = \frac{1}{9}\int \frac{dx}{\frac{1}{9} + x^2} = \frac{1}{9}\cdot\frac{1}{\frac{1}{3}}\tan^{-1}\frac{x}{\frac{1}{3}} + c = \tfrac{1}{3}\tan^{-1} 3x + c$

(e) $\displaystyle\int \frac{dx}{\sqrt{4 - 9x^2}} = \frac{1}{3}\int \frac{dx}{\sqrt{\frac{4}{9} - x^2}} = \tfrac{1}{3}\cdot\sin^{-1}\frac{3x}{2} + c$

(f) $\displaystyle\int \frac{dx}{5 + 4x^2} = \frac{1}{4}\int \frac{dx}{\frac{5}{4} + x^2} = \frac{1}{4}\cdot\frac{2}{\sqrt{5}}\tan^{-1}\frac{2x}{\sqrt{5}} + c = \frac{1}{2\sqrt{5}}\tan^{-1}\frac{2x}{\sqrt{5}} + c$

$$\frac{d}{d\theta}(\cot \theta) = -\operatorname{cosec}^2 \theta$$

$$\frac{d}{d\theta}(\sec \theta) = \sec \theta \tan \theta$$

$$\frac{d}{d\theta}(\operatorname{cosec} \theta) = -\operatorname{cosec} \theta \cot \theta$$

Example 24 Evaluate $\displaystyle\int_0^{\pi/6} \sin^2 \theta \, d\theta$.

Although you can differentiate $\sin^2 \theta$ you cannot integrate $\sin^2 \theta$. You need to put it in terms of a function that you can integrate.

$$\cos 2\theta = 1 - 2\sin^2 \theta$$

Hence

$$\sin^2 \theta = \tfrac{1}{2}(1 - \cos 2\theta) \quad \text{and you can integrate } \cos 2\theta.$$

$$\therefore \int_0^{\pi/6} \sin^2 \theta \, d\theta = \frac{1}{2}\int_0^{\pi/6} (1 - \cos 2\theta) \, d\theta$$

$$= \frac{1}{2}\left[\theta - \frac{\sin 2\theta}{2} \right]_0^{\pi/6}$$

$$= \frac{1}{2}\left[\frac{\pi}{6} - \frac{1}{2}\sin \frac{\pi}{3} - (0 - 0) \right]$$

$$= \frac{\pi}{12} - \frac{\sqrt{3}}{8}$$

Note: This a very important technique.

- Exercise 8.7   See page 162.

8.8 Revision of Unit

- Miscellaneous Ex 8A   See page 163.

8.9 + A level Questions

- Miscellaneous Ex 8B   See page 164.

Unit 8 EXERCISES

Exercise 8.1   (All angles are in radians).

1   Differentiate $\sin 2x$ from first principles

2   Differentiate w.r.t. θ:

(a) $3 \sin \theta$ (b) $4 \cos \theta$ (c) $\dfrac{\tan \theta}{5}$ (d) $2 \sin \theta - \cos \theta$

(e) $3 \sin \theta - \tan \theta$ (f) $1 - 2 \cos \theta$ (g) $\dfrac{\tan \theta + 3 \cos \theta}{2}$

3   Evaluate:

(a) $\displaystyle\int 2 \sin x \, dx$ (b) $\displaystyle\int \dfrac{\cos x}{3} \, dx$ (c) $\displaystyle\int 3 \sec^2 x \, dx$

(d) $\displaystyle\int (3 \sin x - \cos x) \, dx$ (e) $\displaystyle\int (\sin x + 2 \sec^2 x) \, dx$

4   Evaluate:

(a) $\displaystyle\int_0^{\pi/2} \cos x \, dx$ (b) $\displaystyle\int_0^{\pi/2} (3 \sin x - 2 \cos x) \, dx$

(c) $\displaystyle\int_0^{\pi/4} \sec^2 x \, dx$ (d) $\displaystyle\int_{\pi/2}^{\pi} (5 \cos x - 2 \sin x) \, dx$

(e) $\displaystyle\int_{3\pi/2}^{2\pi} (3 \cos x + 4 \sin x) \, dx$

5   Find $\dfrac{d^2 s}{dt^2}$ if:

(a) $s = 5 \cos t$ (b) $s = 4 \cos t - 5 \sin t$ (c) $s = 1 - 3 \sin t$

6   Find the rate of increase of y w.r.t. x for each of the following functions when $x = \pi/3$

(a) $y = 2 \sin x + 5 \cos x$ (b) $y = 4 \tan x - \cos x + 3 \sin x$

7   Find the gradients of the following curves where $x = 0$:

(a) $y = \cos x$ (b) $y = \tan x$ (c) $y = 4 \sin x - 2 \cos x$

8   Find the areas contained between the x-axis and:

(a)   the curve $y = \sin x$ and the ordinates $x = \pi/6$ and $x = \pi/3$
(b)   the curve $y = 3 \sec^2 x$ and the ordinates $x = 0$ and $x = \pi/4$

9   Find the volume of revolution formed when the part of the curve $y = 3\sqrt{\cos x}$, between 0 and $\pi/4$ is rotated about the x-axis

*10   Find the equations of the tangent and normal to $y = 3 \sin x + \cos x$ at the point where $x = \pi/4$

*11   Find the values of $\tan x$ for which $\dfrac{dy}{dx}$ is zero in each of the following:

(a) $y = \sin x + \cos x$ (b) $y = 2 \sin x - \cos x$
(c) $y = 3 \cos x - 4 \sin x$

*12   Use the results $\cos x = 1 - 2 \sin^2 \dfrac{x}{2} = 2 \cos^2 \dfrac{x}{2} - 1$ to find the following

(a) $\dfrac{d}{dx}\left(\sin^2 \dfrac{x}{2} \right)$ (b) $\dfrac{d}{dx}\left(\cos^2 \dfrac{x}{2} \right)$

(c) $\displaystyle\int 2 \sin^2 \dfrac{x}{2}\, dx$ (d) $\displaystyle\int \cos^2 \dfrac{x}{2}\, dx$

Exercise 8.2

1   Differentiate w.r.t. x:

(a) $\sin 6x$ (b) $\cos \dfrac{x}{2}$ (c) $\tan 3x$ (d) $\cos (0.2x)$

(e) $\sin \dfrac{x}{5}$ (f) $2 \tan 10x$ (g) $3 \cos 5x$ (h) $\tfrac{1}{2} \sin 2x$

(i) $4 \tan \dfrac{x}{2}$ (j) $\tfrac{1}{3} \sin \dfrac{x}{3}$

2   Integrate w.r.t. x:

(a) $\sin 6x$ (b) $\cos \dfrac{x}{2}$ (c) $\sec^2 3x$ (d) $\sin \dfrac{3x}{2}$

(e) $2 \sec^2 \dfrac{x}{2}$ (f) $4 \cos 4x$ (g) $\tfrac{1}{3} \sin \dfrac{x}{5}$ (h) $\tfrac{1}{3} \sin 3x$

(i) $8 \sec^2 4x$ (j) $0.3 \cos (0.3x)$

3   Differentiate w.r.t. x:

(a) $\sin (x + 1)$ (b) $\cos (2x - 1)$ (c) $2 \sin (3x - 1)$

(d) $\tfrac{1}{3} \cos (3x + 2)$ (e) $2 \sin (1 - 2x)$ (f) $4 \cos \left(\dfrac{x}{2} + 1 \right)$

(g) $3 \tan (1 - 6x)$ (h) $\sin (\pi x - \pi/4)$ (i) $\cos \left(\dfrac{\pi x}{2} + \dfrac{\pi}{3} \right)$

(j) $\pi \tan (2 - \pi x)$

4 Integrate w.r.t. t:

(a) $\sin(3t + 2)$ (b) $\cos(2t - 1)$ (c) $\sec^2\left(\dfrac{t}{2} + 3\right)$

(d) $2\sin(4 - t)$ (e) $\cos\left(\pi t + \dfrac{\pi}{4}\right)$

5 Differentiate w.r.t. x:

(a) $\sin^2 x$ (b) $\cos^3 x$ (c) $\tan^5 x$ (d) $3\sin^4 x$

(e) $\tfrac{1}{2}\cos^4 x$ (f) $\sin^2 2x$ (g) $\cos^3 4x$ (h) $\tan^2 3x$

(i) $3\sin^4 \dfrac{x}{4}$ (j) $\sqrt{\sin x}$ (k) $\sqrt{\cos x}$ (l) $\sqrt{\tan x}$

(m) $\dfrac{1}{\sin^2 x}$ (n) $\dfrac{1}{\cos^3 2x}$ (o) $\sqrt{\tan^3 4x}$

6 Evaluate:

(a) $\displaystyle\int_0^{\pi/4} \cos 2x\, dx$ (b) $\displaystyle\int_0^{\pi/3} \sin 3x\, dx$

(c) $\displaystyle\int_0^{\pi} \dfrac{1 - \cos 2x}{2}\cdot dx$ (d) $\displaystyle\int_0^{\pi/4} \sec^2 3x\, dx$

(e) $\displaystyle\int_{-\pi/2}^{\pi/2} (2\sin 2x + 1)\, dx$ (f) $\displaystyle\int_{\pi/2}^{\pi} (3\sin 2x - 4\cos 2x)\, dx$

7 Find the gradient of the curve $y = 4\sin 2x$ at the point where $x = \pi/6$. Find the equation of the tangent at this point

8 If $r = \sin 3t + 2\cos 2t$ find the value of $\dfrac{d^2 r}{dt^2}$ when $r = \pi^c/4$

9 Find the area enclosed between the curve $y = 2\cos 2x$ and the ordinates $x = \pi/4$, $x = \pi/2$

10 If $r = \sin^2 \theta$, show that $\dfrac{dr}{d\theta} = \sin 2\theta$ and determine the value of $\dfrac{d^2 r}{d\theta^2}$ when $\theta = \pi/6$

11 Find the volume swept out when the part of the curve $y = \sqrt{\sin 4x}$ between $x = 0$ and $x = \pi/4$ is rotated about the x-axis

*12 Find the equation of the normal to the curve $y = 1 - 2\sin^2 x$ at the point where $x = \pi/8$

*13 Find values of θ, $0 \leqslant \theta \leqslant \pi$, for which $\dfrac{dr}{d\theta}$ vanishes where $r = \cos 2\theta - 2\cos \theta$

*14　Evaluate:

(a) $\displaystyle\int_0^1 2\sin\left(\frac{\pi t}{2}+\frac{\pi}{4}\right)dt$　　(b) $\displaystyle\pi\int_{-\pi/2}^{\pi/2}\cos\left(t-\frac{\pi}{3}\right)dt$

*15　Evaluate the integrals

(a) $\displaystyle\int_0^{\pi/2}2\sin^2 x\,dx$　　(b) $\displaystyle\int_0^{\pi/2}2\cos^2 x\,dx$

using the results $\cos 2x = 2\cos^2 x - 1 = 1 - 2\sin^2 x$

Exercise 8.3

Differentiate w.r.t. x:

1　$x\cos x$　　　　　　　**2**　$x^2\sin x$　　　　　　**3**　$x\tan x$

4　$\dfrac{2x}{1+x}$　　　　　　**5**　$\dfrac{3x^2-1}{1-x}$　　　　**6**　$\sin x\cdot\cos x$

7　$\dfrac{\tan x}{x}$　　　　　　**8**　$\dfrac{x}{\cos x}$　　　　　**9**　$\dfrac{\cos x}{2x^3}$

10　$\dfrac{1}{(1-2x)^2}$　　　**11**　$x(1-\cos x)$　　**12**　$x\sin 2x$

13　$x^2\cos x - x$　　**14**　$3x - x^3\tan 2x$　　**15**　$(x+3)(2\sin x - \cos x)$

16　By putting $\tan\theta = \dfrac{\sin\theta}{\cos\theta}$ and differentiating as a quotient show that the derivative of $\tan\theta$ is $\sec^2\theta$

17　Differentiate (a) $\cot\theta$　　(b) $\csc\theta$

18　Find the values of x for which the gradients of the following curves are zero:

(a) $y = x^2(1-x)$　　(b) $y = \dfrac{1-x}{x^2}$

*19　Find $\dfrac{d^2r}{dt^2}$ when:

(a) $r = \dfrac{t}{2+t}$　　(b) $r = 3t\sin t$　　(c) $r = \dfrac{\cos t}{t}$

*20　Given that $\theta = \pi + x\cos\theta$, find $\dfrac{dx}{d\theta}$ when $\theta = \pi/4$

In questions 21–25 find $\dfrac{dy}{dx}$:

21 $\quad y = \dfrac{2\sin 2x}{x}$ **22** $\quad y = 2\sin 2x \cdot \cos 5x$

23 $\quad y = x\cos^2 x$ **24** $\quad y = (2+x)\cos^3 x$

*25 $\quad y = \dfrac{x}{\sqrt{x-1}}$

*26 If $y = \dfrac{(1+x)\sin x}{x+3}$ show that $\dfrac{dy}{dx} = \dfrac{(x+3)(x+1)\cos x + 2\sin x}{(x+3)^2}$

**27 Find and classify the turning points of $y = \dfrac{x}{(x-1)(x-4)}$

Exercise 8.4

Find $\dfrac{dy}{dx}$ in each of the following question 1–10:

1 $\quad x^2 - 3y^2 = 5$ **2** $\quad y^3 - x^3 = 3$

3 $\quad x^2 + y^2 + 4x - 1 = 0$ **4** $\quad y^2 = 8x$

5 $\quad y^3 = 3(x-1)$ **6** $\quad xy = 3$

7 $\quad 2x^2 - xy + y^2 + 5y - 4 = 0$ **8** $\quad \dfrac{1}{x} + \dfrac{1}{y} = 1$

9 $\quad x^3 y^2 - 3x = 0$ **10** $\quad \sqrt{x} + \sqrt{y} = 3$

11 Find $\dfrac{dr}{d\theta}$ if:

(a) $r^2 \cos\theta = 3$ (b) $r\{\sin 2\theta + \cos 2\theta\} = r + r\sin\theta + 1$

12 Find the gradients of each of the following curves:
(a) $y^2 = 8x$ at $(2,4)$
(b) $x^2 + 2y^2 = 9$ at $(1,2)$

*13 If $x^2 + 2xy + 3y^2 = 3$, find $\dfrac{dy}{dx}$ and prove that $\dfrac{d^2 y}{dx^2} = \dfrac{-6}{(x+3y)^3}$

*14 Find the value of $\dfrac{d^2 s}{dt^2}$ when $s = -2$ if $2s^2 t^3 + 2s + 3 = 0$

*15 Find the gradients of the curves $y^2 = 2x$ and $x^2 = 2y$ at their points of intersection. Hence find the angles at which the curves cut

*16 Find the points of intersection of the curves $xy = 1$, and $x^2 - 2y^2 + 1 = 0$ and determine the angles at which the curves cut each other

*17 Find the equations of the tangent and the normal to the curve $4x^2 - 25y^2 = 100$ at the point $(5\sqrt{2}, 2)$

*18 Find the maximum and minimum values of y if

$$x^2 - 2y^2 + 6x - 3y + 18 = 0$$

**19 If $r^2 = 9\sin 2\theta$, find the values of θ for which r is a maximum or a minimum

Exercise 8.5

1 Find the coordinates of the points on the curve

$$x = 2\sin\phi, \quad y = 3\cos\phi$$

where ϕ has values 0, $\pi/4$, $\pi/2$, $5\pi/6$, $-\pi/3$ giving your answers in surd form or as a rational number

2 Find the length of the chord joining the points $t = -1$, $t = 2$ on the locus $(4t^2, 8t)$

3 Find the gradient of the chord joining the points on the curve $x = 2 + \sin\theta$, $y = 3\cos\theta$ for which θ has values 0 and $\pi/2$

4 Find the coordinates of the points on the locus $(2\sin^3 t, 4\cos^3 t)$ where

$$t = \frac{\pi}{6}, \frac{\pi}{4}, \frac{\pi}{3}, \frac{-\pi}{2}, \text{ giving your answer in surd or rational form}$$

5 Find the gradient of the curve $x = 2t - t^3$, $y = 2 - t^2$ at the point with parameter 3

6 Find the point where the line $y + x = 3$ intersects the locus $(3t + 1, 2 - t)$

7 Prove that $y = x + 2$ is a tangent to the curve $x = 2t^2$, $y = 4t$

Find the Cartesian equation of each of the following curves:

8 $x = 3t^2$, $y = 6t$ **9** $x = 3t$, $y = \dfrac{3}{t}$

10 $x = 2\sin\phi$, $y = 3\cos\phi$ **11** $x = 5\cos\phi$, $y = 1 + 3\sin\phi$

12 $x = 2 - p$, $y = p^3 + 3$

Find the Cartesian equations of each of the following loci:

13 $(2 - 3t, 3 + 2t)$ **14** $(3 + 2\cos\theta, 1 - \sin\theta)$

*15 $\left(\dfrac{3t}{1+t}, \dfrac{2-t}{1+t}\right)$ ***16** $\left(2t + \dfrac{1}{t}, 2t - \dfrac{1}{t}\right)$

*17 $(\cos\theta, \cos 2\theta)$ ***18** $(\sin^3\theta, \cos^3\theta)$

19 Find the equation of the normal to the curve $x = 2t^2$, $y = 4t$ at the point with parameter t

20 Find the equation of the tangent to the curve $xy = c^2$ at the point $\left(ct, \dfrac{c}{t}\right)$

21 Prove that the equation of the tangent to the parabola $y^2 = 4ax$ at the point P $(at^2, 2at)$ is $ty - x = at^2$. Write down, or obtain, the equation of the normal to the parabola at P. The tangent and the normal meet the x-axis at T and N respectively. Express $PT^2 + PN^2$ in terms of a and t

(J)

Exercise 8.6

1 Without using tables or calculators, find the principal value of the following inverse functions:

(a) $\sin^{-1}\tfrac{1}{2}$ (b) $\tan^{-1}(-1)$ (c) $\cos^{-1} 0.5$

(d) $\sin^{-1}\dfrac{\sqrt{3}}{2}$ (e) $\sin^{-1}\left(-\dfrac{1}{\sqrt{2}}\right)$ (f) $\cos^{-1}\left(-\dfrac{\sqrt{3}}{2}\right)$

(g) $\sin^{-1}(-0.5)$ (h) $\sec^{-1} 2$ (i) $\cot^{-1} 1/\sqrt{3}$

2 If $y = \sin^{-1} 3x$ show that $\dfrac{dy}{dx} = \dfrac{3}{\sqrt{1 - 9x^2}}$

3 If $y = \tan^{-1}\dfrac{x}{2}$ show that $\dfrac{dy}{dx} = \dfrac{2}{4 + x^2}$

In questions 4–19 differentiate w.r.t. x:

4 $\sin^{-1} 2x$ **5** $\tan^{-1}\dfrac{x}{3}$ **6** $\cos^{-1} 4x$

7 $\sin^{-1}\sqrt{x}$ **8** $\cos^{-1} x^2$ **9** $\tan^{-1}\dfrac{1}{x}$

10 $\cot^{-1}\dfrac{2}{x}$ **11** $\sin^{-1}(x - 1)$ **12** $\cos^{-1}\dfrac{1}{x - 1}$

13 $\operatorname{cosec}^{-1} x$ ***14** $\tan^{-1}\dfrac{2 + x}{1 - 2x}$ ***15** $\sin^{-1}\left(\dfrac{2 - x}{2 + x}\right)$

16 $\sec^{-1} 2x$ **17** $\tan^{-1}\cos x$ **18** $\sin^{-1}\cos x$

19 $\sin^{-1}(2 \cdot \sin x)$

20 Given that $y = \sin^{-1}\sqrt{x}$, prove that $\dfrac{dy}{dx} = \dfrac{1}{\sin 2y}$ (J)

21 Prove that $\dfrac{d}{dx}(\sin^{-1}x) = \dfrac{1}{(1-x^2)^{1/2}}$

Given that the variables x and y satisfy the equation

$$\sin^{-1}2x + \sin^{-1}y + \sin^{-1}(xy) = 0$$

find $\dfrac{dy}{dx}$ when $x = y = 0$ (J)

22 Given that $-1 < x < 1$ and $0 < \cos^{-1}x < \pi$, and that $(1-x^2)^{1/2}$ denotes the positive square root of $(1-x^2)$, find the derivative of the function

$$f(x) = \cos^{-1}x - x(1-x^2)^{1/2}$$

expressing your answer as simply as possible

Prove that, as x increases in the interval $-1 < x < 1$, $f(x)$ decreases, and sketch the graph of $f(x)$ in this interval   (J)

Exercise 8.7

Evaluate the following integrals:

1 $\displaystyle\int \dfrac{dx}{\sqrt{4-x^2}}$ **2** $\displaystyle\int \dfrac{dx}{\sqrt{3-x^2}}$ **3** $\displaystyle\int \dfrac{dx}{4+x^2}$

4 $\displaystyle\int_0^1 \dfrac{dx}{\sqrt{2-x^2}}$ **5** $\displaystyle\int \dfrac{dx}{\sqrt{1-16x^2}}$ **6** $\displaystyle\int \dfrac{dx}{\sqrt{1-25x^2}}$

7 $\displaystyle\int \dfrac{dx}{\sqrt{1-2x^2}}$ **8** $\displaystyle\int_0^3 \dfrac{dx}{9+x^2}$ **9** $\displaystyle\int \dfrac{dx}{16+x^2}$

10 $\displaystyle\int \dfrac{dx}{16+9x^2}$ **11** $\displaystyle\int \dfrac{dx}{9+25x^2}$ **12** $\displaystyle\int \dfrac{dx}{1+4x^2}$

13 $\displaystyle\int \dfrac{dx}{1+3x^2}$ **14** $\displaystyle\int \dfrac{dx}{\sqrt{16-9x^2}}$ **15** $\displaystyle\int \dfrac{dx}{\sqrt{2-3x^2}}$

16 $\displaystyle\int_0^{\pi/4} \cos^2 x \, dx$ **17** $\displaystyle\int_0^{\pi} \sin^2 2x \, dx$ **18** $\displaystyle\int_{\pi/4}^{\pi/3} \tan^2 x \, dx$

19 $\displaystyle\int_0^{\pi/4} (1+\sin^2 x)\,dx$ **20** $\displaystyle\int_{-\pi/6}^{\pi/6} (\cos^2 3x - 2)\,dx$

Differentiate w.r.t. x

21 $\cot x + \operatorname{cosec} x$ **22** $x \sec x$ **23** $\tan x - \operatorname{cosec} x$

24 $\cot^2 x$ **25** $\sec^3 x$ **26** $\dfrac{1+\cot x}{1-\cot x}$

Evaluate:

*27 $\displaystyle\int_{-2}^{-1} \frac{dx}{1+(x+2)^2}$ ***28** $\displaystyle\int_{5}^{7} \frac{dx}{\sqrt{16-(x-3)^2}}$

**29 $\displaystyle\int_{1}^{1+1/\sqrt{5}} \frac{dx}{3+5(x-1)^2}$

**30   Using $\cos^4 x = (\cos^2 x)^2$ and $\cos 2x = 2\cos^2 x - 1$

show that $\displaystyle\int_{0}^{\pi/4} \cos^4 x \, dx = \frac{3\pi}{32} + \frac{1}{4}$

Miscellaneous Exercise 8A

1   Differentiate $\cos \dfrac{x}{2}$ from first principles

2   Find $\dfrac{dy}{dx}$:

   (a) $y = (1 + 3\sin x)^2$ (b) $y = \sin^2 2x + \cos^3 3x$

   (c) $y = \sqrt[2]{\sin^3 x}$ (d) $y = \dfrac{\cos^2 2x - \sin^2 2x}{\cos^2 2x}$

3   If $y = C\sin(\alpha t + \beta)$ where C, α, β are constants, show that $\dfrac{d^2 y}{dt^2} + \alpha^2 y = 0$

4   Evaluate:

   (a) $\displaystyle\int_{0}^{\pi} \sec^2 \frac{x}{4} \, dx$ (b) $\displaystyle\int_{-\pi/2}^{\pi/2} (\cos x + 2\cos 2x) \, dx$

   (c) $\displaystyle\int_{0}^{\pi/4} (1 + \sin x)^2 \, dx$

5   Find the maximum and minimum values of $2x - \tan x$ for $0 \leqslant x \leqslant \pi$

6   If $x = \sin 2t$ and $y = 2\cos t$ find $\dfrac{dx}{dt}$ and $\dfrac{dy}{dt}$

   Hence find $\dfrac{dy}{dx}$ in terms of t. For what values of t is the tangent to the curve parallel to the x-axis?

7   Find the equation of the tangent to the curve
$$y = 2(\theta - \sin \theta), \quad x = 2(1 - \cos \theta)$$
at the point where $\theta = \pi/2$

8 A particle moves in a straight line. The distance from a fixed point, s, at a time t seconds is given by

$$s = 8t + 4\cos 2t$$

Find the velocity and acceleration after $\pi/2$ seconds. At what time is the particle first at rest?

9 Sketch the curve $y = 2\cos x/2$ for values of x between 0 and π. Find the area contained between this part of the curve and the x-axis

10 Find the area below the x-axis between the curve $y = \frac{1}{2}\sin 2x - \sin x$ and the x-axis, for $0 \leqslant x \leqslant \pi$

11 If $p(r + 1) = 3$ find the value of $\dfrac{dp}{dr}$ when $r = 2$

12 Differentiate (a) $x\sin x \cdot \cos x$ (b) $\dfrac{x\sin x}{x - 1}$

13 If $x = \dfrac{2t}{1 + t^2}$, $y = \dfrac{1 - t^2}{1 + t^2}$ express $\dfrac{dy}{dx}$ in terms of t

*14 If $y = 1 - \cos x$ is rotated about the x-axis for $0 \leqslant x \leqslant \pi/2$, find the volume of revolution formed

*15 Find the maximum and minimum values of $\dfrac{x^2 - 4x + 3}{4x - 13}$

Miscellaneous Exercise 8B

1 Given that x and y are related by $x^3 + y^3 = 3xy$, find $\dfrac{dy}{dx}$ in terms of x and y (L: 8.4)

2 Given that $y = \dfrac{\sin x - \cos x}{\sin x + \cos x}$ show that $\dfrac{dy}{dx} = 1 + y^2$

Prove that $\dfrac{d^2y}{dx^2}$ is zero only when $y = 0$ (J: 8.3, 8.4)

3 Show that $\dfrac{d}{dx}\left(\dfrac{x}{1 + x}\right) = \dfrac{1}{(1 + x)^2}$

A curve is described by the equation $\dfrac{y}{1 + y} + \dfrac{x}{1 + x} - x^2y^3 = 0$. Find the equation of the tangent to the curve at the point $(1, 1)$ (J: 3.3, 8.4)

4 Evaluate $\dfrac{dy}{dx}$ when $y = 1$, given that

(a) $y(x + y) = 3$ (b) $x = \dfrac{1}{(4 - t)^2}$, $y = \dfrac{t}{4 - t}$, $0 < t < 4$ (L: 8.4, 8.5)

5 Determine the maximum and the minimum values of the function

$$y = \sqrt{3}\sin 2x + \cos 2x$$

and find the values of x in the interval $0 \leqslant x \leqslant \pi$ for which they occur. Find also the points in the interval $0 \leqslant x \leqslant \pi$ at which $y = 0$ and sketch the graph of y in this interval.

Calculate, in terms of π, the volume swept out when the region bounded by the curve

$$y = \sqrt{3}\sin 2x + \cos 2x$$

the line $x = \dfrac{\pi}{4}$ and the coordinate axes, is rotated about the x-axis through an angle of $2\pi^c$ (J: 5.5, 7.2, 8.1)

6 Given that $x = \theta - \sin\theta$, $y = 1 - \cos\theta$, show that

$$\frac{dy}{dx} = \cot\frac{\theta}{2} \quad \text{and} \quad \frac{d^2y}{dx^2} + \frac{1}{y^2} = 0 \qquad\qquad \text{(J: 4.3, 8.5)}$$

7 Given that $y = \dfrac{x^2 - 1}{2x^2 + 1}$, find $\dfrac{dy}{dx}$ and state the set of values of x for which $\dfrac{dy}{dx}$ is positive.

Find the greatest and least values of y for $0 \leqslant x \leqslant 1$ (L: 8.3)

8 Given that $y = \sec x \cdot \tan^2 x$

show that $y\dfrac{dy}{dx} = a\tan^7 x + b\tan^5 x + c\tan^3 x$

where a, b and c are constants to be determined   (J: 1.6, 8.7)

9 Given that $x = \sec\theta + \tan\theta$ and $y = \operatorname{cosec}\theta + \cot\theta$ show that

$$x + \frac{1}{x} = 2\sec\theta \quad \text{and} \quad y + \frac{1}{y} = 2\operatorname{cosec}\theta$$

Find $\dfrac{dx}{d\theta}$ and $\dfrac{dy}{d\theta}$ in terms of θ and hence show that

$$\frac{dy}{dx} = -\left(\frac{1 + y^2}{1 + x^2}\right) \qquad\qquad \text{(J: 8.5, 8.7)}$$

10 Given that $x = \sin\theta$ and $y = \sin n\theta$, where $0 < \theta < \pi/2$ and n is a constant

(a) find $\dfrac{d}{d\theta}\left(\dfrac{dy}{dx}\right)$, in terms of θ and n

(b) prove that $(1 - x^2)\dfrac{d^2y}{dx^2} = x\dfrac{dy}{dx} - n^2y$ (A: 8.5)

11　Given that $y^2 - 5xy + 8x^2 = 2$, prove that $\dfrac{dy}{dx} = \dfrac{5y - 16x}{2y - 5x}$.

The distinct points P and Q on the curve $y^2 - 5xy + 8x^2 = 2$ each have x-coordinate 1. The normals to the curve at P and Q meet at the point N. Calculate the coordinates of N　　(A: 3.3, 8.4)

12　A curve joining the points $(0, 1)$ and $(0, -1)$ is represented parametrically by the equations $x = \sin\theta$, $y = (1 + \sin\theta)\cos\theta$ where $0 \leqslant \theta \leqslant \pi$. Find $\dfrac{dy}{dx}$ in terms of θ and determine the x, y coordinates of the points on the curve at which the tangents are parallel to the x-axis. Sketch the curve.

The region in the quadrant $x \geqslant 0$, $y \geqslant 0$ bounded by the curve and the coordinate axes is rotated about the x-axis through an angle of 2π. Show that the volume swept out is given by

$$V = \pi \int_0^1 (1 + x)^2 (1 - x^2)\, dx$$

Evaluate V, leaving the result in terms of π　　(J: 5.5, 8.5)

13　Show that $\dfrac{d}{dx}\left(\dfrac{\sin 2x}{\cos 2x - \sin 2x}\right) = \dfrac{2}{(\cos 2x - \sin 2x)^2}$

The area between the curve $y = \dfrac{2}{\cos 2x - \sin 2x}$, the x-axis and the lines $x = \dfrac{\pi}{4}$ and $x = \dfrac{5\pi}{12}$ is rotated completely about the x-axis. Show that the volume generated may be written in the form $\pi(3 - \sqrt{3})$

(A: 8.3, 5.3, 5.5)

*14　State the derivatives of $\sin x$ and $\cos x$ and use these results to show that the derivative of $\tan x$ is $\sec^2 x$. Show further that $\dfrac{d}{dx}(\tan^{-1} x) = \dfrac{1}{1 + x^2}$

A vertical rod AB of length 3 units is held with its lower end B at a distance 1 unit vertically above a point O. The angle subtended by AB at a variable point P on the horizontal plane through O is θ. Show that

$$\theta = \tan^{-1} x - \tan^{-1}\frac{x}{4} \qquad \text{where } x = \text{OP}$$

Prove that, as x varies, θ is a maximum when $x = 2$ and that the maximum value of θ can be expressed as $\tan^{-1}\frac{3}{4}$　　(J: 3.6, 8.3, 8.6)

*15　Investigate the nature of the turning points of the curve

$$y = (x - 1)(x - a)^2$$

in the cases of (a) $a > 1$, (b) $a < 1$. Sketch on separate diagrams the curves $y = (x - 1)(x - a)^2$ when $a > 1$ and when $0 < a < 1$　　(L: 8.3)

*16   Prove that $\dfrac{d}{dx}(\sin 2x - 2x\cos 2x) = 4x\sin 2x$

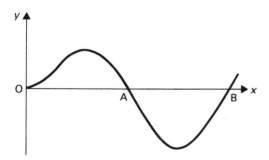

The curve shown in the figure is part of the graph of the function $y = x\sin 2x$. Write down the coordinates of the points A and B

Show that the area of the region bounded by the curve and the line segment AB is three times that of the area of the region bounded by the curve and the line segment OA   (A: 8.3)

*17   The points P $(ap^2, 2ap)$ and Q $(aq^2, 2aq)$ on the parabola $y^2 = 4ax$ are such that the tangent to the parabola at P is parallel to the chord OQ, where O is the origin. Show that $q = 2p$

The tangents to the parabola at P and Q meet at the point R. Find the coordinates of R in terms of a and p. Find the equation of the perpendicular bisector of the line PQ and obtain an expression in terms of a and p for the perpendicular distance of R from this bisector

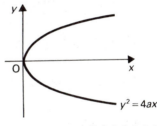

(J: 2.1, 2.3, 8.5)

*18   In the triangle ABC, angle $ACB = \theta$ where $0 < \theta < \dfrac{\pi}{2}$ and AC has length d. An ant starts at A and crawls directly to the line CB, then crawls directly to the line CA, then directly to the line CB and so on until it ultimately reaches C. Show that the total distance travelled is s where

$s = \dfrac{d\sin\theta}{1 - \cos\theta}$. Show that s increases as θ decreases.   (A: 6.2, 8.1)

UNIT 9

EXPONENTIAL AND LOGARITHMIC FUNCTIONS

9.1 Logarithms

Logarithm is another word for index or power.

$$2^3 = 8$$

⇔ 3 is the power to which the base 2 must be raised to give 8
⇔ 3 is the logarithm which, with a base 2, gives 8
⇔ $3 = \log_2 8$

> If $y = a^x$ then $x = \log_a y$

Example 1 Express in logarithmic form: (a) $5^3 = 125$ (b) $10^{-1} = 0.1$

(a) $5^3 = 125 \Rightarrow \underline{\log_5 125 = 3}$

(b) $10^{-1} = 0.1 \Rightarrow \underline{\log_{10} 0.1 = -1}$

Example 2 Express $\log_{1/5} 25 = -2$ in index form.

$$\log_{1/5} 25 = -2 \Rightarrow \underline{(1/5)^{-2} = 25}$$

Example 3 Evaluate (a) $\log_{10} 100$ (b) $\log_2 64$.

(a) $10^2 = 100$ ∴ $\log_{10} 100 = \underline{2}$

(b) $2^6 = 64$ ∴ $\log_2 64 = \underline{6}$

Laws of logarithms

1 $\log_a b + \log_a c = \log_a bc$

2 $\log_a b - \log_a c = \log_a b/c$

3 $\log_a \dfrac{1}{x} = -\log_a x$

4 $\log_a x^k = k \log_a x$

5 $\log_a (a^x) = x$

6 $a^{\log_a x} = x$

Example 4 Express $\log (p^2/q^3)$ in terms of $\log p$ and $\log q$.

$$\log \frac{p^2}{q^3} = \log p^2 - \log q^3 \qquad \text{(Law 2)}$$

$$\therefore \log \frac{p^2}{q^3} = 2 \log p - 3 \log q \qquad \text{(Law 4)}$$

Example 5   Simplify (a) $\log 100 - 2\log 5$ (b) $2 + \log_3 5$

(a) $\log 100 - 2\log 5 = \log 100 - \log 25$
$$= \log(100/25)$$
$\therefore\ \log 100 - 2\log 5 = \log 4$

(b) $2 + \log_3 5 = \log_3 9 + \log_3 5$
$\therefore 2 + \log_3 5 = \log_3 45$

Note: If the base of a log is not included then it is not essential to the question.

● **Exercise 9.1**   See page 179.

9.2 Solving Exponential and Logarithmic Equations

A function of the form a^x is called an **exponential function**. Exponential functions are inverses of log functions.

Example 6   Solve $5^x = 103$

$$\log 5^x = \log 103$$
$$x\log 5 = \log 103$$
$$x = \frac{\log 103}{\log 5} = 2.880 \quad \text{(3.d.p)}$$

Example 7   Solve $2^{2x} + 3 \cdot 2^x - 4 = 0$

This is a quadratic in 2^x
i.e.
$$(2^x)^2 + 3(2^x) - 4 = 0$$
$$(2^x + 4)(2^x - 1) = 0$$
$$2^x = -4 \quad \text{and} \quad 2^x = 1$$
$$\downarrow \qquad\qquad\qquad \downarrow$$
$$\text{no solution} \qquad x = 0$$

Changing the Base of a Logarithm

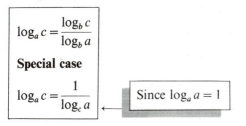

$$\log_a c = \frac{\log_b c}{\log_b a}$$

Special case

$$\log_a c = \frac{1}{\log_c a}$$ Since $\log_a a = 1$

Proof

Let
$$\log_a c = x$$
$$\Rightarrow \quad c = a^x$$
$$\Rightarrow \log_b c = \log_b a^x \longleftarrow \boxed{\text{Taking logarithms to base } b}$$
$$= x \log_b a$$
$$= \log_a c \cdot \log_b a$$
$$\Rightarrow \log_a c = \frac{\log_b c}{\log_b a}$$

Example 8   Evaluate $\log_3 7$ using a calculator.

$$\log_3 7 = \frac{\log_{10} 7}{\log_{10} 3} = \underline{1.771} \quad \text{(3.d.p.)}$$

Example 9   Solve $\log_3 x - 3\log_x 3 + 2 = 0$.

$$\log_x 3 = \frac{1}{\log_3 x}$$

∴ equation becomes
$$\log_3 x - \frac{3}{\log_3 x} + 2 = 0$$

or
$$(\log_3 x)^2 + 2\log_3 x - 3 = 0$$

$$(\log_3 x - 1)(\log_3 x + 3) = 0$$
$$\log_3 x = 1 \text{ and } \log_3 x = -3$$
$$x = 3 \text{ and } \qquad x = 3^{-3} = \tfrac{1}{27}$$

Example 10   Solve for x and y the equations $xy = 270$ and $\log_{10} x - 2\log_{10} y = 1$.

$$xy = 270 \quad (1) \qquad \text{and} \qquad \log_{10} x - 2\log_{10} y = 1 \qquad (2)$$

$$\Rightarrow \log_{10}\frac{x}{y^2} = 1$$

$$\Rightarrow \qquad \frac{x}{y^2} = 10^1 = 10$$

$$\Rightarrow \qquad x = 10y^2 \qquad (3)$$

Substituting (3) into (1)

$$\Rightarrow \qquad 10y^3 = 270$$
$$\Rightarrow \qquad y^3 = 27$$
$$\Rightarrow \qquad \underline{y = 3, x = 90}$$

- Exercise 9.2   See page 180.

9.3 Graph and Derivatives

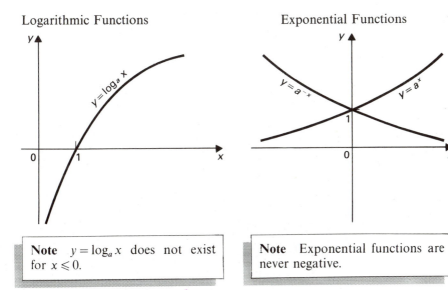

Logarithmic Functions

Exponential Functions

Note $y = \log_a x$ does not exist for $x \leqslant 0$.

Note Exponential functions are never negative.

Differentiating Log Functions

$$\frac{d}{dx}(\log_a x) = \frac{1}{x}\cdot\log_a e$$

$$\text{where } e = \lim_{t \to 0}(1 + t)^{1/t} \approx 2.718$$

Natural (or Naperian) Logarithms

$$\frac{d}{dx}(\log_e x) = \frac{1}{x}\cdot\log_e e = \frac{1}{x}$$

$\log_e x$ is the most commonly used logarithm, but it is more generally written as $\ln x$.

$$\boxed{\frac{d}{dx}(\ln x) = \frac{1}{x}}$$

Example 11
Differentiate w.r.t. x: (a) $\ln(2x + 1)$ (b) $\ln(x^2 - 1)$ (c) $\ln \sin x$

(a) $\dfrac{d}{dx}(\ln(2x + 1)) = \dfrac{2}{2x + 1}$

(b) $\dfrac{d}{dx}(\ln(x^2 - 1)) = \dfrac{2x}{x^2 - 1}$

(c) $\dfrac{d}{dx}(\ln \sin x) = \dfrac{\cos x}{\sin x} = \cot x$

Example 12   Differentiate w.r.t. x (a) $\ln(x-1)^2$ (b) $\ln\sqrt{\dfrac{1+x}{1-x}}$

> These functions can be differentiated using the techniques of Example 11, but it is much easier if you use the properties of logs to simplify first.

(a) $\dfrac{d}{dx}(\ln(x-1)^2) = \dfrac{d}{dx}(2\ln(x-1)) = \dfrac{2}{x-1}$

(b) $\dfrac{d}{dx}\left(\ln\sqrt{\dfrac{1+x}{1-x}}\right) = \dfrac{d}{dx}\left[\tfrac{1}{2}(\ln(1+x) - \ln(1-x))\right] = \dfrac{1}{2}\left[\dfrac{1}{1+x} - \dfrac{(-1)}{1-x}\right]$

$$= \dfrac{1}{1-x^2}$$

Differentiating the Exponential Function

e^x is known as **the exponential function**

$$\dfrac{d}{dx}(e^x) = e^x \text{ and hence } \int e^x\,dx = e^x + c$$

Example 13   Differentiate w.r.t. x (a) e^{3x} (b) e^{x^2} (c) $e^{\sin x}$
(d) $e^{(x^3 + x)}$

(a) $\dfrac{d}{dx}(e^{3x}) = \underline{3e^{3x}}$

(b) $\dfrac{d}{dx}(e^{x^2}) = \underline{2x\,e^{x^2}}$

(c) $\dfrac{d}{dx}(e^{\sin x}) = \underline{\cos x \cdot e^{\sin x}}$

(d) $\dfrac{d}{dx}(e^{(x^3 + x)}) = \underline{(3x^2 + 1)e^{(x^3 + x)}}$

- Exercise 9.3   See page 181.

9.4 Integration

$1/x$ exists for positive and negative values of x, but $\ln x$ does *not* exist for negative values of x. Hence:

if $x > 0$ $\displaystyle\int \dfrac{1}{x}\,dx = \ln x + c$

and if $x < 0$ $\displaystyle\int \dfrac{1}{x}\,dx = \ln(-x) + c$

$\Rightarrow \boxed{\displaystyle\int \dfrac{1}{x}\,dx = \ln|x| + c}$

$$\int \frac{dx}{ax+b} = \frac{1}{a}\int \frac{a}{ax+b}\,dx = \frac{1}{a}\ln|ax+b| + c$$

$$\int \frac{f'(x)}{f(x)}\,dx = \ln|f(x)| + c$$

Example 14   Integrate (a) $\dfrac{3}{2x-4}$ (b) $\dfrac{x^2}{x^3+1}$

(a) $\displaystyle\int\frac{3}{2x-4}\,dx = \frac{3}{2}\int\frac{2}{2x-4}\,dx = \tfrac{3}{2}\ln|2x-4| + c$

> derivative of the denominator

(b) $\displaystyle\int\frac{x^2}{x^3+1}\,dx = \frac{1}{3}\int\frac{3x^2}{x^3+1}\,dx = \tfrac{1}{3}\ln|x^3+1| + c$

> $\displaystyle\int f'(x)e^{f(x)}\,dx = e^{f(x)} + c$

Example 15   Integrate   (a) e^{2x} (b) e^{-x} (c) $3xe^{x^2}$
(d) $\sec^2 2x\, e^{\tan 2x}$

(a) $\displaystyle\int e^{2x}\,dx = \frac{1}{2}\int 2e^{2x}\,dx = \tfrac{1}{2}e^{2x} + c$

(b) $\displaystyle\int e^{-x}\,dx = -\int -e^{-x}\,dx = -e^{-x} + c$

(c) $\displaystyle\int 3x\,e^{x^2}\,dx = \frac{3}{2}\int (2x)e^{x^2}\,dx = \tfrac{3}{2}e^{x^2} + c$

(d) $\displaystyle\int \sec^2 2x\, e^{\tan 2x}\,dx = \frac{1}{2}\int(2\sec^2 2x)e^{\tan 2x}\,dx = \tfrac{1}{2}e^{\tan 2x} + c$

> **Note**   Always check by differentiation

Example 16   Evaluate (a) $\int \cot x\,dx$ (b) $\int \tan x\,dx$

(a) $\displaystyle\int \cot x\,dx = \int\frac{\cos x}{\sin x}\,dx = \ln|\sin x| + c$

(b) $\displaystyle\int \tan x\,dx = -\int\frac{(-\sin x)}{\cos x}\,dx = -\ln|\cos x| + c \text{ or } \ln|\sec x| + c$

Example 17   Evaluate:

(a) $\displaystyle\int_{-2}^{-1}\frac{dx}{1-3x} = -\frac{1}{3}\int_{-2}^{-1}\frac{-3\,dx}{1-3x} = -\tfrac{1}{3}\big[\ln|1-3x|\big]_{-2}^{-1}$

$$= -\tfrac{1}{3}[\ln 4 - \ln 7]^{-2}$$

$$= \tfrac{1}{3}\ln 7 - \tfrac{1}{3}\ln 4 \qquad = \tfrac{1}{3}\ln \tfrac{7}{4}$$

(b) $\displaystyle \int_1^2 \frac{dx}{x-5} = [\ln|x-5|]_1^2 = \ln|-3| - \ln|-4| = \ln\tfrac{3}{4}$

(c) $\displaystyle \int_{-\pi/2}^{\pi/2} \frac{\cos x}{2+\sin x}\,dx = [\ln|2+\sin x|]_{-\pi/2}^{\pi/2} = \ln 3 - \ln 1 = \ln 3$

(d) $\displaystyle \int_0^{1/2} e^{4x}\,dx = [\tfrac{1}{4}e^{4x}]_0^{1/2} = \tfrac{1}{4}(e^2 - e^0) = \tfrac{1}{4}(e^2 - 1)$

Example 18   Integrate $\left(1 + \dfrac{1}{x}\right)^3$.

$$\int\left(1+\frac{1}{x}\right)^3 dx = \int\left(1 + \frac{3}{x} + \frac{3}{x^2} + \frac{1}{x^3}\right)dx$$

$$= \int\left(1 + 3\left(\frac{1}{x}\right) + 3\cdot x^{-2} + x^{-3}\right)dx$$

$$= x + 3\ln|x| - \frac{3}{x} - \frac{1}{2x^2} + c$$

• **Exercise 9.4**   See page 182.

9.5 Series Expansions

$e^x = 1 + x + \dfrac{x^2}{2!} + \dfrac{x^3}{3!} + \cdots + \dfrac{x^r}{r!} + \cdots$	Valid for all x
$\ln(1+x) = x - \dfrac{x^2}{2} + \dfrac{x^3}{3} - \dfrac{x^4}{4} + \cdots + \dfrac{(-1)^{r+1}}{r}x^r + \cdots$	Valid for $-1 < x \leqslant 1$
$\ln(1-x) = -x - \dfrac{x^2}{2} - \dfrac{x^3}{3} - \dfrac{x^4}{4} - \cdots - \dfrac{x^r}{r} - \cdots$	Valid for $-1 \leqslant x < 1$
$\cos x = 1 - \dfrac{x^2}{2!} + \dfrac{x^4}{4!} - \dfrac{x^6}{6!} + \cdots + \dfrac{(-1)^r x^{2r}}{(2r)!} + \cdots$	
$\sin x = x - \dfrac{x^3}{3!} + \dfrac{x^5}{5!} - \dfrac{x^7}{7!} + \cdots + \dfrac{(-1)^r x^{2r+1}}{(2r+1)!} + \cdots$	Valid for all x (in radians)
$(1+x)^n = 1 + nx + \dfrac{n(n-1)x^2}{2!} + \cdots + \dfrac{n!}{r!(n-r)!}x^r + \cdots$	
$(1+x)^{-1} = 1 - x + x^2 - x^3 + \cdots + (-1)^r x^r + \cdots$	Valid for $|x| < 1$
$(1-x)^{-1} = 1 + x + x^2 + x^3 + \cdots + x^r + \cdots$	

All these expansions may be quoted.

Example 19　Expand $\ln\left[(2+x)/(1-3x)^2\right]$ as a series of ascending powers of x up to, and including, the term in x^3. Give the general term and the range of values of x for which the expansion is valid.

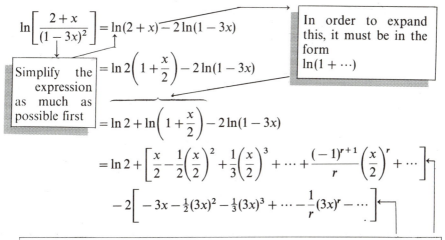

$$= \ln 2 + \tfrac{13}{2}x + \tfrac{71}{8}x^2 + \tfrac{433}{24}x^3 + \cdots$$

General term $= \left[\dfrac{(-1)^{r+1}}{2^r} + 2.3^r\right]\dfrac{x^r}{r}$

$\ln\left(1+\dfrac{x}{2}\right)$ is valid for $-1 < \dfrac{x}{2} \leqslant 1$ or $-2 < x \leqslant 2$

$\ln(1-3x)$ is valid for $-1 \leqslant 3x < 1$ or $-\tfrac{1}{3} \leqslant x < \tfrac{1}{3}$

$\Rightarrow$ the total expansion is valid for $-\tfrac{1}{3} \leqslant x < \tfrac{1}{3}$

> These are the values for which both expansion are valid.

Example 20　Find the first four terms of the expansion of $\tfrac{1}{2}(e^x + e^{-x})$.

$$\tfrac{1}{2}(e^x + e^{-x}) = \frac{1}{2}\left[1 + x + \frac{x^2}{2!} + \frac{x^3}{3!} + \frac{x^4}{4!} + \cdots + \left(1 - x + \frac{x^2}{2!} - \frac{x^3}{3!} + \frac{x^4}{4!} + \cdots\right)\right]$$

$$\Rightarrow \tfrac{1}{2}(e^x + e^{-x}) = 1 + \frac{x^2}{2!} + \frac{x^4}{4!} + \frac{x^6}{6!} + \cdots$$

> In sums of expansions, you should consider more than the requisite number of terms, since one or more terms may disappear.

Example 21　If x is so small that fifth and higher powers can be neglected show that $\sin^2 x \approx x^2 - \tfrac{1}{3}x^4$.

$$\sin x = x - \frac{x^3}{3!} + \cdots$$

$$\therefore \sin^2 x = \left(x - \frac{x^3}{3!} + \cdots\right)\left(x - \frac{x^3}{3!} + \cdots\right)$$

$$\therefore \sin^2 x \approx \left(x^2 - \frac{x^4}{3!} - \frac{x^4}{3!}\right) \longleftarrow \boxed{\text{Ignoring terms in } x^5 \text{ and above}}$$

$$\therefore \sin^2 x \approx x^2 - \tfrac{1}{3}x^4 \quad \text{Q.E.D.}$$

Example 22   Expand $\ln(1 - 3x + x^2)$ as far as the term in x^4.

$$\ln(1 - 3x + x^2)$$
$$= \ln[1 - (3x - x^2)]$$
$$= -(3x - x^2) - \tfrac{1}{2}(3x - x^2)^2 - \tfrac{1}{3}(3x - x^2)^3 - \tfrac{1}{4}(3x - x^3)^4 - \cdots$$
$$\approx -3x + x^2 - \tfrac{1}{2}(9x^2 - 6x^3 + x^4) - \tfrac{1}{3}(27x^3 - 27x^4) - \tfrac{1}{4}(81x^4)$$

$$\boxed{\text{Omitting any powers above } x^4}$$

$$\therefore \ln(1 - 3x + x^2) \approx -3x + x^2$$
$$\left.\begin{array}{l} -\tfrac{9}{2}x^2 \quad +3x^3 - \tfrac{1}{2}x^4 \\ \qquad\qquad -9x^3 + 9x^4 \\ \qquad\qquad\qquad -\tfrac{81}{4}x^4 \end{array}\right\} \longleftarrow \boxed{\begin{array}{l}\text{stacking terms leads}\\\text{to fewer errors}\end{array}}$$

$$\therefore \ln(1 - 3x + x^2) \approx -3x - \tfrac{7}{2}x^2 - 6x^3 - \tfrac{51}{4}x^4$$

- **Exercise 9.5**   See page 183.

9.6 Reduction of a Relationship to a Linear Form

Linear Relationships

A linear relationship is a relationship of the form

$$Y = mX + c$$

where $m = $ gradient of the line
and $c = $ intercept the vertical axis

Non-Linear Relationships

When variables are connected by a non-linear law, it is often possible to find related variables which are connected by a linear law.

Example 23   The following values of x and y have been found experimentally:

x	1	2	3	4
y	1.26	1.59	1.87	2.16

It is known that they are connected by an equation of the form $y^3 = ax^2 + b$. Plot y^3 against x^2 and hence find equation connecting x and y.

$X(= x^2)$ 1   4   9   16

$Y(= y^3)$ 2   4   6.5   10.1

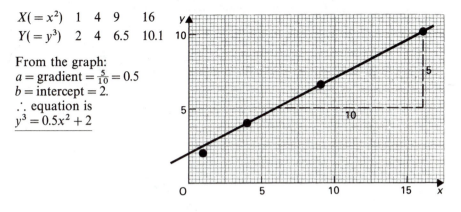

From the graph:
$a = $ gradient $= \frac{5}{10} = 0.5$
$b = $ intercept $= 2$.
∴ equation is
$y^3 = 0.5x^2 + 2$

Relationships of the form $\dfrac{1}{y} + \dfrac{1}{x} = \dfrac{1}{a}$

Plotting X as $\dfrac{1}{x}$ and Y as $\dfrac{1}{y}$ gives the vertical intercept as $\dfrac{1}{a}$.

Relationships of the form $y = pq^x$

If $\qquad y = pq^x$

then $\log y = \log(pq^x) \Rightarrow \log y = (\log q) \times x + \log p$

Compare with $\qquad\qquad\qquad Y \;=\; m\; X \qquad c$

Relationships of the form $y = ax^k$

Here $\qquad\qquad \log y = \log(ax^k) \Rightarrow \log y = k \log x + \log a$

$\qquad\qquad\qquad\qquad\qquad Y = m\; X \;+\; c$

Example 24   The following values of x and y are believed to obey a law of the form $y = ax^k$

x 20   40   60   80   100

y 110   300   560   860   1200

Show graphically that there is a relationship of the form $y = ax^k$ and find approximate values for a and k.

$$\log_{10} y = k \log_{10} x + \log_{10} a$$

$$Y \;= m \quad X \;+ \quad c$$

$X \;(= \log_{10} x)$ 1.301   1.602   1.778   1.903   2.000

$Y \;(= \log_{10} y)$ 2.041   2.477   2.748   2.934   3.079

To make the values obtained as accurate as possible, the scales must be as large as possible. This means that the graph scales will be truncated and hence the intercept c cannot be found from the graph

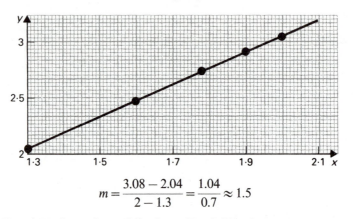

$$m = \frac{3.08 - 2.04}{2 - 1.3} = \frac{1.04}{0.7} \approx 1.5$$

Hence X and Y obey a law of the form $Y = 1.5X + \log_{10} a$.
Substituting $X = 2$, $Y = 3.08 \Rightarrow \log_{10} a \approx 0.08 \Rightarrow a \approx 1.20$.

$$\therefore \ \log_{10} y \approx 1.5 \log_{10} x + \log_{10}(1.2)$$
$$\Rightarrow \log_{10} y \approx \log_{10} x^{1.5} \times 1.2$$
$$\therefore \qquad y \approx 1.2x^{1.5}$$

●   **Exercise 9.6**   See page 184.

9.7 Finding the Method of Reduction: Revision of Unit

1   Reduce the equation to three terms.

2   Make one term constant.

3   Remove the unknown constant(s) from the coefficient of one of the variable terms.

Example 25
Reduce each of the following to the form $Y = mX + C$.
(a) $ay = x^3 + x + d$ (b) $p \ln y = x^2 - qx$

(c) $y = \dfrac{1}{(x - r)(x - s)}$

(a) $ay = x^3 + x + d$
     $\Rightarrow ay = (x^3 + x) + d$ (1) and (2) are satisfied.

$$\Rightarrow \quad y = \underbrace{\frac{1}{a}}(x^3 + x) + \frac{d}{a} \quad (3)$$

$$\updownarrow \quad \updownarrow \quad \updownarrow \qquad \updownarrow$$
$$Y = m \quad X \quad + c$$

(b) $p \ln y = x^2 - qx$ (1) is satisfied

$$\Rightarrow p \frac{\ln y}{x} = x - q \qquad (2)$$

$$\Rightarrow \qquad x = p \left(\frac{\ln y}{x} \right) + q \quad (3)$$

$$\begin{array}{cccc} \updownarrow & \updownarrow & \updownarrow & \updownarrow \\ Y = & m & X & + c \end{array}$$

(c) $y = \dfrac{1}{(x - r)(x - s)}$

$$\Rightarrow \frac{1}{y} = (x - r)(x - s) = x^2 - (r + s)x + rs$$

$$\Rightarrow x^2 - \underbrace{\frac{1}{y}}_{\;} = \underbrace{(r + s)}x - rs$$

$$\begin{array}{cccc} \updownarrow & & \updownarrow\;\updownarrow\;\updownarrow \\ Y & = & m\;X + c \end{array}$$

- Miscellaneous Exercise 9A   See page 186.

9.8 + A Level Questions

- Miscellaneous Exercise 9B   See page 187.

Unit 9   EXERCISES

Exercise 9.1

1 Express in logarithmic form:

(a) $3^2 = 9$ (b) $10^4 = 10000$ (c) $4^{3/2} = 8$
(d) $512 = 8^3$ (e) $1728 = (144)^{3/2}$ (f) $\frac{1}{2} = 8^{-1/3}$
(g) $6^0 = 1$ (h) $1/25 = 5^{-2}$ (i) $9 = 27^{2/3}$
(j) $p = q^3$ (k) $x^y = z$ (l) $m^n = 3$

2 Express in index form:

(a) $\log_5 125 = 3$ (b) $\log_{10} 100 = 2$ (c) $\log_3 81 = 4$
(d) $\log_3 1 = 0$ (e) $\log_{1/2} 8 = -3$ (f) $\log_{125} 5 = \frac{1}{3}$
(g) $\log_p 1 = 0$ (h) $\log_q p = 3$ (i) $\log_5 x = y$
(j) $\log_a 3 = c$ (k) $\log_p q = r$ (l) $q = \log_r p$

3 Express in terms of $\log p$, $\log q$ and $\log r$:

(a) $\log pr$ (b) $\log p/q$ (c) $\log pqr$
(d) $\log pq/r$ (e) $\log q^2$ (f) $\log p^2/q$
(g) $\log q^3 p^2$ (h) $\log 1/p$ (i) $\log \sqrt{p}$
(j) $\log \sqrt{q/r}$ (k) $\log r\sqrt{p}$ (l) $\log p^4/r\sqrt{q}$

4 Simplify:

(a) $\log 3 + \log 5$ (b) $\log 8 - \log 2$
(c) $\log 2 + \log 15 - \log 5$ (d) $3 \log 5$
(e) $2 \log 5 + \log 2$ (f) $3 \log 4 - 4 \log 2$
(g) $\frac{1}{2} \log 9$ (h) $2 \log 8 + 3 \log 4 - 3 \log 16$
(i) $\frac{1}{3} \log 27 - 3 \log 2$ (j) $\frac{1}{2} \log 81 - 2 \log 3$

5 Evaluate:

(a) $\log_{10} 1000$ (b) $\log_2 16$ (c) $\log_3 9$
(d) $\log_9 3$ (e) $\log_8 1$ (f) $\log_{125} 25$
(g) $\log_{10} 0.1$ (h) $\log_2 2^3$ (i) $\log_{1/2} 4$
(j) $\log_{11} 121$ (k) $\log_{144} 12$ (l) $\log_8 \frac{1}{2}$

6 Express in terms of $\log_{10} a$ and $\log_{10} b$:

(a) $\log_{10} 10a$ (b) $\log_{10} \dfrac{100}{b^2}$ (c) $\log_{10} \dfrac{\sqrt{b}}{1000}$

*7 Simplify:

(a) $1 - \log_{10} 5$ (b) $2 \log_{10} 2 + 2$
(c) $\frac{1}{2} \log_{10} 16 - \frac{1}{3} \log_{10} 125 + 1$ (d) $2 \log_4 3 + 1$
(e) $2 - 2 \log_6 3$ (f) $3 - \log_4 8$

*8 Simplify:

(a) $\dfrac{\log 9}{\log 3}$ (b) $\dfrac{\log 125}{\log 5}$ (c) $\log(x^2 - 1) - \log(x + 1)$

(d) $3 \log_a 5 + 2 \log_a 2 - \log_a 50 - \log_a 10$

*9 Evaluate:

(a) $\log_a a^3$ (b) $\log_p p^q$ (c) $\log_{10} 10$
(d) $10^{\log_{10} 5}$ (e) $3^{\log_3 9}$

Exercise 9.2

1 Solve the equations: (answers to 3 d.p).

(a) $5^x = 20$ (b) $3^x = 7$
(c) $2^x = 15$ (d) $2^{3x} = 21$
(e) $4^{x-1} = 12$ (f) $3^{2x-1} = 4$

2 Evaluate, using a calculator, to 3 d.p:

(a) $\log_5 13$ (b) $\log_3 5$ (c) $\log_{15} 2$ (d) $\log_{0.2} 6$

3 Solve the equations:

(a) $3.3^{2x} - 4.3^x + 1 = 0$ (b) $2^{2x+1} - 5.2^x + 2 = 0$
(c) $2.5^{2x} - 12.5^x + 10 = 0$ (d) $4^x - 5.2^x - 24 = 0$

4 Solve $(\log_2 x)^2 - 5 \log_2 x + 6 = 0$

5 If $\log_3 x + \log_x 3 = 2$ find x

6 Solve the simultaneous equations $\log_x y = 2$, $xy = 27$

7 Solve simultaneously $2^y = 8^x$ and $2 \log y = \log x + \log 3$

8 Solve $2 \log_5 x = 2 \log_5 3 + \log_5 (x - 2)$

*9 Solve $\log_4 (2x - 1) = \log_2 x$

*10 Find the positive value of x that satisfies the equation

$$\log_3 x = \log_9 (x + 6)$$

*11 Solve the simultaneous equations $\log(y + x) = 2 \log x$ and $\log y = \log 2 + \log(x - 1)$

**12 Solve $2^{x+1} + 7 - 4.2^{-x} = 0$

Exercise 9.3

1 On one diagram, sketch and label the graphs of $y = 2^x$, $y = 2^{-x}$, $y = 3^x$ and $y = -3^x$

2 On three separate diagrams sketch the graphs of $y = \log x$, $y = 2 \log x$ and $y = 2 + \log x$

3 Differentiate w.r.t. x:

   (a) $\ln 3x$ (b) $\ln(x + 5)$ (c) $\ln(3x - 1)$
   (d) $\ln(2 - x)$ (e) $\ln(x^2 - x + 1)$ (f) $\ln(1 - x^3)$
   (g) $\ln \cos x$ (h) $\ln \sin 2x$ (i) $\ln \tan x$

4 Differentiate w.r.t. x:

   (a) $3e^x$ (b) e^{-x} (c) e^{-5x}
   (d) $e^{(1/2)x}$ (e) e^{x^3} (f) $e^{\tan x}$

   (g) e^{e^x} (h) $e^x + \dfrac{1}{e^x}$ (i) $\left(e^x - \dfrac{1}{e^x}\right)^2$

5 Differentiate each of the following after first simplifying the log function:

   (a) $\ln \sqrt{2 + x}$ (b) $\ln\left(\dfrac{2 + x}{2 - x}\right)$ (c) $\ln(3x - 1)^4$

   (d) $\ln \sin^2 x$ (e) $\ln \dfrac{1}{x}$ (f) $\ln \sqrt{\dfrac{2 + x}{1 + 3x}}$

   (g) $\ln \dfrac{x^3}{x + 1}$ (h) $\ln(\sqrt[3]{x^3 - 1})$ (i) $\ln \sec^2 x$

6 Find dy/dx if $y = x^2 \ln x$

7 If $y = (\ln x)/x$ prove that dy/dx is zero when $x = e$

8 Find the gradient of the curve $y = (3x^2 - 1)\ln x$ at the point where $x = 1$

9 If $y = x^3 e^x$ show that dy/dx vanishes when $x = 0$ and $x = -3$

10 Find the stationary value of the function te^{-t} in terms of e and determine whether it is a maximum or a minimum

*11 If $x = e^{-t}\sin t$ verify that $\dfrac{d^2x}{dt^2} + \dfrac{2dx}{dt} + 2x = 0$

*12 Differentiate w.r.t. x:

(a) $(2x - 1)\ln(2x - 1)$ (b) $\ln(1 + e^x)$

(c) $e^x \ln 3x$ (d) $e^{-3x}\cos 3x$

(e) $\ln\left(\dfrac{1 + \sin x}{1 - \sin x}\right)$ (f) $\ln\left(\dfrac{1 - x}{\sqrt{1 + x^2}}\right)$

(g) $\ln\left(\dfrac{1 + \sqrt{x}}{1 - \sqrt{x}}\right)$ (h) $e^{x\sin x}$

Exercise 9.4

1 Integrate w.r.t. x:

(a) $\dfrac{4}{x}$ (b) $\dfrac{1}{4x}$ (c) $1 - \dfrac{1}{x}$ (d) $\dfrac{1}{x + 2}$

(e) $\dfrac{1}{2x - 3}$ (f) $\dfrac{1}{3x + 1}$ (g) $\dfrac{1}{1 - 2x}$ (h) $\dfrac{2}{3(x - 4)}$

(i) $\dfrac{3}{5x - 2}$ (j) $\dfrac{1}{3 - x}$ (k) $\dfrac{2}{5 - 2x}$ (l) $\dfrac{4}{1 - 3x}$

2 Integrate w.r.t. x:

(a) e^{3x} (b) e^{-3x} (c) $1/e^x$ (d) $3e^x$

(e) $4e^{2x}$ (f) $3e^{-4x}$ (g) $3x^2 e^{x^3}$ (h) $\cos x\, e^{\sin x}$

3 Integrate w.r.t. x:

(a) $\left(1 - \dfrac{1}{x}\right)^2$ (b) $\left(x + \dfrac{1}{x}\right)^2$ (c) $\left(e^x + \dfrac{1}{e^x}\right)^2$

4 Evaluate:

(a) $\displaystyle\int \dfrac{2x}{x^2 - 1}\,dx$ (b) $\displaystyle\int \dfrac{3x^2 - 2x}{x^3 - x^2 + 1}\,dx$ (c) $\displaystyle\int \dfrac{x + 1}{x^2 + 2x + 3}\,dx$

(d) $\displaystyle\int \dfrac{x^3}{x^4 - 3}\,dx$ (e) $\displaystyle\int \dfrac{\sin x}{1 + \cos x}\,dx$ (f) $\displaystyle\int \dfrac{x^5}{x^6 + 2}\,dx$

(g) $\displaystyle\int \tan 4x\,dx$ (h) $\displaystyle\int \cot 3x\,dx$ (i) $\displaystyle\int \dfrac{\sec^2 x}{1 + \tan x}\,dx$

5 Evaluate:

(a) $\displaystyle\int_0^2 e^{-x}dx$ (b) $\displaystyle\int_0^1 e^{(1-2x)}dx$ (c) $\displaystyle\int_0^{1/2}\left(1-\frac{e^{-x}}{2}\right)dx$

(d) $\displaystyle\int_0^1 \frac{dx}{3x+1}$ (e) $\displaystyle\int_{1/4}^{1/2}\frac{x}{1-x^2}dx$ (f) $\displaystyle\int_0^{\pi/6}\tan 2x\,dx$

6 Find the area included between the curve $y(x-1)=2$, the x-axis and the ordinates $x=-2$ and $x=-3$

7 If $pv=3$, evaluate $\int_1^3 p\,dv$

*8 Evaluate $\int dx/(\sin x\cdot\cos x)$ by multiplying the numerator and denominator by $\sec^2 x$

*9 Evaluate $\int \sec x\,dx$ by multiplying the numerator and denominator by $\sec x+\tan x$

**10 Find the maximum and minimum points on the curve $y=xe^{-x^2}$

Exercise 9.5

Expand each of the following up to, and including, the term in x^3. Give the general term and the range of values of x for which the expansion is valid.

1 e^{5x} **2** $\ln\left(\dfrac{1-x}{1+x}\right)^3$ **3** $\ln(3-x)$

Give the first three non-zero terms in each of the following expansions:

4 $x\cos 2x$ **5** $(1-x)e^x$ **6** $\frac{1}{2}(e^x-e^{-x})$ **7** $\dfrac{x+e^x}{e^x}$

8 Find, in ascending powers of x, the first four non-zero terms of the expansions of (a) $\sqrt{1/(1-2x)}$ (b) $e^x\cdot e^{x^2}$ giving each term in its simplest form.

If x is so small that powers higher than x^3 are negligible, show that

$$\sqrt{\frac{1}{1-2x}}-e^x\cdot e^{x^2}\approx kx^3$$

and find the value of k

9 Find the first three non-zero terms of the expansions in ascending powers of x of (a) $\sin px$ (b) xe^{-qx} where p and q are non-zero constants

If, further

$$xe^{-qx}\sin px\approx-2x^2+12x^3+rx^4$$

evaluate p, q and r

10 Show that $\ln\left(1+\dfrac{1}{y}\right) - \ln\left(1-\dfrac{1}{y}\right) = \ln\left(\dfrac{y+1}{y-1}\right)$, for $y > 1$.

Hence, show that, if $y > 1$

$$y\ln\left(\frac{y+1}{y-1}\right) \approx 2 + \frac{a}{y^2} + \frac{b}{y^4}$$

where a and b are constants, giving their values. Use this result to calculate the values of $\ln\{(11/9)^{10}\}$ to 5 decimal places

Exercise 9.6

In this exercise, where graphs are required to be drawn, they should be at least A5 size (half A4) for accuracy.

1 Find the relationship connecting the variables in each of the following cases:

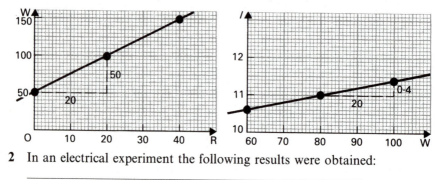

2 In an electrical experiment the following results were obtained:

V	85	80	74	60	48	35	20
R	292	280	268	232	204	175	139

By drawing a graph of R against V find an equation giving R in terms of V

3 In a cooling experiment, the temperature $\theta°\,C$ was recorded for various values of T minutes, the time elapsed.

T	0	10	25	35	50	60
θ	8.2	4.3	−1.9	−5.8	−11.9	−15.8

Find the law connecting T and θ in the form $\theta = T_0 + kT$

4 The following values of p and q have been found by experiment:

p	2	3	4	5	6
q	0	2	5.5	10	15.5

It is known that they are connected by an equation of the form $q = ap^2 + b$. Plot q against p^2. Which experimental data pair is obviously in error. Find approximate values of a and b.

5 The variables u and v are connected by

$$\frac{1}{u} + \frac{1}{v} = \frac{1}{f}$$

where f is a constant. By plotting the reciprocal of v against the reciprocal of u, estimate the value of f

u	15	20	25	40	50
v	30	20	17	13	12.5

6 Given the following values of x and y

x	1	2	3	4	5
y	-3	6	39	108	225

by plotting y/x against x^2, verify that x and y are connected by an equation of the form $y = \alpha x^3 + \beta x$ and find approximate values for α and β

7 A student performs as experiment and records the following data for two variables y and x.

x	0.5	2	4	7	8
y	1.4	2.8	4.0	5.3	5.7

He believes that there is a relationship between y and x of the form $y = ax^b$. By plotting $\ln y$ against $\ln x$ show that this relationship holds and find a and b to 2 significant figures

After completing the experiment, the student realised that he should have measured y when $x = 6$. Obtain from your graph, an estimate to 2 s.f. of the value of y when $x = 6$ (L)

8

x	1	2	3	4	5
y	14.1	15.8	17.8	19.9	22.4

The values x and y obtained in an experiment are believed to satisfy a relationship of the form $y = ab^x$ where a and b are constants. Verify the relationship graphically and calculate approximate values of a and b

9

x	1.21	2.55	3.46	4.91	5.71
y	10.0	79.3	189.6	521.7	809.9

By means of a straight line graph verify that x and y are related by a law of the form $y = ax^3 + bx^2$ where a and b are constants. Use the graph to estimate the values of a and b

Miscellaneous Exercise 9A

Reduce each of the relationships to the form $Y = mX + c$. In each case give the functions corresponding to X and Y and the constants corresponding to m and c.

1 $\dfrac{1}{y} = ax^2 + b$ **2** $y^3 = ax^2 + abx$

3 $\dfrac{a}{x} + \dfrac{b}{y} = 2$ **4** $ay = b^x$

5 $s = ab^{-t}$ **6** $y(y - p) = x - q$

7 $ae^y = x(x - b)$ **8** $y = (x - a)(x - b)$

9 Express in logarithmic form:
    (a) $5^3 = 125$ (b) $32 = 8^{5/3}$

10 Express in index form:
    (a) $\log_2 16 = 4$ (b) $\log_9 1 = 0$

11 Express $\log xy^3 / \sqrt{z}$ in terms of $\log x$, $\log y$ and $\log z$

12 Simplify (a) $\frac{1}{3}\log 27$ (b) $\log_a a^3$ (c) $\ln e^x$

13 Evaluate (a) $\log_2 64$ (b) $\log_{100} 10$ (c) $\ln e^2$

14 Solve (a) $4^{2x} = 27$ (b) $3^{x-1} \cdot 4^x = 109$

15 Solve $5^{2x} + 2.5^x - 3 = 0$

16 If $\log_5 x - 2\log_x 5 = 1$ find x

17 Differentiate w.r.t. x:
    (a) $\ln(x^2 + 3)$ (b) $\ln \sin x$ (c) e^{5x} (d) $e^{(-1/2)x}$
    (e) $\frac{1}{2}e^{x^2}$ (f) $\ln(\sec x + \tan x)$

18 Differentiate, giving your answer in its simplest form:
    (a) $\ln(2x + 1)^3$ (b) $\ln 1/\sqrt{x - 1}$

19 Integrate w.r.t. x:

    (a) $\dfrac{2}{x}$ (b) $2e^{-x}$ (c) $\dfrac{3}{x} - \dfrac{2}{x^2}$ (d) $\dfrac{1}{x + 1}$

    (e) $3e^{2x}$ (f) $\dfrac{3x + 1}{3x^2 + 2x - 1}$ (g) $3x^2 e^{x^3}$

20 Evaluate:

    (a) $\displaystyle\int_0^4 e^{(1/2)x}\, dx$ (b) $\displaystyle\int_0^3 \dfrac{2x^2}{1 + x^3}\, dx$ (c) $\displaystyle\int_2^3 \dfrac{dx}{1 - x}$

Miscellaneous Exercise 9B

Differentiate with respect to x.

1 (a) $\dfrac{\sin x}{e^x}$ (b) $\ln(1 + \tan^2 x)$ (L: 8.2, 8.3, 9.3)

2 (a) $\dfrac{e^{-x}}{x}$ (b) $\ln(1 + \sin x)$ (L: 8.2, 8.3, 9.3)

3 (a) $x^3 \ln x$ (b) $\dfrac{1 + \cos x}{x}$ (L: 8.3, 9.3)

4 (a) $\sin^2 3x$ (b) $\ln(1 + \sec x)$ (L: 8.2, 9.3)

5 If $2 \log_y x + 2 \log_x y = 5$, show that $\log_y x$ is either $\frac{1}{2}$ or 2. Hence find all pairs of values of x and y which satisfy simultaneously the equation above and the equation $xy = 27$ (J: 9.2)

6 Given that $y = 6^x$, find, to 2 decimal places, the value of x when $y = 0.5$ (L: 9.2)

7 Evaluate $\displaystyle\sum_{r=1}^{20} \log_{10}(1.1)^r$ giving your answer correct to 2 s.f.   (J: 6.1, 9.1)

8 Given that $y = A \cos(\log_e x) + B \sin(\log_e x)$, where A and B are constants and $x > 0$, show that

$$x^2 \frac{d^2 y}{dx^2} + x \frac{dy}{dx} + y = 0 \qquad \text{(A: 8.2, 9.3)}$$

9 Show that the tangent at the point of inflexion on the curve

$$y = 2x - \frac{1}{x} - 2 \ln x$$

passes through the origin   (L: 3.3, 3.5, 9.3)

10 Find, from first principles, a formula for the sum of the integers $1, 2, 3, \ldots, n$.

Prove that $\displaystyle\sum_{r=0}^{n} \log_a (2a^r) = \frac{1}{2}(n + 1)(n + \log_a 4)$ (J: 6.1, 6.3, 9.1)

11 Given that $y = e^{-x} \sin(x\sqrt{3})$, prove that

$$\frac{dy}{dx} = -2e^{-x} \sin\left(x\sqrt{3} - \frac{\pi}{3}\right)$$

Show also that $\dfrac{d^3 y}{dx^3} = ky$ for some constant k and state the value of k (A: 8.3, 9.3)

12 Simplify $\ln\left[\dfrac{(1+x)e^{-2x}}{1-x}\right]^{1/2}$ and show that its derivative is $\dfrac{x^2}{1-x^2}$

Hence, or otherwise, evaluate $\dfrac{dy}{dx}$ at $x=0$ for the function

$y=\left[\dfrac{(1+x)e^{-2x}}{1-x}\right]^{1/2}$ (J: 9.3, 8.4)

13 The equation of a curve is $6y^2=e^xy^3-2e^{4x}$. Given that (a, b) is the point on the curve at which $dy/dx=0$, show that $b=2e^a$.

By substituting the coordinates into the equation of the curve, obtain a further relation and hence find the values of a and b (A: 8.4, 9.3)

14 Find the four values of x for which the curve $y=\sqrt{2}\,e^{-x}\cos x$ has turning points in the range $0\leqslant x\leqslant 4\pi$, writing your values in ascending order. Show that the corresponding y values form the first four terms of a geometric progression. Show that the sum to infinity of the geometric progression is $-e^{\pi/4}/(e^\pi+1)$ (A: 1.5, 6.2, 8.3, 9.3)

15 Starting from first principles, prove that the sum of the first n terms of a G.P., whose first term is a and whose common ratio is r, is $a(1-r^n)/(1-r)$

State the range of values of r for which the sum to infinity exists.

Determine the range of values of x for which the sum to infinity of the series

$$1-e^x+e^{2x}-e^{3x}+\cdots+(-1)^m e^{mx}+\cdots$$

exists.

Write down (or find) an expression for the sum to infinity in this case
                                                                                    (J: 6.2, 9.3)

16 Solve the simultaneous equations

$$\log_x y+2\log_y x=3$$
$$\log_9 y+\log_9 x=3$$ (J: 9.2)

17 (a) Find the values of x which satisfy the equation

$$4\log_3 x=9\log_x 3$$

(b) By taking $\log_{10} 5\approx 0.7$, obtain an estimate of the root of the equation

$$10^{y-5}=5^{y+2}$$

giving your answer to the nearest integer   (A: 9.2)

18 Rewrite the following equations in suitable form to display a linear relationship in each case between two of the variables, x, $\ln x$, y, and $\ln y$.

(a) $\dfrac{2}{x}+\dfrac{3}{y}=\dfrac{4}{xy}$

(b) $5x=6^y$ (L: 9.7)

19 Given that $p = \log_y x$ and $q = \log_z x$, where $x \neq 1$, express, in terms of p and q

(a) $\log_x (yz)$
(b) $\log_z y$

If $y + z = 1$, prove that $\dfrac{1}{p} = \log_x (1 - x^{1/q})$ (A: 9.1, 9.2)

20 Use the series expansion for $\sin x$ to write down in ascending powers of x the first three non-zero terms in the series expansion of $\sin x / x$

Write down in ascending powers of y the first three non-zero terms in the series expansion of $\log_e (1 - y)$

Use these two series to show that if x is small enough for terms in x^6 and higher powers of x to be neglected, then

$$\log_e \sin x - \log_e x = Ax^2 + Bx^4$$

for constants A and B whose numerical values are to be found

(A: 9.5)

21 Express $\log_9 xy$ in terms of $\log_3 x$ and $\log_3 y$. Without using tables, solve for x and y the simultaneous equations

$$\log_9 xy = \tfrac{5}{2}$$
$$\log_3 x \log_3 y = -6$$

expressing your answers as simply as possible   (J: 9.2)

22 An infinite geometric series has first term a and common ratio r, where $a > 0$ and $0 < r < 1$. Show that D, the difference between the sum to infinity of the series and the sum of the first n terms of the series, is given by

$$D = \frac{ar^n}{1 - r}$$

Deduce that $\log_{10} D = n \log_{10} r + \log_{10}\left(\dfrac{a}{1 - r}\right)$

The following table gives values of D, rounded to the nearest 100, corresponding to some values of n.

n	4	6	10	14	16
D	4200	3000	1600	800	600

By drawing an appropriate linear graph, estimate to 2 significant figures, a value for r and a value for a (A: 6.2, 9.6)

23 Without using tables, solve each of the following equations for x, expressing your answers as simply as possible:

(a) $9 \log_x 5 = \log_5 x$ (b) $\log_8\left(\dfrac{x}{2}\right) = \dfrac{\log_8 x}{\log_8 2}$ (J: 9.2)

UNIT 10 ALGEBRAIC TECHNIQUES

10.1 Polynomials

A function of the form

$$P(x) = a_n x^n + a_{n-1} x^{n-1} + \cdots + a_0 \qquad (a_n \neq 0)$$

where n is a positive integer, is called a **polynomial of degree n**

The real numbers $a_n, a_{n-1}, a_{n-2}, \ldots, a_0$ are called the **coefficients of** x^n, x^{n-1}, x^{n-2}, etc.

For instance
$x^5 - 4x^2 + 3x$ is a polynomial of degree 5. A **linear function**, such as $ax + b$, is a polynomial of degree 1. A **quadratic function** is a polynomial of degree 2. A polynomial of degree 3 is often called a **cubic**.

Product of Two Polynomials

Example 1 Multiply $5x^2 - x + 3$ by $2x^3 + x^2 + 2$.

$$(5x^2 - x + 3)(2x^3 + x^2 + 2) = 10x^5 + 5x^4 \qquad\quad + 10x^2$$
$$- 2x^4 - \ x^3 \qquad\quad - 2x$$
$$+ 6x^3 + \ 3x^2 \qquad + 6$$
$$\therefore (5x^2 - x + 3)(2x^3 + x^2 + 2) = 10x^5 + 3x^4 + 5x^3 + 13x^2 - 2x + 6$$

> The system of stacking terms minimises errors.

Example 2 Find the coefficient of x^4 in the product of $x^3 - 3x^2 + x - 1$ and $2x^3 + x^2 - 3x + 2$.

$$\overbrace{(x^3 - 3x^2 + \underbrace{x}_{} - 1)(2x^3 + x^2 - 3x + 2)}$$

Term in x^4 is

$$(x^3)(-3x) + (-3x^2)(x^2) + (x)(2x^3) = -3x^4 + (-3x^4) + 2x^4 = -4x^4$$
$$\therefore \text{ coefficient of } x^4 = -4$$

Using Factors of a Polynomial to Simplify Algebraic Fractions

Example 3   Simplify $\dfrac{2x}{x^2 - 4} + \dfrac{7}{2x^2 + x - 6}$.

$$\frac{2x}{x^2 - 4} + \frac{7}{2x^2 + x - 6} = \frac{2x}{(x - 2)(x + 2)} + \frac{7}{(2x - 3)(x + 2)}$$

$$= \frac{2x(2x - 3) + 7(x - 2)}{(x - 2)(x + 2)(2x - 3)}$$

$$= \frac{4x^2 + x - 14}{(x - 2)(x + 2)(2x - 3)}$$

$$= \frac{(4x - 7)(x + 2)}{(x - 2)(x + 2)(2x - 3)}$$

$$= \frac{4x - 7}{(x - 2)(2x - 3)}$$

Division of Polynomials

Example 4   Divide $x^4 - 3x + 10$ by $x - 2$.

$$
\begin{array}{r}
x^3 + 2x^2 + 4x + 5 \\
x - 2 \enclose{longdiv}{x^4 \qquad\qquad - 3x + 10} \\
\underline{x^4 - 2x^3} \qquad\qquad\quad \\
2x^3 \qquad\qquad \\
\underline{2x^3 - 4x^2} \qquad\quad \\
4x^2 - 3x \quad \\
\underline{4x^2 - 8x} \quad \\
5x + 10 \\
\underline{5x - 10} \\
20
\end{array}
$$

When $x^4 - 3x + 10$ is divided by $x - 2$ the **quotient** is $x^3 + 2x^2 + 4x + 5$ and the **remainder** is 20, or $x^4 - 3x + 10 \equiv (x - 2)(x^3 + 2x^3 + 4x + 5) + 20$

●   Exercise 10.1   See page 206.

10.2 The Factor Theorem

> If, for a given function $f(x)$, $f(a) = 0$, then $x - a$ is a factor of $f(x)$

Example 5   Factorise completely $x^4 - 3x^3 + 4x^2 - 8$.

Let $f(x) \equiv x^4 - 3x^3 + 4x^2 - 8$
Then $f(1) = 1 - 3 + 4 - 8 \neq 0$ $\therefore x - 1$ is not a factor of $f(x)$
   $f(-1) = 1 + 3 + 4 - 8 = 0$ $\therefore x + 1$ *is* a factor of $f(x)$

$$
\begin{array}{r}
x^3 - 4x^2 + 8x - 8 \\
x + 1\overline{\smash{\big)}\ x^4 - 3x^3 + 4x^2 \qquad\quad - 8} \\
x^4 + x^3 \\
\hline
-4x^3 + 4x^2 \\
-4x^3 - 4x^2 \\
\hline
8x^2 \\
8x^2 + 8x \\
\hline
-8x - 8 \\
-8x - 8 \\
\hline
-
\end{array}
$$

Now let $g(x) \equiv x^3 - 4x^2 + 8x - 8$
Then $g(-1) = -1 - 4 - 8 - 8 \neq 0$ $\therefore x + 1$ is not a factor of $g(x)$
   $g(2) =$ $8 - 16 + 16 - 8 = 0$ $\therefore x - 2$ is a factor of $g(x)$

Using long division again

$$x^3 - 4x^2 + 8x - 8 = (x - 2)(x^2 - 2x + 4)$$

Also $x^2 - 2x + 4$ has no linear factors.
$\therefore x^4 - 3x^3 + 4x^2 - 8 \equiv (x + 1)(x - 2)(x^2 - 2x + 4)$

> **Notes 1** The factors of 8 are $1, 2, 4, 8$ so the values we must choose for a must belong to the set $\{\pm 1, \pm 2, \pm 4, \pm 8\}$.
>
> **2** After having taken out the first factor from $f(x)$ it must be tried again as a possible factor of $g(x)$ since repeated factors occur frequently.

• Exercise 10.2   See page 207.

10.3 The Remainder Theorem

> When a polynomial $f(x)$ is divided by $(x - a)$ the remainder is $f(a)$
> $$f(x) = (x - a)\ \underset{\uparrow}{Q(x)}\ +\ \underset{\uparrow}{f(a)}$$
> $$\text{quotient}\quad\text{remainder}$$

Proof

$$\frac{f(x)}{x - a} = Q(x) + \frac{r}{x - a} \longleftarrow$$

The remainder is always at least one degree less than the divisor and it must therefore, in this case, be a constant.

Hence $f(x) = (x - a)Q(x) + r$

When $x = a$, $f(a) = 0 \cdot Q(a) + r$.
$\therefore$ remainder $= r = f(a)$

> **Note:** When a polynomial $f(x)$ of degree n is divided by a polynomial of degree r, then the quotient will be of degree $(n - r)$ and the remainder, at most, of degree $(r - 1)$.

Example 6   When the polynomial $f(x) = x^3 + ax^2 + bx - 3$ is divided by $(x - 1)$ and $(x - 2)$ its remainders are -3 and -1 respectively. Find the values of a and b.

When $f(x)$ is divided by $(x - 1)$ it has remainder -3

$$\Rightarrow f(1) = -3$$
$$\Rightarrow 1 + a + b - 3 = -3$$
$$\Rightarrow a + b = -1 \qquad (1)$$

When $f(x)$ is divided by $(x - 2)$ it has remainder -1

$$\Rightarrow f(2) = -1$$
$$\Rightarrow 8 + 4a + 2b - 3 = -1$$
$$\Rightarrow 4a + 2b = -6 \qquad (2)$$

Solving (1) and (2) simultaneously $\Rightarrow a = -2, b = 1$

Example 7   Find the remainder when $2x^3 + x^2 - 9x + 8$ is divided by $(x - 1)(x + 3)$.

Since the divisor is a quadratic, the remainder must at most be of degree 1. Let the remainder be $Ax + B$.

Let $\qquad\qquad f(x) \equiv 2x^3 + x^2 - 9x + 8 \qquad (1)$
then $\qquad\qquad f(x) = (x - 1)(x + 3)Q(x) + Ax + B$
Hence $\qquad\qquad f(1) = A + B \overset{(1)}{=} 2$
and $\qquad\qquad f(-3) = -3A + B \overset{(1)}{=} -10$ $\Bigg\} \Rightarrow A = 3, \quad B = -1$

$\therefore$ remainder $= 3x - 1$

Repeated Factors

> If $f(x)$ has a repeated factor $(x - a)$ then $f'(a)$ has a factor $x - a$

Proof
If $f(x)$ has a repeated factor $(x - a)$

then $\qquad\qquad f(x) = (x - a)^2 h(x)$
Hence $\qquad\qquad f'(x) = 2(x - a)h(x) + (x - a)^2 h'(x)$
$\qquad\qquad \therefore f'(x) = (x - a)[2h(x) + (x - a)h'(x)]$
$\qquad\qquad \Rightarrow f'(x)$ has a factor $(x - a)$

but a linear factor of $f'(x)$ is **not necessarily** a factor of $f(x)$.

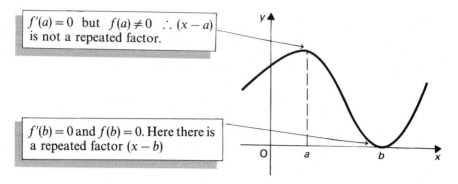

$f'(a) = 0$ but $f(a) \neq 0$ ∴ $(x - a)$ is not a repeated factor.

$f'(b) = 0$ and $f(b) = 0$. Here there is a repeated factor $(x - b)$

Example 8   Find any repeated factors of $x^4 + 4x^3 - 2x^2 - 12x + 9$.

Let $f(x) \equiv x^4 + 4x^3 - 2x^2 - 12x + 9$
then $f'(x) = 4x^3 + 12x^2 - 4x - 12$
Hence $f'(x) = 4(x - 1)(x + 1)(x + 3)$

Using the factor theorem

and, checking the values of $f(x)$ for $x = \pm 1, -3$ we get

$$f(1) = 0, \quad f(-1) \neq 0, \quad f(-3) \neq 0$$

∴ $(x + 1)$ and $(x + 3)$ are not factors of $x^4 + 4x^3 - 2x^2 - 12x + 9$ and $(x - 1)$ is the only repeated factor

Example 9   Use the remainder theorem to find the remainder when $x^4 - 6x^3 + 10x^2 - 2x$ is divided by $(x - 3)^2$.

Let $f(x) \equiv x^4 - 6x^3 + 10x^2 - 2x$ (1)
and let $f(x) \equiv (x - 3)^2 Q(x) + Ax + B$ (2)
                    $f(3) = 3A + B = 3$ (3)

But, no other equation linking A and B can be found directly from equation (2) without involving $Q(x)$. So, we consider the derived function.

$(1) \Rightarrow f'(x) = 4x^3 - 18x^2 + 20x - 2 \Rightarrow f'(3) = 4$
and $(2) \Rightarrow f'(x) = 2(x - 3)Q(x) + (x - 3)^2 Q'(x) + A \Rightarrow f'(3) = A$ $\Big\} \Rightarrow A = 4$
and $(3) \Rightarrow B = -9$.

∴ remainder $= 4x - 9$

- **Exercise 10.3**   See page 208.

10.4 Relationships between the Roots and the Coefficients of a Quadratic Equation

Let α and β be the roots of the equation $ax^2 + bx + c = 0$.

and
$$\left. \begin{array}{r} \therefore \ (x - \alpha)(x - \beta) = 0 \\ ax^2 + bx + c = 0 \end{array} \right\} \text{ have the same solution}$$

and
$$\left. \begin{array}{r} \Rightarrow x^2 - (\alpha + \beta)x + \alpha\beta = 0 \\ x^2 + \left(\dfrac{b}{a}\right)x + \left(\dfrac{c}{a}\right) = 0 \end{array} \right\} \text{ have the same solution}$$

$$\therefore \quad \boxed{\begin{array}{c} \alpha + \beta = -\dfrac{b}{a} \\[2mm] \text{and} \quad \alpha\beta = \dfrac{c}{a} \end{array}}$$

The equation may also be written as

$$x^2 - (\text{sum of roots})x + (\text{product of roots}) = 0$$

Note: Roots have the same sign $\Rightarrow c/a > 0$
But $c/a > 0 \Rightarrow$ roots have the same sign only if $b^2 \not< 4ac$

Example 10   Find the sum and product of the roots of the equation $2x^2 - 5x + 1 = 0$.

$$\text{Sum of roots} = \tfrac{5}{2}$$
$$\text{Product of roots} = \tfrac{1}{2}$$

Example 11   Find the equation given that the sum of its roots is -5 and the product of its roots is $\tfrac{2}{3}$.

The equation is
$$x^2 + 5x + \tfrac{2}{3} = 0$$
$$\Rightarrow 3x^2 + 15x + 2 = 0$$

Example 12   The roots of the equation $3x^2 + 5x + 6 = 0$ are α and β. Find the values of $\dfrac{1}{\alpha} + \dfrac{1}{\beta}$ and $\dfrac{1}{\alpha\beta}$. Hence write down the equation whose roots are $\dfrac{1}{\alpha}$ and $\dfrac{1}{\beta}$.

Since $3x^2 + 5x + 6 = 0$ has roots α and β,

then
$$\alpha + \beta = -\tfrac{5}{3} \tag{1}$$

and
$$\alpha\beta = 2. \tag{2}$$

Hence
$$\frac{1}{\alpha} + \frac{1}{\beta} = \frac{\alpha + \beta}{\alpha\beta} \overset{(1)(2)}{=} \frac{-\tfrac{5}{3}}{2} = -\frac{5}{6}$$

and
$$\frac{1}{\alpha\beta} \overset{(2)}{=} \frac{1}{2}$$

$\therefore$ new equation is $x^2 + \tfrac{5}{6}x + \tfrac{1}{2} = 0$
or $6x^2 + 5x + 3 = 0$

Example 13   If α and β are the roots of $x^2 + 5x - 1 = 0$ find the equation whose roots are α^3 and β^3.

Since $x^2 + 5x - 1 = 0$ has roots α and β,

$$\alpha + \beta = -5 \tag{1}$$

and

$$\alpha\beta = -1 \tag{2}$$

If the new equation has roots α^3 and β^3 then sum of new roots is

$$\alpha^3 + \beta^3 = (\alpha + \beta)(\alpha^2 - \alpha\beta + \beta^2)$$
$$= (\alpha + \beta)[(\alpha + \beta)^2 - 3\alpha\beta]$$
$$\overset{(1)(2)}{=\!=\!=}(-5)[(-5)^2 - 3(-1)] = -140$$

and product of new roots is

$$\alpha^3\beta^3 = (\alpha\beta)^3 \overset{(2)}{=} (-1)^3 = -1$$

$\therefore$ new equation is $\underline{x^2 + 140x - 1 = 0}$

- Exercise 10.4   See page 209.

10.5 Inequalities

1   Any number may be added to, or taken away from, both sides of an inequality.

2   Both sides of an inequality may be multiplied or divided by the same **positive** number.

3   Both sides of an inequality may be multiplied or divided by the same **negative** number, provided the inequality sign is reversed.

Example 14   Solve (a) $\frac{2}{3}x - 3 > 4$ (b) $3 - 5x \geqslant 10$

(a) $\frac{2}{3}x - 3 > 4$
$\Rightarrow 2x - 9 > 12$
$\quad 2x > 21$
$\quad \underline{x > \frac{21}{2}}$

(b) $3 - 5x \geqslant 10$
$\quad -5x \geqslant 7$
$\quad \underline{x \leqslant -\frac{7}{5}}$

Example 15   For what values of x is $(x - 1)(x + 2) < 0$?

$$(x - 1)(x + 2) < 0$$

either $+$ $-$ $\Rightarrow$ $x > 1$ *and also* $x < -2 \Rightarrow$ no solution

or $-$ $+$ $\Rightarrow$ $x < 1$ *and also* $x > -2 \Rightarrow -2 < x < 1$

$\therefore$ the inequality holds for $\underline{-2 < x < 1}$

Example 16   For what values of x is $2x^2 > x + 3$?

$$2x^2 > x + 3$$
$$\Rightarrow \quad 2x^2 - x - 3 > 0$$

$$(2x - 3)(x + 1) > 0$$

| either | + | + | $\Rightarrow$ $x > \frac{3}{2}$ *and also* $x > -1 \Rightarrow x > \frac{3}{2}$ |
| or | − | − | $\Rightarrow$ $x < \frac{3}{2}$ *and also* $x < -1 \Rightarrow x < -1$ |

$\therefore$ the inequality holds for $x < -1$ and $x > \frac{3}{2}$

But what if the quadratic expression does not factorise?

Example 17   For what values of x is $2x^2 + 2 > 3x$?

$$2x^2 + 2 > 3x$$
$$\Rightarrow 2x^2 - 3x + 2 > 0$$

This will not factorize. Solve by 'completing the square'.

$$\Rightarrow \quad x^2 - \tfrac{3}{2}x + 1 > 0$$
$$\Rightarrow (x - \tfrac{3}{4})^2 + 1 - \tfrac{9}{16} > 0$$
$$\Rightarrow \qquad (x - \tfrac{3}{4})^2 > -\tfrac{7}{16}$$

This is true for all real values of x and consequently the inequality is satisfied for all real values of x

Example 18   For what values of x is $2x^2 - 4x + 1 \geqslant 0$?

$$2x^2 - 4x + 1 \geqslant 0$$

This does not factorise, so 'completing the square',

$$x^2 - 2x + \tfrac{1}{2} \geqslant 0$$
$$(x - 1)^2 - 1 + \tfrac{1}{2} \geqslant 0$$
$$(x - 1)^2 \geqslant \tfrac{1}{2}$$
$$x - 1 \geqslant \frac{1}{\sqrt{2}} \quad \text{and} \quad x - 1 \leqslant -\frac{1}{\sqrt{2}}$$
$$\therefore x \geqslant 1 + \frac{1}{\sqrt{2}} \quad \text{and} \quad x \leqslant 1 - \frac{1}{\sqrt{2}}$$

Example 19   Solve $\dfrac{x - 1}{x} \leqslant 0$.

$$\frac{x - 1}{x} \leqslant 0$$

| either +/− | $\Rightarrow$ | $x \geqslant 1$ | *and* | $x < 0$ | $\Rightarrow$ | no solution |
| or −/+ | $\Rightarrow$ | $x \leqslant 1$ | *and* | $x > 0$ | $\Rightarrow$ | $0 < x \leqslant 1$ |

$\therefore$ solution is $0 < x \leqslant 1$ | **Note** $x \neq 0$ since you cannot divide by zero

Example 20   Solve $\dfrac{x+1}{x} > 4$.

> Inequations must always be rearranged to give ··· > 0 or < 0

$$\therefore \qquad \frac{x+1}{x} > 4$$

$$\Rightarrow \quad \frac{x+1}{x} - 4 > 0$$

$$\Rightarrow \frac{(x+1) - 4x}{x} > 0$$

$$\Rightarrow \quad \frac{1 - 3x}{x} > 0$$

either $+/+ \;\Rightarrow\; 1 - 3x > 0$ *and* $x > 0 \;\Rightarrow\; x < \frac{1}{3}$ *and* $x > 0 \;\Rightarrow\; 0 < x < \frac{1}{3}$

or $\quad -/- \;\Rightarrow\; 1 - 3x < 0$ *and* $x < 0 \;\Rightarrow\; x > \frac{1}{3}$ *and* $x < 0 \;\Rightarrow\;$ no solution

$\therefore$ solution is $0 < x < \frac{1}{3}$

● Exercise 10.5   See page 210.

10.6 Further Inequalities

The Modulus of x

The modulus of x is written $|x|$.

$$|x| = x \quad \text{if} \quad x \geqslant 0 \quad \textbf{\textit{and}} \quad |x| = -x \quad \text{if} \quad x \leqslant 0$$

The statement 'x is numerically less than 1' can be written

$$|x| < 1 \quad \text{or} \quad -1 < x < 1$$

The Modulus of a Function

$$|f(x)| \leqslant a \Leftrightarrow -a \leqslant f(x) \leqslant a$$

Example 21   Express in modulus form (a) $-3 < x < 3$
(b) $-3 < 4x < 5$ (c) $(2x - 1)^2 > 4$.

(a) $-3 < x < 3 \Rightarrow |x| < 3$

(b) $-3 < 4x < 5$
$\Rightarrow -4 < 4x - 1 < 4$
$\Rightarrow \qquad |4x - 1| < 4$

(c) $(2x - 1)^2 > 4 \Rightarrow |2x - 1| > 2$

Example 22   Solve $|2x - 3| > 5$.

$$|2x - 3| > 5 \Rightarrow 2x - 3 > 5 \quad \text{and} \quad 2x - 3 < -5$$
$$\Rightarrow 2x \quad\;\; > 8 \quad \text{and} \quad 2x \qquad < -2$$
$$\Rightarrow x \qquad > 4 \quad \text{and} \quad x \qquad < -1$$

Example 23 For what values of p has the equation $4x^2 + 8x - 8 = p(4x - 3)$ no real roots?

The equation can be written as

$$4x^2 + x(8 - 4p) + (3p - 8) = 0$$

The roots are imaginary if '$b^2 < 4ac$'
i.e. if

$$(8 - 4p)^2 < 4 \cdot 4(3p - 8)$$

$(\div 16)$ $\Rightarrow$ $(2 - p)^2 < 3p - 8$

$\Rightarrow p^2 - 7p + 12 < 0$

$\Rightarrow (p - 3)(p - 4) < 0$

either $+ \quad -$ $\Rightarrow p > 3$ *and* $p < 4$ $\Rightarrow$ no solution

or $\quad - \quad +$ $\Rightarrow p < 3$ *and* $p > 4$ $\Rightarrow 3 < p < 4$

$\therefore$ solution is $3 < p < 4$

Example 24 For what values of x is $\dfrac{(2x - 1)(x + 2)}{(x - 2)}$ positive?

Method 1

$\therefore$ solutions are $-2 < x < \frac{1}{2}$ and $x > 2$ $\qquad$ or $x \in] - 2, \frac{1}{2}[\cup]2, \infty[$

Method 2

$\begin{matrix} + & + \\ & + \end{matrix}$ $\Rightarrow x > \frac{1}{2}$ and $x > -2$ and $x > 2 \Rightarrow x > 2$

$\begin{matrix} - & - \\ & + \end{matrix}$ $\Rightarrow x < \frac{1}{2}$ and $x < -2$ and $x > 2 \Rightarrow$ no solution

$\begin{matrix} + & - \\ & + \end{matrix}$ $\Rightarrow x > \frac{1}{2}$ and $x < -2$ and $x > 2 \Rightarrow$ no solution

$\begin{matrix} - & + \\ & - \end{matrix}$ $\Rightarrow x < \frac{1}{2}$ and $x > -2$ and $x < 2 \Rightarrow -2 < x < \frac{1}{2}$

$\therefore$ Solutions are $-2 < x < \frac{1}{2}$ and $x > 2$

Example 25 Find the set of possible values of y given that $y = \dfrac{x^2 + 8}{x - 1}$ and x is real.

$$y = \frac{x^2 + 8}{x - 1} \Rightarrow \qquad y(x - 1) = x^2 + 8$$

$$\Rightarrow x^2 - yx + (y + 8) = 0$$

For real values of x $b^2 \geqslant 4ac$ ⟵⎯⎯ $\boxed{1.8}$

$$\Rightarrow \qquad\qquad y^2 \geqslant 4(y + 8)$$
$$\Rightarrow y^2 - 4y - 32 \geqslant 0$$
$$\Rightarrow (y + 4)(y - 8) \geqslant 0$$

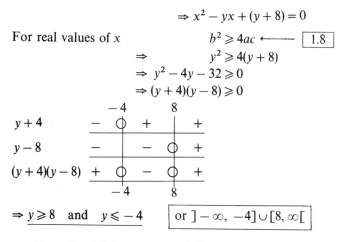

$$\Rightarrow y \geqslant 8 \quad \text{and} \quad y \leqslant -4 \qquad \boxed{\text{or} \;]-\infty, \; -4] \cup [8, \infty[}$$

● **Exercise 10.6**   See page 210.

10.7 Partial Fractions

> **Rule 1**   Before a fractional function can be expressed directly in partial fractions, the numerator must be of at least one degree less than the denominator.

Example 26   Write the fraction $6x^3 + x^2 + 8/(3x - 1)(x + 2)$ in a form suitable for expressing in partial fraction form.

$(3x - 1)(x + 2) = 3x^2 + 5x - 2$ is of lower order than the numerator so we divide it into the numerator.

$$
\begin{array}{r}
2x - 3 \\
3x^2 + 5x - 2 \,{\overline{\smash{\big)}\,6x^3 + x^2 + 8}} \\
\underline{6x^3 + 10x^2 - 4x } \\
-9x^2 + 4x \\
\underline{-9x^2 - 15x + 6} \\
19x + 2
\end{array}
$$

$$\therefore \frac{6x^3 + x^2 + 8}{(3x - 1)(x + 2)} \equiv 2x - 3 + \frac{19x + 2}{(3x - 1)(x + 2)}$$

> **Rule 2**   For each linear factor $ax + b$ in the denominator of a fractional function, there is a partial fraction of the form $\dfrac{A}{ax + b}$ where A is a constant.

Example 27   Express $3x/(x - 1)(2x + 1)(x + 2)$ in partial fractions.

Let $$\frac{3x}{(x - 1)(2x + 1)(x + 2)} \equiv \frac{A}{x - 1} + \frac{B}{2x + 1} + \frac{C}{x + 2}$$
$$\therefore \; 3x \equiv A(2x + 1)(x + 2) + B(x - 1)(x + 2)$$
$$+ C(x - 1)(2x + 1)$$

Let $x = 1$ LHS $= 3$ RHS $= 9A \Rightarrow A = \tfrac{1}{3}$

Let $x = -2$ LHS $= -6$ RHS $= 9C \Rightarrow C = -\frac{2}{3}$

Let $x = -\frac{1}{2}$ LHS $= -\frac{3}{2}$ RHS $= -\dfrac{9B}{4} \Rightarrow B = -\frac{2}{3}$

$$\therefore \frac{3x}{(x-1)(2x+1)(x+2)} \equiv \frac{1}{3(x-1)} - \frac{2}{3(2x+1)} - \frac{2}{3(x+2)}$$

Rule 3   For each linear factor $ax + b$ repeated r-times in the denominator, there will be r partial fractions of the form

$$\frac{A_1}{ax+b}, \frac{A_2}{(ax+b)^2}, \frac{A_3}{(ax+b)^3} \cdots \frac{A_r}{(ax+b)^r}$$

Example 28   Express $\dfrac{2x^2 - 1}{(x-1)^3(x+1)}$ in partial fractions.

Let
$$\frac{2x^2 - 1}{(x-1)^3(x+1)} \equiv \frac{A}{x-1} + \frac{B}{(x-1)^2} + \frac{C}{(x-1)^3} + \frac{D}{x+1}$$

$$\therefore 2x^2 - 1 \equiv A(x-1)^2(x+1) + B(x-1)(x+1)$$
$$+ C(x+1) + D(x-1)^3$$

Let $x = 1$ $\Rightarrow 1 = 2C$ $\Rightarrow C = \frac{1}{2}$
Let $x = -1 \Rightarrow 1 = -8D \Rightarrow D = -\frac{1}{8}$

Equating coefficient of $x^3 : A + D = 0 \Rightarrow A = \frac{1}{8}$
Equating coefficient of $x^0 : A - B + C - D = -1$
$$\Rightarrow \frac{1}{8} - B + \frac{1}{2} + \frac{1}{8} = -1 \Rightarrow B = \frac{7}{4}.$$

$$\therefore \frac{2x^2 - 1}{(x-1)^3(x+1)} \equiv \frac{1}{8(x-1)} + \frac{7}{4(x-1)^2} + \frac{1}{2(x-1)^3} - \frac{1}{8(x+1)}$$

Rule 4   For each quadratic factor in the denominator there will be a partial fraction of the form $\dfrac{Ax + B}{ax^2 + bx + C}$

Example 29   Express $\dfrac{x^3 + 3}{x^4 - 1}$ in partial fractions.

Let
$$\frac{x^3 + 3}{x^4 - 1} \equiv \frac{A}{x-1} + \frac{B}{x+1} + \frac{Cx + D}{x^2 + 1}$$

$$\therefore x^3 + 3 \equiv A(x+1)(x^2+1) + B(x-1)(x^2+1) + (Cx+D)(x^2-1)$$

Let $x = 1$ $\Rightarrow 4 = 4A$ $\therefore A = 1$ ⟵ | These techniques have
Let $x = -1 \Rightarrow 2 = -4B$ $\therefore B = -\frac{1}{2}$ | been met earlier in 1.9

Equating coefficient of $x^3 : A + B + C = 1 \Rightarrow C = \frac{1}{2}$
Equating coefficient of $x^2 : A + B + D = 0 \Rightarrow D = -\frac{1}{2}$

$$\therefore \frac{x^3 + 3}{x^4 - 1} \equiv \frac{1}{2}\left\{ \frac{2}{x-1} - \frac{1}{x+1} + \frac{x-1}{x^2+1} \right\}$$

Exercise 10.7   See page 211.

10.8 The Expansion of Rational Algebraic Fractions

Example 30   Expand the fraction $\dfrac{x}{(1-x)(2+x)}$ in ascending powers of x as far as the term in x^3 and state the range of values of x for which the expansion is valid

$$\frac{x}{(1-x)(2+x)} \equiv \frac{1}{3(1-x)} - \frac{2}{3(2+x)}$$ ⟨using partial fraction techniques⟩

$$= \tfrac{1}{3}[(1-x)^{-1} - 2(2+x)^{-1}]$$

$$= \frac{1}{3}\left[(1-x)^{-1} - 2 \cdot 2^{-1}\left(1 + \frac{x}{2}\right)^{-1}\right]$$

$$= \frac{1}{3}\left[(1 + x + x^2 + x^3 + \cdots) - \left(1 - \frac{x}{2} + \frac{x^2}{4} - \frac{x^3}{8} + \cdots\right)\right]$$

$$= \frac{1}{3}\left[\frac{3x}{2} + \frac{3x^2}{4} + \frac{9x^3}{8} + \cdots\right]$$

$$= \frac{x}{2} + \frac{x^2}{4} + \frac{3x^3}{8} + \cdots$$

$(1-x)^{-1}$ is valid for $|x| < 1$

$\left(1 + \dfrac{x}{2}\right)$ is valid for $\left|\dfrac{x}{2}\right| < 1$ i.e. $|x| < 2$ ⟩ ⟹ The expansion is valid for $|x| < 1$

Example 31   Find the first three terms of the expansion of $\dfrac{2}{x^2-4}$. For what values of x is the expansion valid?

$$\frac{2}{x^2-4} \equiv \frac{1}{2(x-2)} - \frac{1}{2(x+2)} = \tfrac{1}{2}[(x-2)^{-1} - (x+2)^{-1}]$$

$$= \tfrac{1}{2}[-(2-x)^{-1} - (2+x)^{-1}]$$

$$= -\frac{1}{4}\left[\left(1 - \frac{x}{2}\right)^{-1} + \left(1 + \frac{x}{2}\right)^{-1}\right]$$

$$= -\frac{1}{4}\left[\left(1 + \frac{x}{2} + \frac{x^2}{4} + \frac{x^3}{8} + \cdots\right) + \left(1 - \frac{x}{2} + \frac{x^2}{4} - \frac{x^3}{8} + \cdots\right)\right]$$

$$= -\frac{1}{4}\left[2 + \frac{2x^2}{4} + \frac{2x^4}{16}\right] + \cdots$$

$$= -\frac{1}{2}\left[1 + \frac{x^2}{4} + \frac{x^4}{16} + \cdots\right]$$

Valid for $|x| < 2$

Example 32   Find the coefficient of x^n in the expansion of $\dfrac{2}{(1+2x)(1-x)}$.

$$\frac{1}{(1+2x)(1-x)} \equiv \frac{4}{3(1+2x)} + \frac{2}{3(1-x)}$$

$$= \tfrac{4}{3}[1 - 2x + (2x)^2 - (2x)^3 + (2x)^4 - \cdots + (-1)^n(2x)^n]$$
$$+ \tfrac{2}{3}[1 + x + x^2 + x^3 + \cdots + x^n]$$

The term in $x^n = \tfrac{4}{3}(-1)^n(2x)^n + \tfrac{2}{3}x^n$

$\therefore$ coefficient of $x^n = \tfrac{1}{3}[2 + (-1)^n 2^{n+2}]$

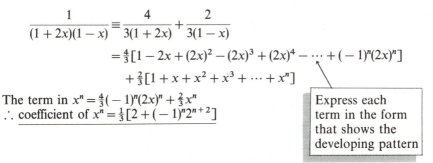

Express each term in the form that shows the developing pattern

Example 33 Expand $\dfrac{2}{3x-1}$ as a series of **descending** powers of x. Find the coefficient of x^{-n}.

$$\frac{2}{3x-1} = 2(3x-1)^{-1}$$

$$= 2(3x)^{-1}\left(1 - \frac{1}{3x}\right)^{-1}$$

$$= \frac{2}{3x}\left[1 + \frac{1}{3x} + \frac{1}{(3x)^2} + \frac{1}{(3x)^3} + \cdots + \frac{1}{(3x)^n} + \cdots\right]$$

$$\therefore \frac{2}{3x-1} = \frac{2}{3x} + \frac{2}{(3x)^2} + \frac{2}{(3x)^3} + \cdots + \frac{2}{(3x)^n} + \cdots$$

$\therefore$ coefficient of $x^{-n} = \dfrac{2}{3^n}$

Note When you want to find the general term, it is essential that each term in the expansion is not fully simplified but left in the form that best shows the developing pattern in the expansion.

e.g. $1 + 2(3x) + 3(3x)^2 + 4(3x)^3 + 5(3x)^4 + \cdots$

rather than $1 + 6x + 27x^2 + 108x^3 + 405x^4 + \cdots$

It is easy to see from the first line that the term in x^n is $(n+1)(3x)^n$.

- Exercise 10.8 See page 212.

10.9 Integration of Rational Algebraic Fractions

Example 34 Find $\displaystyle\int \frac{1}{x(x^2-1)} \, dx$.

$$\int \frac{1}{x(x^2-1)} \, dx = \int \left(-\frac{1}{x} + \frac{1}{2(x+1)} + \frac{1}{2(x-1)}\right) dx$$

$$= -\ln|x| + \tfrac{1}{2}\ln|x+1| + \tfrac{1}{2}\ln|x-1| + c$$

$$= \tfrac{1}{2}\ln\left|\frac{x^2-1}{x^2}\right| + c$$

Using partial fraction techniques

Example 35   Evaluate $\displaystyle\int_0^{1/\sqrt{3}} \frac{2}{(1+x^2)(1-x)}\,dx.$

$\boxed{\text{Using partial fraction techniques}}$

$$\int_0^{1/\sqrt{3}} \frac{2}{(1+x^2)(1-x)}\,dx = \int_0^{1/\sqrt{3}} \left(\frac{x+1}{1+x^2} + \frac{1}{1-x}\right)dx$$

$\boxed{\text{You cannot integrate this as it is, so you split it thus:}}$

$$= \int_0^{1/\sqrt{3}} \left(\frac{x}{1+x^2} + \frac{1}{1+x^2} + \frac{1}{1-x}\right)dx$$

$$= [\tfrac{1}{2}\ln|1+x^2| + \tan^{-1}x - \ln|1-x|]_0^{1/\sqrt{3}}$$

$$= \left[\tfrac{1}{2}\ln\tfrac{4}{3} + \tan^{-1}\frac{1}{\sqrt{3}} - \ln\left(1 - \frac{1}{\sqrt{3}}\right)\right]$$

$$\quad - [\tfrac{1}{2}\ln 1 + \tan^{-1}0 - \ln 1]$$

$$= \tfrac{1}{2}\ln\tfrac{4}{3} + \frac{\pi}{6} - \ln\left(\frac{\sqrt{3}-1}{\sqrt{3}}\right)$$

$$= \tfrac{1}{2}\ln\tfrac{4}{3} + \frac{\pi}{6} - \tfrac{1}{2}\ln\left(\frac{(\sqrt{3}-1)^2}{3}\right)$$

$$= \tfrac{1}{2}\ln\frac{4}{4-2\sqrt{3}} + \frac{\pi}{6}$$

$$= \tfrac{1}{2}\ln\frac{2}{2-\sqrt{3}} + \frac{\pi}{6}$$

● **Exercise 10.9**   See page 212.

10.10 The Method of Differences

Example 36   Find the sum $\displaystyle S_n = \sum_{r=1}^{n} \frac{4r}{(2r-1)(2r+1)(2r+3)}.$

Expressing the general term in partial fractions gives

$$\frac{4r}{(2r-1)(2r+1)(2r+3)} \equiv \frac{1}{4(2r-1)} + \frac{1}{2(2r+1)} - \frac{3}{4(2r+3)}$$

Hence $S_n = \tfrac{1}{4} + \tfrac{1}{6} - \tfrac{3}{20}$
$\qquad\qquad + \tfrac{1}{12} + \tfrac{1}{10} - \tfrac{3}{28}$
$\qquad\qquad + \tfrac{1}{20} + \tfrac{1}{14} - \tfrac{3}{36}$
$\qquad\qquad + \cdots$
$\qquad\qquad \vdots$

$\boxed{\text{Sufficient terms should be included at the beginning and end to give a clear picture of the pattern of cancelling}}$

$$+\frac{1/}{4(2n-5)}+\frac{1/}{2(2n-3)}-\frac{3/}{4(2n-1)}$$

$$+\frac{1/}{4(2n-3)}+\frac{1/}{2(2n-1)}-\frac{3}{4(2n+1)}$$

$$+\frac{1/}{4(2n-1)}+\frac{1}{2(2n+1)}-\frac{3}{4(2n+3)}$$

$$\therefore S_n=\frac{1}{4}+\frac{1}{6}+\frac{1}{12}+\frac{1}{2(2n+1)}-\frac{3}{4(2n+1)}-\frac{3}{4(2n+3)}$$

$$\Rightarrow S_n=\frac{1}{2}-\frac{4n+3}{2(2n+1)(2n+3)}$$

Example 37   Find the sum of the series $\dfrac{1}{1.3}+\dfrac{1}{2.4}+\dfrac{1}{3.5}+\cdots+\dfrac{1}{n(n+2)}$

Since the last term is $\dfrac{1}{n(n+2)}$ the general term is $\dfrac{1}{r(r+2)}$ sum is

$$\sum_1^n\frac{1}{r(r+2)}=\sum_1^n\left(\frac{1}{2r}-\frac{1}{2(r+2)}\right)\longleftarrow \boxed{\text{Using partial fraction techniques}}$$

$$=\frac{1}{2}\left[\frac{1}{1}-\frac{1/}{3}\right.$$

$$+\frac{1}{2}-\frac{1/}{4}$$

$$+\frac{1/}{3}-\frac{1/}{5}$$

$$+\cdots$$
$$\vdots$$

$$+\frac{1/}{n-1}-\frac{1}{n+1}$$

$$\left.+\frac{1/}{n}-\frac{1}{n+2}\right]$$

$$\therefore \text{Sum}=\frac{1}{2}\left[1+\frac{1}{2}-\frac{1}{n+1}-\frac{1}{n+2}\right]=\frac{3}{4}-\frac{2n+3}{2(n+1)(n+2)}$$

● **Exercise 10.10**   See page 213.

10.11 A Short Cut: Revision of Unit: A Level Questions

Example 38
The line $2y+6x=9$ cuts the curve $y=x^2-2x+4$ at P and Q. Find the coordinates of the mid point of PQ.

The line is $y = \frac{9}{2} - 3x$ (1)

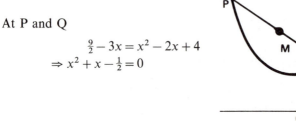

At P and Q

$$\frac{9}{2} - 3x = x^2 - 2x + 4$$
$$\Rightarrow x^2 + x - \frac{1}{2} = 0$$

The roots of this equation are x_P and x_Q and the sum of the roots $x_P + x_Q = -1$.

But

$$x_M = \frac{x_P + x_Q}{2} = -\frac{1}{2}$$

x_M lies on (1) so

$$y_M = \frac{9}{2} - 3(-\frac{1}{2}) = 6$$

$\therefore$ M is $(-\frac{1}{2}, 6)$

Example 39   Any point $P(t^2, 2t)$ lies on the curve $y^2 = 4x$ for all values of t. The line $y = 3x - 1$ intersects the curve at Q $(t_1^2, 2t_1)$ and R $(t_2^2, 2t_2)$. Find the values of $t_1 + t_2$ and $t_1 t_2$

AT Q and R

$$y = 3\left(\frac{y^2}{4}\right) - 1$$
$$\Rightarrow 3y^2 - 4y - 4 = 0$$

The roots of this equation are y_Q and y_R so the sum of the roots is

$$y_Q + y_R = 2t_1 + 2t_2 = \frac{4}{3}$$
$$\therefore t_1 + t_2 = \frac{2}{3}$$

and the product of roots is

$$y_Q + y_R = 4t_1 t_2 = -\frac{4}{3}$$
$$\therefore t_1 t_2 = -\frac{1}{3}$$

● Miscellaneous Exercise 10.B   See page 213.

Unit 10 EXERCISES

Exercise 10.1

1   Find the product of:
(a) $2x^2 - x + 5$ and $x + 3$
(b) $3x^3 - 2x^2 + 5x - 2$ and $2x - 1$
(c) $x^2 + 5x - 2$ and $3x^2 - x + 1$
(d) $5x^3 - 2x + 4$ and $x^2 - 1$
(e) $3x^4 - 2x^3 + 6x - 3$ and $3x + 2$
(f) $x^5 + x^4 + x^3 + x^2 + x + 1$ and $x - 1$

2 Find the coefficients of the given terms in the following products.

(a) $(2x - 3)(x^3 - x^2 + 2x - 5)$; x, x^3
(b) $(x^2 - 2x + 1)(3x^2 + 2x - 1)$; x, x^2
(c) $(x^3 + x - 1)(2x^2 - 3x + 1)$; x^2, x^4
(d) $(3x^3 - x^2 - 2x + 5)(x^3 + 2x^2 - x - 3)$; x^2; x^5

3 Simplify the following expressions:

(a) $\dfrac{4}{x^2 - 1} - \dfrac{7}{2x^2 - 3x - 5}$

(b) $\dfrac{1}{x + 3} + \dfrac{9x}{x^3 + 27}$

(c) $\dfrac{x^4 - x}{2x - 3x^2} \times \dfrac{3x^2 + x - 2}{1 - x^2}$

(d) $\dfrac{x^4 - 16}{x^2 - x - 6} \div \dfrac{2x^2 - 5x + 2}{2x^2 - 7x + 3}$

4 Find the remainder when:

(a) $5x^2 + 3x - 6$ is divided by $x - 2$
(b) $x^3 - 2x^2 + 6x - 2$ is divided by $x - 5$
(c) $x^4 - 3x^3 + 2x^2 - x + 5$ is divided by $x + 1$
(d) $x^5 + 15$ is divided by $x + 2$

5 Find the quotient and remainder when:

(a) $3x^2 - 2x + 5$ is divided by $x - 1$
(b) $4x^3 - 4x^2 + 5x + 3$ is divided by $2x - 3$
(c) $x^3 + 9$ is divided by $x - 3$
(d) $3x^5 - x^4 - 6x^3 + 11x^2 + 1$ is divided by $3x - 1$
(e) $x^4 - 2x^3 + 5x - 4$ is divided by $x^2 - x - 1$
(f) $2x^4 - x^3 + 3x^2 - 3$ is divided by $x^2 - 1$
(g) $3x^2 - 2ax - 3a^2$ is divided by $x + a$
(h) $a^3 - 3a^2b + 2b^3$ is divided by $a - b$

*6 Find the remainder when:

(a) $x^4 - 3x^3 + 3x$ is divided by $2x - 1$
(b) $9x^5 - 2x^2$ is divided by $3x + 1$
(c) $x^3 + 2x^2 + 5$ is divided by $x + a$
(d) $x^2 + ax + b$ is divided by $x - c$

Exercise 10.2

1 Determine whether the linear functions are factors of the given polynomials:

(a) $x^3 - 7x + 6$; $x - 1$
(b) $2x^2 + 3x - 5$; $x + 1$
(c) $x^3 - 5x^2 + 3x + 6$; $x - 3$
(d) $x^3 - 8$; $x - 2$
(e) $6x^3 + x^2 - 4x + 1$; $3x - 1$
(f) $x^3 + ax^2 + a^2x - a^3$; $x + a$

2 Factorise completely:

(a) $x^3 + x^2 - 5x + 3$
(b) $x^4 - 3x^3 + 6x^2 - 12x + 8$
(c) $x^4 + x^3 - 8x^2 - 9x - 9$
(d) $2x^3 - 3x^2 - 3x + 2$
(e) $3x^4 - 2x^3 + 5x^2 - 4x - 2$
(f) $x^3 - y^3$
(g) $8x^3 - 27$

3   If $x^2 - 3x + a$ has a factor $(x - 1)$, find a

4   If $x^2 + 6x + p$ has a factor $(x + 2)$, find the other factor

5   If 3 is root of $x^2 + tx - 6 = 0$, find t

6   If one root of $x^2 - 5x + a = 0$ is 2, find the other root

*7   If $(x - a)$ is a factor of the expression
$$ax^3 - 3x^2 - 5ax - 9$$

Find the possible values of a and factorise the expression for each of these values

Exercise 10.3

1   Find the remainder when $x^5 + 2x^3 - 3x + 1$ is divided by $x - 1$

2   Find the remainder when $8x^4 - 3x^2 + 2x + 4$ is divided by $2x + 1$

3   The remainder when $x^5 - px^2 + 2$ is divided by $x + 1$ is 4. Find the value of p

4   When the polynomial $f(x) = x^3 + x^2 + px + q$ is divided by $x + 1$ and $x - 2$ its remainders are -4 and 11 respectively. Find the values of p and q

5   Find the remainder when $x^4 + 2x^3 - 7x^2 + 3x - 6$ is divided by $(x - 2)(x + 4)$

6   When $x^5 - 4x^4 + 4x^3 + 7$ is divided by $(x + 1)(x - 1)(x - 3)$ the remainder is $ax^2 + bx + c$. Find the values of a, b and c

7   Determine whether the following functions have repeated factors and, if they have, find them:
   (a) $f(x) = x^4 - 8$
   (b) $f(x) = 2x^3 - 3x^2 - 12x + 20$
   (c) $f(x) = x^4 - 8x^2 + 16$

8   Find the remainder when $x^3 - 5x^2 - 3x$ is divided by $(x + 1)^2$

9   It is known that $x^4 + 4x^3 - 8x^2 - 48x + k$ has two equal roots. Find all the possible values of k

*10   When a polynomial $P(x)$ is divided by $(x - 1)$ the remainder is 3. When $P(x)$ is divided by $(x - 2)$ the remainder is 5. Find, by writing
$$P(x) \equiv (x - 1)(x - 2)Q(x) + ax + b$$
the remainder when $P(x)$ is divided by $(x - 1)(x - 2)$

*11   Find the constants m and n such that $(x - 3)$ is a common factor of $x^3 - 3x^2 + mx - n$ and $x^3 - mx^2 - 7x + n$

Exercise 10.4

1 Write down the sum and product of the roots of each of the following equations:

(a) $x^2 - 5x + 3 = 0$ (b) $3x^2 + 2x - 5 = 0$ (c) $x(x-2) = x+1$

(d) $(x+3)(x-2) = 4$ (e) $\dfrac{(x+4)}{3} = \dfrac{5}{(x-1)}$ (f) $2x + 3 = \dfrac{4}{x}$

2 Write down the equation, given that the sum and product of the roots are:

(a) $4, 5$ (b) $-3, \frac{1}{2}$ (c) $\frac{1}{4}, -\frac{3}{5}$ (d) $-\frac{1}{2}, 0$

(e) $0, \frac{2}{3}$ (f) k, k^2 (g) $-(k+2), 3k^2$ (h) $\dfrac{p}{q}, \dfrac{q^2}{r}$

3 State whether each of the following is true or false:

(a) If $x^2 - px + q = 0$ has two positive roots then $q > 0$
(b) If $x^2 - px + q = 0$ has two negative roots then $q < 0$
(c) If the roots of $x^2 + rx + s = 0$ have opposite signs then $s < 0$
(d) If $s < 0$ then the roots of $x^2 + rx + s = 0$ have opposite signs.
(e) If the graph of $x^2 + rx + s = 0$ is symmetrical about the y-axis then $s = -k^2$ for some real number k.

4 The roots of the equation $2x^2 - 3x + 5 = 0$ are α and β. Find the value of:

(a) $2\alpha + 2\beta$ (b) $\alpha^2\beta^2$ (c) $\dfrac{1}{\alpha} + \dfrac{1}{\beta}$ (d) $(\alpha+2)(\beta+2)$

(e) $\alpha^2 + \beta^2$ (f) $\alpha^2\beta + \alpha\beta^2$ (g) $(\alpha + \beta)^2$ (h) $(\alpha - \beta)^2$

5 The roots of $x^2 - 3x + 1 = 0$ are α and β. Find the equation whose roots are:

(a) $(\alpha + 1, \beta + 1)$ (b) $\dfrac{1}{\alpha}, \dfrac{1}{\beta}$ (c) α^2, β^2

6 Find the equation whose roots are double those of $3x^2 - 5x - 1 = 0$, without solving the given equation

7 Find the equation whose roots are the reciprocals of those of $5x^2 + 4x + 2 = 0$, without solving the given equation

8 Find the equation whose roots are two more than those of $x^2 - 3x - 2 = 0$, without solving the given equation

9 Find the equation whose roots are one less than those of $x^2 + 7x + 3 = 0$, without solving the given equation

10 Find the value of k if one root of $2x^2 + kx - 3 = 0$ is minus the other root

*11 Find the value of m if one root of $8x^2 + mx + 27 = 0$ is the square of the other root

*12   If the roots of $2x^2 - 5x + 2 = 0$ are α and β, find the value of:

(a) $(\alpha + 1)(\beta + 1)$ (b) $\dfrac{1}{\alpha + 1} + \dfrac{1}{\beta + 1}$

(c) $(\alpha - \beta)^2$ (d) $\dfrac{\alpha}{\beta} + \dfrac{\beta}{\alpha}$

*13   The roots of the equation $x^2 - 5x + 3 = 0$ are α and β. Find the equation whose roots are $\dfrac{1}{2\alpha + \beta}$ and $\dfrac{1}{2\beta + \alpha}$

*14   The roots of the equation $3x^2 + 6x - 1 = 0$ are α and β. Find the equation whose roots are $\dfrac{1}{\alpha^2 + 1}$ and $\dfrac{1}{\beta^2 + 1}$

Exercise 10.5

Solve:

1 $3x + 1 > 5$

2 $4x - 2 > x + 5$

3 $3(1 - 2x) \leqslant x$

4 $-3x < 6$

5 $\frac{2}{5}x \geqslant 3(1 - x)$

6 $\frac{3}{4}(x - 1) < \frac{1}{2}(1 - 2x) + 3$

7 $x(x - 2) \leqslant 0$

8 $(x - 3)(x + 2) > 0$

9 $(1 - 3x)(2x + 1) > 0$

10 $2x > x^2 - 3$

11 $2x^2 \geqslant 3x + 2$

12 $\dfrac{x - 2}{x + 1} < 0$

13 $x^2 > x + 6$

14 $(x + 8)(x - 3) \leqslant 3x$

15 $(2x - 1)(x - 2) > 5$

16 $(x + 3)^2 + 1 > 0$

17 $(x - 1)^2 < 0$

18 $(x - 2)^2 \leqslant 0$

19 $x^2 + x + 1 > 0$

20 $2x^2 + 7 \leqslant 4x$

21 $2x^2 - 8x + 7 > 0$

22 $3x^2 + 18x + 25 \leqslant 0$

23 $\dfrac{x}{x - 1} \geqslant 0$

24 $\dfrac{x}{x - 1} \geqslant 3$

25 $x^3 > 4x$

26 $2x^2 + 7x + 5 > 0$

27 $2x^2 + 7x + 5 \leqslant 0$

28 $10x^2 - 10x + 1 < 0$

*29 $\dfrac{x + 5}{x + 1} < 4$

*30 $\dfrac{x(x - 1)}{x + 1} > x$

Exercise 10.6

1   Express in modulus form:
(a) x is numerically less than 4
(b) $-2 < x < 2$
(c) $-2 < x < 6$

(d) $x > 3$ and $x < -3$
(e) $-3 < 2x < 7$
(f) $-4 < 3x < 0$
(g) $(x+1)^2 > 9$
(h) $(x-2)^2 \leqslant 5$
(i) $(x-2)^2 > 5$
(j) $(x+\frac{3}{2})^2 - \frac{5}{4} < 0$

2 Express in the form $a < x < b$

(a) $|x| < 5$
(b) $|x-1| < 3$
(c) $|3x+1| < 5$

3 Solve:

(a) $|x+3| < 2$ (b) $|4x-1| > 5$
(c) $|5-2x| < 3$ (d) $|1-3x| > 7$

4 Find the condition that $x^2 + 2(a+2)x + 9a = 0$ has no real roots

5 For what values of k has $x^2 + 2x + 7 = k(2x+1)$ real roots?

6 Prove that $x^2 - 2px + p^2 - q^2 - r^2 = 0$ has real roots

In questions (7)–(15) solve:

7 $(x-1)(x-2)x > 0$ **8** $\dfrac{(2x+3)(x-4)}{x-1} > 0$

9 $\dfrac{x^2-x-2}{2x-3} < 0$ **10** $\dfrac{x^2(x+1)}{x-1} > 0$

11 $x+1 < \dfrac{6}{x}$ **12** $\dfrac{x}{2} > \dfrac{18}{x}$

13 $x+2 > \dfrac{30}{x+1}$ **14** $3x+8 < \dfrac{3}{x}$

**15 Given that $y = \dfrac{x^2+k}{x+2}$ and x is real, find

(a) the set of possible values of y where $k = 5$
(b) the set of values of k for which y can take all real values

Find the values of m for which the line $y = mx$ touches the curve $y = \dfrac{2x^2+1}{2(x+2)}$ and show that the acute angle between the tangets from the origin to this curve is $\tan^{-1} 3$ (A)

Exercise 10.7

Express in partial fractions:

1 $\dfrac{x}{(2-x)(1+x)}$ **2** $\dfrac{3x-1}{(3x+1)(x-2)}$ **3** $\dfrac{2x}{(x-1)(x+3)}$

4 $\dfrac{3}{(x-2)^2(x+2)}$　　　**5** $\dfrac{2x^2-3}{x(x^2+2)}$　　　**6** $\dfrac{x^2+2x}{x^2-9}$

7 $\dfrac{5}{x(x-1)^2}$　　　**8** $\dfrac{2}{x(x^2+3)}$　　　**9** $\dfrac{x^2-x-1}{x^2-x-2}$

10 $\dfrac{2x-3}{x^3(x+2)}$　　　**11** $\dfrac{x^2}{(x-3)^2}$　　　**12** $\dfrac{13-5x}{(x+1)(x-2)^2}$

13 $\dfrac{3x^2+x}{(x+1)(x^2+4)}$　　　**14** $\dfrac{6}{1-x^3}$　　　**15** $\dfrac{x^2+2}{x(x^2-1)}$

16 $\dfrac{x+1}{(x-2)(x-1)(x-3)}$　　　***17** $\dfrac{4x^3}{(1+x^2)(1-x)^2}$　　　***18** $\dfrac{x^3+1}{x^3+2x}$

*19 $\dfrac{x}{(x+1)(3-x^2)}$　　　***20** $\dfrac{2}{x(x^2+4)^2}$

Exercise 10.8

Find the first three terms in each of the following expansions. For what values of x is each expansion valid?

1 $\dfrac{2x-4}{(1-2x)(1+x)}$　　　**2** $\dfrac{x}{(1+2x)(1+3x)}$　　　**3** $\dfrac{x-1}{(2-x)(3-x)}$

4 $\dfrac{1}{x^2-x-2}$　　　**5** $\dfrac{2x}{3-2x-x^2}$　　　**6** $\dfrac{x+1}{(1-x)^3}$

7 $\dfrac{1+x^2}{(1-x^2)(3+x)}$　　　**8** $\dfrac{1}{(1-x)(1-2x)}$　　　**9** $\dfrac{2x-1}{(x+1)^2(1-3x)}$

10 Show that the coefficient of x^n in the expansion of $\dfrac{2-3x}{1-3x+2x^2}$ is $1+2^n$

11 Expand $\dfrac{1}{x^2-4x+3}$ in ascending powers of x. Find the coefficient of x^{n+1}

*12 Expand $\dfrac{1}{x+2}$ as a series of descending powers of x. Find the coefficient of x^{-n}. For what values of x is the expansion valid.

**13 Find the first three terms of the expansion of $\dfrac{1}{(1+x)(2-x^2)}$

*14 Find the coefficient of x^n in the expansion of $\dfrac{x-2}{(x+1)^2(x-1)}$

Exercise 10.9

Evaluate:

1 $\displaystyle\int \dfrac{dx}{x^2-1}$　　　**2** $\displaystyle\int_5^6 \dfrac{dx}{(x-3)(x-4)}$　　　**3** $\displaystyle\int_4^6 \dfrac{dx}{x^2(x-2)}$

4 $\displaystyle\int_1^3 \dfrac{dx}{x(x^2+1)}$　　　***5** $\displaystyle\int_1^2 \dfrac{4x-1}{(2x-1)^2(x+5)}dx$　　　**6** $\displaystyle\int_1^2 \dfrac{x^2}{(2x+1)(x+2)}dx$

7 $\displaystyle\int_0^1 \frac{x-1}{x^2-7x+12}\,dx$ ***8** $\displaystyle\int \frac{dx}{2-x^2}$ **9** $\displaystyle\int_1^2 \frac{3dx}{x^3(x^2+2)}$

10 $\displaystyle\int_2^3 \frac{dx}{(x-1)^2(x^2+1)}$ **11** $\displaystyle\int_{-1}^0 \frac{x^2}{(x-2)^2}\,dx$ **12** $\displaystyle\int \frac{3x^2+24}{(x+2)(x^2+5)}\,dx$

Exercise 10.10

1 Find the value of $\displaystyle\sum_1^n \frac{1}{r(r+1)(r+2)}$

2 Show that $\displaystyle\sum_3^n \frac{1}{(r+1)(r+2)} = \frac{n-2}{4(n+2)}$

3 Show that $\displaystyle\sum_1^{2n} \frac{1}{r(r+1)} = \frac{2n}{2n+1}$

4 Find the sum of $\displaystyle\frac{1}{2\cdot4} + \frac{1}{3\cdot5} + \frac{1}{4\cdot6} + \cdots + \frac{1}{(n+1)(n+3)}$

5 Show that $\displaystyle\frac{1}{3\cdot5} + \frac{1}{4\cdot12} + \frac{1}{5\cdot21} + \cdots + \frac{1}{n(n^2-4)} = \frac{1}{8}\left[\frac{11}{12} - \frac{2}{n^2-1} - \frac{2}{n(n+2)}\right]$

*6 Show that $\displaystyle\sum_1^n \frac{1}{(2r-1)(2r+1)} = \frac{n}{2n+1}$. Find $\displaystyle\lim_{n\to\infty} \frac{n}{2n+1}$ (L)

*7 Show that $\displaystyle\frac{1}{r!} - \frac{1}{(r+1)!} = \frac{r}{(r+1)!}$

and find the corresponding expression for

$$\frac{1}{r!} + \frac{1}{(r+1)!}$$

Given that S_n and T_n are the sum of the first n terms of the series whose rth terms are

$$\frac{r}{(r+1)!} \quad \text{and} \quad \frac{(-1)^r(r+2)}{(r+1)!}$$

respectively, find S_{2n} and T_{2n} and show that $T_{2n} = -S_{2n}$ (J)

Miscellaneous Exercise 10B

1 PQ is a chord of gradient m ($\neq 0$) of the parabola $y^2 = 4x$. Find the equation of PQ given that it passes through $(1,0)$. Find the coordinates of the mid point M of PQ in terms of m. Deduce that M lies on the curve whose equation is $y^2 = 2(x-1)$ (J: 10.11)

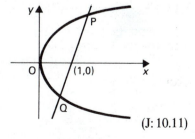

2 Given that $2x-1$ is a factor of the polynomial

$$8x^3 - 12x^2 + ax - 1$$

find the value of the constant a (J: 10.2)

Hwk

3   The roots of the quadratic equation

$$x^2 + 5x + 3 = 0$$

are α and β. Find a quadratic equation with roots α^2 and β^2 (L: 10.4)

4   Find (a) $\displaystyle\int \frac{x+1}{x(2x+1)}\,dx$ (b) $\displaystyle\int \frac{x(2x+1)}{x+1}\,dx$ (J: 10.9)

5   Given that a, b and h are real constants, show that the roots of the quadratic equation

$$(x-a)(x-b) = h^2$$ (L: 1.8)

are real

6   Given that $f(x) = x^3 - 3x^2 + 4$, show that $f(2) = f'(2) = 0$, where

$$f'(x) = \frac{d}{dx}[f(x)]$$

Hence factorise $f(x)$ completely.

Sketch the graph of the curve

$$y = x^3 - 3x^2 + 4$$

marking on your sketch the coordinates of the points at which the curve meets the coordinate axes   (A: 3.4, 10.2)

7   Find the two values of k for which the equation

$$\frac{1}{x+1} - \frac{k}{x} + \frac{4}{x-1} = 0$$

has equal roots. Solve the equation for each of these values of k (J: 1.8)

8   Given that $y = \log_3 x$, express

$$k\log_3 x = \log_x 3 - k$$

as a quadratic equation in y. Show that the sum of the roots of this quadratic is independent of k.

Find the possible values of x when $k = \frac{1}{6}$ (A: 9.2, 10.4)

9   The curve with equation $y = x^3 + ax^2 + bx - 20$ has a minimum point at $(4, -4)$. Find the values of a and b. Express y as a product of three linear factors   (A: 1.1, 3.4, 10.2)

10   Given that $f(x) \equiv a - 2x - x^2$, where a is a constant, find

(a)   the value of a for which the roots of the equation $f(x) = 0$ differ by 3
(b)   the set of values of a for which $f(x) < 0$ for all values of x
   (L: 1.8, 10.4, 10.5)

11 Find the set of values of x for which

$$\frac{x}{x+1} > 2 \qquad \text{(L: 10.6)}$$

12 When the polynomial $P(x)$, where

$$P(x) \equiv ax^3 + 5x^2 + 2x + b$$

is divided by $(x-1)$ the remainder is 2. When $P(x)$ is divided by $(x-2)$ the remainder is 5. Determine the constants a and b (L: 10.3)

13 Express the function $y = \dfrac{2x^2}{(2x-1)(x+1)}$ in partial fractions.

Hence, or otherwise, find the value of $\dfrac{d^2y}{dx^2}$ when $x = 1$ (J: 10.7)

14 Find the real values of k for which the equation $x^2 + (k+1)x + k^2 = 0$ has

(a) real roots
(b) one root double the other (A: 1.8, 10.4)

15 The arithmetic mean of two numbers p and q is 39 and their geometric mean is 15. Write down a quadratic equation whose roots are p and q Hence, or otherwise, find the values of p and q (L: 6.10, 10.4)

16 Express the function $\dfrac{1 + 3x^2}{(1-x)^2(1+x)}$ as the sum of three partial fractions.

Hence, or otherwise, find the first three terms in the expansion of the function in ascending powers of x.

17 The roots of the equation $x^2 + (x+1)^2 = k$ are α and β.

(a) Prove that $\alpha^3 + \beta^3 = \frac{1}{2}(1 - 3k)$
(b) Find in terms of k, a quadratic equation in x whose roots are α^3 and β^3. (A: 10.4)

18 Express the polynomial

$$f(x) = 2x^4 + x^3 - x^2 + 8x - 4$$

as the product of two linear factors and a quadratic factor $q(x)$.

Prove that there are no real values of x for which $q(x) = 0$ (A: 10.2, 1.8)

19 Find the set of values of x for which

$$\frac{2}{x-2} < \frac{1}{x+1} \qquad \text{(L: 10.6)}$$

20 Find the values of the constants p and q so that the polynomial

$$px^3 - 11x^2 + qx + 4$$

is divisible by $(x-1)$, and has remainder 70 when divided by $(x-3)$.

Express the polynomial as a product of 3 linear factors (L: 10.2, 10.3)

21 Express $\alpha^3 + \beta^3$ as the product of a linear and a quadratic factor

Given that $\qquad \alpha + \beta = 2 \quad$ and $\quad \alpha^3 + \beta^3 = 32$

find the value of $\alpha\beta$.

Write down a quadratic equation with numerical coefficients and roots α, β. Hence calculate the exact values of α and β (J: 1.9, 10.4)

22 Given that
$$f(x) \equiv \frac{1}{x(x+2)}$$

express $f(x)$ in partial fractions.

Hence, or otherwise, find (a) $\dfrac{d^4 f(x)}{dx^4}$ (b) $\displaystyle\int_1^3 f(x)\,dx$ (L: 10.6, 10.8)

23 Find the set of values of x for which $x > |3x - 8|$ (L: 10.6)

24 Express the function $f(x) = \dfrac{1}{(1+x)(1-3x)}$ in partial fractions. Find the first four terms in the expansion of $f(x)$ in ascending powers of x, and obtain the coefficient of x^n (J: 10.8)

25 (a) Prove that $\displaystyle\int_1^2 \frac{6}{(3+x)(3-x)}\,dx = \ln\frac{5}{2}$

(b) Evaluate, in terms of π, $\displaystyle\int_0^2 \frac{x^2}{x^2+4}\,dx$ (J: 10.9)

26 (a) Find the values of k for which the equation $x^2 + (k+4)x + 5k = 0$ has equal roots

(b) The roots of the equation $x^2 + (k+4)x + 5k = 0$ are α and β. Given that $k \neq 0$, form a quadratic equation, with the coefficients expressed in terms of k, whose roots are α/β and β/α (A: 1.8, 10.4)

27 Find the ranges of values of x for which the sum of the infinite geometric series
$$1 + \frac{5}{3x-4} + \left(\frac{5}{3x-4}\right)^2 + \cdots$$

exists   (J: 6.2, 10.6)

28 Given that
$$f(x) \equiv \frac{1}{(x+1)(x+3)}$$

express $f(x)$ in partial fractions and find
$$\sum_{r=1}^{n} \frac{1}{(2r+1)(2r+3)} \qquad \text{(L: 10.10)}$$

29 (a) Find the constants A, B, C in the identity
$$\frac{5x^2 + 4x - 20}{(x+2)(x^2+4)} \equiv \frac{A}{x+2} + \frac{Bx+C}{x^2+4}$$

(b) Find $\displaystyle\int \frac{x}{x^2+4}\,dx$ and $\displaystyle\int \frac{1}{x^2+4}\,dx$

(c) Using the answers to parts (a) and (b) above show that

$$\int_0^2 \frac{5x^2+4x-20}{(x+2)(x^2+4)}\,dx = a\ln 2 - b\pi$$

where a and b are positive integers, and find a and b (J: 10.9)

30 Find the constants p and r so that the polynomial x^3+px+r has a remainder -9 when it is divided by $(x+1)$ and a remainder -1 when it is divided by $(x-1)$ (L: 10.3)

31 Show that, if x is real, $2x^2+6x+9$ is always positive. Hence, or otherwise, solve the inequality

$$\frac{(x+1)(x+3)}{x} > \frac{x+6}{3}$$

(J: 1.8, 10.6)

32 Given that α and β are the roots of the equation

$$x^2 - 6x + 1 = 0$$

state the values of $\alpha+\beta$ and $\alpha\beta$. Evaluate $\alpha^2+\beta^2$ and $\alpha^3+\beta^3$.

Given that $e^x = 1 + x + \dfrac{x^2}{2!} + \dfrac{x^3}{3!} + \cdots$,

find the numerical values of the coefficients of the first four terms in the expansion in ascending powers of t of

$$e^{\alpha t} + e^{\beta t}$$

(J: 10.3)

*33 Prove that the equation of the tangent at the point $P(2ap, ap^2)$ on the parabola $4ay = x^2$, where $a > 0$, is $px - y - ap^2 = 0$

Given that this tangent meets the parabola $y^2 = 4ax$ at the points $Q(at_1^2, 2at_1)$ and $R(at_2^2, 2at_2)$, express $t_1 + t_2$ and $t_1 t_2$ in terms of p

The midpoint of QR is M, and N is the point $[at_1 t_2, a(t_1+t_2)]$. Show that the distance between M and N is $s = 2a/p^2 + 2ap$

Determine the minimum value of s as p varies in the range $p > 0$ (J: 3.3, 3.6, 10.11)

*34 The equation $x^2 + 6x + c = 0$ has real roots which differ by $2n$. Show that $n^2 = 9 - c$

Given that the roots also have opposite signs, find the set of possible values of n (J: 10.4, 10.6)

11.1 Revision of Calculus Techniques

Example 1   Differentiate w.r.t. x: (a) $\sin^3 x$ (b) $\cos 2x$ (c) $\tan^4 3x$
(d) $\ln(x^3 + 3x)$ (e) e^{x^2}

(a) $\dfrac{d}{dx}(\sin^3 x) = \dfrac{d}{dx}((\sin x)^3) = \underline{3\sin^2 x \cdot \cos x}$

(b) $\dfrac{d}{dx}(\cos 2x) = \underline{-2\sin 2x}$

(c) $\dfrac{d}{dx}(\tan^4 3x) = 4\tan^3 3x \cdot \sec^2 3x \times 3 = \underline{12\tan^3 3x \cdot \sec^2 3x}$

(d) $\dfrac{d}{dx}[\ln(x^3 + 3x)] = \underline{\dfrac{3x^2 + 3}{x^3 + 3x}}$

(e) $\dfrac{d}{dx}[e^{x^2}] = e^{x^2} \times 2x = \underline{2xe^{x^2}}$

Example 2   Differentiate w.r.t. x: (a) $x^2 \ln x$ (b) $\dfrac{e^{3x}}{2x + 1}$
(c) $3x^2 + \sin 5y + 2xy = 4$.

(a) $\dfrac{d}{dx}(\overset{u}{x^2} \overset{v}{\ln x}) = 2x \cdot \ln x + x^2 \cdot \overset{v'}{\dfrac{1}{x}} = \underline{2x\ln x + x}$

overbraces: $u'\ v$, $u\ \frac{1}{x}$

(b) $\dfrac{d}{dx}\left(\dfrac{e^{3x}}{2x + 1}\right) \genfrac{}{}{0pt}{}{u}{v} = \dfrac{\overset{v}{(2x + 1)}\overset{u'}{3e^{3x}} - \overset{u}{e^{3x}} \cdot \overset{v'}{2}}{\underset{v^2}{(2x + 1)^2}} = \underline{\dfrac{e^{3x}(6x + 1)}{(2x + 1)^2}}$

(c) $3x^2 + \sin 5y + 2xy = 4$ is an implicit function. Differentiating implicitly:

| since xy is a product |

$$6x + 5\cos 5y\dfrac{dy}{dx} + 2\left[x \cdot \dfrac{dy}{dx} + y\right] = 0$$

Example 3   If $x = \sin\theta$, $y = 1 - \cos\theta$ find $\dfrac{dy}{dx}$ and $\dfrac{d^2y}{dx^2}$.

$\dfrac{dy}{d\theta} = \sin\theta, \quad \dfrac{dx}{d\theta} = \cos\theta$

$\Rightarrow \dfrac{dy}{dx} = \dfrac{dy}{d\theta} \times \dfrac{d\theta}{dx} = \dfrac{\sin\theta}{\cos\theta}$

$$\Rightarrow \frac{dy}{dx} = \tan \theta$$

$$\frac{d^2y}{dx^2} = \frac{d}{dx}\left(\frac{dy}{dx}\right) = \frac{d}{dx}(\tan \theta) = \frac{d}{d\theta}(\tan \theta) \times \frac{d\theta}{dx} = \sec^2 \theta \times \frac{1}{\cos \theta}$$

$$\Rightarrow \frac{d^2y}{dx^2} = \sec^3 \theta \qquad \boxed{\text{using the chain rule}}$$

Example 4   Evaluate the following integrals:

(a) $\displaystyle\int_1^2 \frac{x^2+1}{x}\,dx$ (b) $\displaystyle\int_0^{\pi/6} \cos 3x\,dx$ (c) $\displaystyle\int_0^1 \frac{3x+1}{3x^2+2x+1}\,dx$

(d) $\displaystyle\int \frac{3}{\sqrt{9-4x^2}}\,dx$ (e) $\displaystyle\int \cos^2 x\,dx$

(a) $\displaystyle\int_1^2 \frac{x^2+1}{x}\,dx = \int_1^2 \left(x+\frac{1}{x}\right)dx = \left[\frac{x^2}{2} + \ln x\right]_1^2 = (2 + \ln 2) - (\tfrac{1}{2} + \ln 1)$

$$= \tfrac{3}{2} + \ln 2$$

(b) $\displaystyle\int_0^{\pi/6} \cos 3x\,dx = \left[\frac{\sin 3x}{3}\right]_0^{\pi/6} = \tfrac{1}{3}[\sin \pi/2 - \sin 0] = \tfrac{1}{3}$

(c) $\displaystyle\int_0^1 \frac{3x+1}{3x^2+2x+1}\,dx = \frac{1}{2}\int_0^1 \frac{6x+2}{3x^2+2x+1}\,dx = [\tfrac{1}{2}\ln(3x^2+2x+1)]_0^1$

$$= \tfrac{1}{2}\ln 6 - \tfrac{1}{2}\ln 1 = \tfrac{1}{2}\ln 6$$

(d) $\displaystyle\int \frac{3}{\sqrt{9-4x^2}}\,dx = \int \frac{3}{2\sqrt{\frac{9}{4}-x^2}}\,dx = \tfrac{3}{2}\cdot \sin^{-1}\frac{x}{3/2} + c = \frac{3}{2}\sin^{-1}\frac{2x}{3} + c$

(e) You cannot integrate $\cos^2 x$ directly.
First you must put it into a form that you can integrate
$$\cos 2x = 2\cos^2 x - 1$$
$$\Rightarrow \cos^2 x = \tfrac{1}{2}(\cos 2x + 1)$$

$$\therefore \int \cos^2 x\,dx = \int \tfrac{1}{2}(\cos 2x + 1)\,dx$$

$$= \frac{1}{2}\left[\frac{\sin 2x}{2} + x\right] + c$$

$$= \tfrac{1}{4}\sin 2x + \frac{x}{2} + c$$

● Exercise 11.1   See page 233.

11.2 Integration By Substitution

Example 5   Integrate $\int (3x-1)^7 \, dx$.

$$\int (3x-1)^7 \, dx \qquad\qquad \text{Let } u = 3x - 1$$

$$= \int u^7 \cdot \frac{du}{3} \qquad\qquad \frac{du}{dx} = 3$$

$$= \frac{1}{3} \cdot \frac{u^8}{8} + c \qquad\qquad \Rightarrow dx = \frac{du}{3}$$

$$= \tfrac{1}{24}(3x-1)^8 + c$$

Solving Indefinite Integrals by Substitution

1   Let $u = f(x)$ be the substitution.

2   Differentiate: $\dfrac{du}{dx} = f'(x) \Rightarrow du = f'(x) \, dx$.

3   If necessary rearrange: $u = f(x) \Rightarrow x = g(u)$.

4   Substitute for x and dx in the integral.

5   Integrate.

6   Reverse the substitution.

Example 6   Evaluate $\displaystyle\int \frac{dx}{\sqrt{x}+1}$.

$$\text{(4)} \qquad \int \frac{dx}{\sqrt{x}+1} = \int \frac{2u \cdot du}{u+1} \qquad\qquad \text{Let } u = \sqrt{x} \qquad \text{(1)}$$

$$\qquad\qquad\qquad u^2 = x$$

$$\text{(5)} \qquad\qquad = \int \left(2 - \frac{2}{u+1}\right) du \qquad\qquad 2u \cdot \frac{du}{dx} = 1$$

$$= 2u - 2\ln(u+1) + c \qquad\qquad \Rightarrow dx = 2u \, du \qquad \text{(2)}$$

$$\text{(6)} \qquad\qquad = 2\sqrt{x} - 2\ln(\sqrt{x}+1) + c$$

Example 7   Evaluate $\displaystyle\int \frac{4x \, dx}{1+x^4}$.

$$\int \frac{4x}{1+x^4} \, dx = \int \frac{2 \, du}{1+u^2} \qquad\qquad \text{Let } u = x^2$$

$$\qquad\qquad\qquad\qquad du = 2x \cdot dx$$

$$= 2\tan^{-1} u + c$$

$$= 2\tan^{-1} x^2 + c$$

Solving Definite Integrals by Substitution

1–4 as for Indefinite Integrals.

5 Use the substitution to change the limits of integration.

6 Integrate.

7 Evaluate, using new limits.

Example 8   Evaluate $\displaystyle\int_0^{\pi/2} \sin^5 x \, dx$.

$$\int_0^{\pi/2} \sin^5 x \, dx$$

$$= \int_0^{\pi/2} -\sin^4 x (-\sin x \, dx)$$

$$= -\int_0^{\pi/2} (1 - \cos^2 x)^2 (-\sin x \, dx)$$

$$= -\int_1^0 (1 - c^2)^2 \, dc$$

$$= -\int_1^0 (1 - 2c^2 + c^4) \, dc$$

$$= -\left[c - \frac{2c^3}{3} + \frac{c^5}{5} \right]_1^0$$

$$= -[0] + [1 - \tfrac{2}{3} + \tfrac{1}{5}] = \tfrac{8}{15}$$

Let $c = \cos x$

$$dc = -\sin x \, dx$$

When $x = 0$, $c = 1$
When $x = \pi/2$, $c = 0$

Example 9   Evaluate $\displaystyle\int_2^4 \frac{x+1}{x\sqrt{x-2}} \, dx$.

$$\int_2^4 \frac{x+1}{x\sqrt{x-2}} \, dx = \int_2^4 \frac{(u^2+3)}{(u^2+2)u} \times 2u \, du$$

$$= \int_0^{\sqrt{2}} \left(2 + \frac{2}{u^2+2} \right) du$$

$$= \left[2u + 2 \cdot \frac{1}{\sqrt{2}} \tan^{-1} \frac{u}{\sqrt{2}} \right]_0^{\sqrt{2}}$$

$$= (2\sqrt{2} + \sqrt{2} \tan^{-1} 1) - (0 + \sqrt{2} \tan^{-1} 0)$$

$$= 2\sqrt{2} + \sqrt{2} \frac{\pi}{4} = 2\sqrt{2} + \frac{\pi}{2\sqrt{2}}$$

Let $x - 2 = u^2$

$$dx = 2u \cdot du$$

$$x = u^2 + 2$$
$$x + 1 = u^2 + 3$$

When $x = 2$, $u = 0$
When $x = 4$, $u = \sqrt{2}$

- Exercise 11.2   See page 234.

11.3 Finding the Substitution I

Finding the best substitution to use is mainly a skill developed through experience but there are some guidelines.

Some Common Types of Substitution

Integral type	Substitution
1 Integrals containing a term of the form $\sqrt{ax+b}$	Let $u^2 = ax + b$
2 Integrals containing a term of the form $(ax+b)^n, n \neq \frac{1}{2}$	Let $u = ax + b$
3 Integrals containing a term of the form $\sqrt{a^2 - x^2}$ or $(\sqrt{a^2 - x^2})^n$	Let $x = a \sin \theta$
4 Integrals containing a term of the form $\sqrt{1 - a^2 x^2}$ or $(\sqrt{1 - a^2 x^2})^n$	Let $ax = \sin \theta$
5 Integrals containing a term of the form $\sqrt{a^2 + x^2}$ or $(\sqrt{a^2 + x^2})^n$	Let $x = a \tan \theta$
6 Integrals containing a term of the form $\sqrt{1 + a^2 x^2}$ or $(\sqrt{1 + a^2 x^2})^n$	Let $ax = \tan \theta$
7 Integrals of odd powers of sine	Let $\cos x = c$
8 Integrals of odd powers of cosine	Let $\sin x = s$
9 Integrals of the form $$\int \frac{dx}{a + b \cos x} \quad \text{or} \quad \frac{dx}{a + b \sin x}$$	Let $\tan \dfrac{x}{2} = t$
10 F.M. only$\}$ Integrals containing a term of the form $\sqrt{x^2 - a^2}$ or $(\sqrt{x^2 - a^2})^n$	Let $x = a \cosh x$

> **Notes 1** This list is not comprehensive. There are other possible substitutions.
>
> **2** In some cases you may feel you have a choice of possible substitutions. Whenever possible, it is usually simpler to use a substitution nearer the top of this list.

In 11.2:   Examples 6 and 9 are of type 1.
Example 5 is of type 2.
Example 8 is of type 7.
However, Example 7 does not fit any of them exactly.

Example 10 Evaluate $\displaystyle\int \frac{dx}{x^2\sqrt{9-x^2}}$, given that $|x| < 3$.

This is type 3.

$$\therefore \int \frac{dx}{x^2\sqrt{9-x^2}} = \int \frac{3\cos\theta\,d\theta}{9\sin^2\theta\;3\cos\theta}$$

$$= \frac{1}{9}\int \csc^2\theta\,d\theta$$

$$= -\tfrac{1}{9}\cot\theta + c$$

$$= -\frac{1}{9}\frac{3\cos\theta}{3\sin\theta} + c$$

$$= -\frac{1}{9}\frac{\sqrt{9-x^2}}{x} + c$$

Let $x = 3\sin\theta$

$dx = 3\cos\theta\cdot d\theta$

$\sqrt{9-x^2} = \sqrt{9-9\sin^2\theta}$
$\qquad\quad = 3\cos\theta$

Example 11 Evaluate $\displaystyle\int (1+25x^2)^{-2}\,dx.$

This is type 6.

$$\therefore \int_0^{1/5} (1+25x^2)^{-2}\,dx = \int_0^{1/5} \frac{dx}{(1+25x^2)^2}$$

$$= \int_0^{\pi/4} \frac{\frac{1}{5}\sec^2\theta\,d\theta}{(\sec^2\theta)^2}$$

$$= \frac{1}{5}\int_0^{\pi/4} \cos^2\theta\,d\theta$$

$$= \frac{1}{5}\int_0^{\pi/4} \frac{(\cos 2\theta + 1)}{2}\,d\theta$$

$$= \frac{1}{10}\left[\frac{\sin 2\theta}{2} + \theta\right]_0^{\pi/4}$$

$$= \frac{1}{10}\left[\frac{1}{2} + \frac{\pi}{4}\right]$$

$$= \frac{2+\pi}{40}$$

Let $5x = \tan\theta$

$5dx = \sec^2\theta\,d\theta$

$1 + 25x^2 = 1 + \tan^2\theta$
$\qquad\qquad = \sec^2\theta.$

When $x = 0$ $\theta = 0$

When $x = \tfrac{1}{5}$ $\theta = \pi/4$

Remember: you cannot integrate $\cos^2\theta$ directly.
$\text{Cos } 2\theta = 2\cos^2\theta - 1$
$\therefore \cos^2\theta = \tfrac{1}{2}(\cos 2\theta + 1)$

● **Exercise 11.3** See page 235.

11.4 Finding the Substitution II

Example 12   Evaluate $\displaystyle\int_0^{\pi/2} \frac{dx}{2 + \cos x}$.

This is type 9.

Let $\tan \dfrac{x}{2} = t$ (1)

Then $\tan x = \dfrac{2t}{1 - t^2}$

$\cos x = \dfrac{1 - t^2}{1 + t^2}$

Differentiating $(1) \Rightarrow \frac{1}{2}\sec^2 \dfrac{x}{2} dx = dt$

$\Rightarrow \frac{1}{2}(1 + t^2)\,dx = dt$

$\Rightarrow \qquad dx = \dfrac{2\,dt}{1 + t^2}$

$\therefore \displaystyle\int_0^{\pi/2} \frac{dx}{2 + \cos x} = \int_0^1 \frac{2\,dt}{(1 + t^2)} \times \frac{1}{\left(2 + \left(\dfrac{1 - t^2}{1 + t^2}\right)\right)}$

When $x = 0$, $t = 0$

When $x = \pi/2$, $t = 1$

$= \displaystyle\int_0^1 \frac{2\,dt}{2(1 + t^2) + (1 - t^2)} = \int_0^1 \frac{2\,dt}{3 + t^2}$

$= 2\left[\dfrac{1}{\sqrt{3}}\tan^{-1}\dfrac{t}{\sqrt{3}}\right]_0^1$

$= \dfrac{2}{\sqrt{3}}\tan^{-1}\dfrac{1}{\sqrt{3}} - 0$

$= \dfrac{2}{\sqrt{3}}\cdot\dfrac{\pi}{6} = \dfrac{\pi}{3\sqrt{3}}$

Example 13   Evaluate $\displaystyle\int \frac{e^x}{e^x - 1}\,dx$.

This fits none of the listed types. So we look back at examples involving e^x that we have met so far (in 11.2). These suggest that we try $u = e^x$ or $u = e^x - 1$.

Method 1

Let $u = e^x - 1$

$du = e^x\,dx$

$\therefore \displaystyle\int \frac{e^x\,dx}{e^x - 1} = \int \frac{du}{u} = \ln u + A$

$= \ln(e^x - 1) + A$

Method 2

Let $u = e^x$

$du = e^x dx$

$$\therefore \int \frac{e^x dx}{e^x - 1} = \int \frac{du}{u - 1} = \ln(u - 1) + B$$

$$= \ln(e^x - 1) + B$$

> So, whichever substitution we chose, we got the same result. But, sometimes the most obvious substitution does not work. So you try another one!

● **Exercise 11.4**   See page 235

11.5 Integration By Parts

$$\int uv' \, dx = uv - \int vu' \, dx$$

Example 14

$$\int \underset{u \ \ v'}{x \sin x} \, dx$$

Let $u = x$ $v' = \sin x$

$u' = 1$ $v = -\cos x$

$$= \underset{u \quad v}{-x \cos x} - \int \underset{u'}{1} \, \underset{v}{(-\cos x)} \, dx$$

$$= -x \cos x + \sin x + c$$

Example 15

$$\int_1^2 \underset{v' \quad u}{x^4 \ln x} \, dx$$

Let $u = \ln x$ $v' = x^4$

$u' = \dfrac{1}{x}$ $v = \dfrac{x^5}{5}$

$$= \left[\frac{x^5}{5} \ln x \right]_1^2 - \int_1^2 \frac{1}{x} \cdot \frac{x^5}{5} \, dx$$

$$= \left[\tfrac{32}{5} \ln 2 - \tfrac{1}{5} \ln 1 \right] - \int_1^2 \frac{x^4}{5} \, dx$$

$$= \tfrac{32}{5} \ln 2 - \left[\frac{x^5}{25} \right]_1^2$$

$$= \tfrac{32}{5} \ln 2 - \left[\tfrac{32}{25} - \tfrac{1}{25} \right]$$

$$= \tfrac{32}{5} \ln 2 - \tfrac{31}{25}$$

Example 16

$$\int_0^1 \underset{u}{x^2} \underset{v'}{e^{-x}} dx \qquad\qquad \text{Let } u = x^2 \qquad v' = e^{-x}$$
$$u' = 2x \qquad v = -e^{-x}$$

$$= [\underset{u}{-x^2 e^{-x}}\underset{v}{}]_0^1 + \int_0^1 \underset{u}{2x}\cdot \underset{v'}{e^{-x}} dx \qquad \text{Let } u = x \qquad v' = e^{-x}$$
$$u' = 1 \qquad v = -e^{-x}$$

$$= -e^{-1} + 2\left\{ [-xe^{-x}]_0^1 + \int_0^1 1\cdot e^{-x} dx \right\}$$

$$= -3e^{-1} + 2[-e^{-x}]_0^1$$

$$= -3e^{-1} - 2e^{-1} + 2e^0$$

$$= \underline{\underline{2 - \frac{5}{e}}}$$

Example 17

$$\int \ln x\, dx = \int \underset{v'}{1}\cdot \underset{u}{\ln x}\, dx \qquad\qquad \text{Let } u = \ln x \qquad v' = 1$$
$$u' = \frac{1}{x} \qquad v = x$$

$$= x\ln x - \int \frac{1}{x}\cdot x\, dx$$

$$= \underline{\underline{x\ln x - x + c}}$$

Example 18

$$\int \tan^{-1} x\, dx = \int \underset{v'}{1}\cdot \underset{u}{\tan^{-1} x}\, dx \qquad \text{Let } u = \tan^{-1} x \qquad v' = 1$$
$$u' = \frac{1}{1+x^2} \qquad v = x$$

$$= x\tan^{-1} x - \int \frac{x}{1+x^2} dx$$

$$= \underline{\underline{x\tan^{-1} x - \tfrac{1}{2}\ln(1+x^2) + c}}$$

Example 19 | A trick to get out of trouble!

$$\int_0^{\pi/2} \underset{v'}{e^x} \underset{u}{\sin x}\, dx \qquad\qquad \text{Let } u = \sin x \qquad v' = e^x$$
$$u' = \cos x \qquad v = e^x$$

$$= [\underset{v}{e^x} \underset{u}{\sin x}]_0^{\pi/2} - \int_0^{\pi/2} \underset{v'}{e^x} \underset{u}{\cos x}\, dx \qquad \text{Let } u = \cos x \qquad v' = e^x$$
$$u' = -\sin x \qquad v = e^x$$

$$= e^{\pi/2} - \left\{ [e^x \cos x]_0^{\pi/2} + \int_0^{\pi/2} e^x \sin x\, dx \right\}$$

$$= e^{\pi/2} + 1 - \int_0^{\pi/2} e^x \sin x\, dx$$

But this is where we started. We are going round in circles!!

Let $\int_0^{/2} e^x \sin x\, dx \equiv I$. We now have

$$I = e^{\pi/2} + 1 - I$$

$$\Rightarrow 2I = e^{\pi/2} + 1$$

or $$I = \tfrac{1}{2}(e^{\pi/2} + 1)$$

$$\Rightarrow \int_0^{\pi/2} \sin x \, dx = \tfrac{1}{2}(e^{\pi/2} + 1)$$

- **Exercise 11.5**   See page 236.

11.6 A Miscellany of Integration Techniques: A Level Questions

- **Exercise 11.6**   See page 237.

11.7 Small Changes and Errors

$$\boxed{\delta y \approx \frac{dy}{dx} \cdot \delta x}$$

Example 20   If $y = x^3 + x^2$ find the approximate increase in y when x increases from 2 to 2.05.

$$y = x^3 + x^2 \Rightarrow \frac{dy}{dx} = 3x^2 + 2x$$

When $x = 2$, $\dfrac{dy}{dx} = 16$

$\therefore \delta y \approx 16 \delta x$ and $\delta x = 0.05$

$\therefore \delta y \approx 16 \times 0.05 \Rightarrow \delta y \approx 0.8$

Example 21   The volume of a sphere is $V = \tfrac{4}{3}\pi r^3$. If a 1% error is made in measuring the radius, estimate the percentage error in the volume.

We know that $\dfrac{\delta r}{r} = 0.01$.

We want to find the value of $\dfrac{\delta V}{V}$.

$$\frac{dV}{dr} = 4\pi r^2 \Rightarrow \delta V \approx 4\pi r^2 \delta r$$

$$\Rightarrow \frac{\delta V}{V} \approx \frac{4\pi r^2 \delta r}{\frac{4}{3}\pi r^3} = \frac{3 \delta r}{r}$$

$$\therefore \frac{\delta V}{V} \approx 3 \times 0.01 = 0.03$$

$\Rightarrow$ Percentage error in V is 3%

Example 22   Use $y = \sqrt{x}$ to estimate $\sqrt{24}$.

$$y = x^{1/2} \Rightarrow \frac{dy}{dx} = \frac{1}{2\sqrt{x}}$$

When $x = 25$, $\dfrac{dy}{dx} = \dfrac{1}{2\sqrt{25}} = \dfrac{1}{10}$ and we let $\delta x = -1$.

$$\therefore \delta y \approx \tfrac{1}{10} \times (-1) = -0.1$$

Also $\delta y = \sqrt{24} - \sqrt{25} = \sqrt{24} - 5$.

$$\therefore \sqrt{24} - 5 \approx -0.1$$
$$\therefore \sqrt{24} \approx 4.9$$

(25, $\sqrt{25}$)

(24, $\sqrt{24}$)

$\delta x = -1$

● **Exercise 11.7**   See page 238.

11.8 First Order Differential Equations with Variables Separable

Equations in which the variables are separable are such that when $\dfrac{dy}{dx}$ is replaced by dy and dx, all the terms in x can be collected on one side and all the terms in y on the other. The general solution is then obtained directly by integration.

Example 23   Find the general solution of $\dfrac{dy}{dx} = xy$.

The equation can be rearranged to give $\dfrac{dy}{y} = x\,dx$.

Hence $\displaystyle\int \frac{dy}{y} = \int x\,dx$.

Integrating $\Rightarrow \ln y = \dfrac{x^2}{2} + C$ is the general solution.

Important note   **The general solution is not unique.**
Other possible general solutions for this D.E. are

$$y = e^{\frac{x^2}{2} + C}, \qquad 2\ln y = x^2 + D, \quad y^2 = e^{x^2 + D}$$

$$y^2 = A e^{x^2}, \quad \ln y + B = \frac{x^2}{2}, \qquad x^2 = 2\ln y + E$$

Example 24   Find the particular solution of $(2 + x)\dfrac{dy}{dx} = y$ given that $y = 6$ when $x = 1$.

This gives $\dfrac{dy}{y} = \dfrac{dx}{x+2} \Rightarrow \int \dfrac{dy}{y} = \int \dfrac{dx}{x+2} \Rightarrow \ln y = \ln(2+x) + A$

But, this is better written as

$$\ln y = \ln(2+x) + \ln C$$

which simplifies to

$$\ln y = \ln C(2+x)$$
$$\Rightarrow \quad y = C(2+x) \longleftarrow \boxed{\text{general solution.}}$$

Given that $y = 6$ when $x = 1$, then $C = 2$.
$\therefore$ the particular solution required is $y = 2(2+x)$

Example 25   Solve $\dfrac{x^2+1}{y+1} = xy\dfrac{dy}{dx}$.

This gives

$$\left(\dfrac{x^2+1}{x}\right)dx = y(y+1)\,dy$$

$$\Rightarrow \int\left(x+\dfrac{1}{x}\right)dx = \int(y^2+y)\,dy$$

$$\Rightarrow \dfrac{x^2}{2} + \ln x \quad = \dfrac{y^3}{3} + \dfrac{y^2}{2} + C$$

Example 26   Find the equation of the curve that satisfies the equation
$\dfrac{dy}{dx} = \dfrac{y}{x^2-1}$ and passes through the point $(0, 1)$.

$$\dfrac{dy}{dx} = \dfrac{y}{x^2-1}$$

$$\Rightarrow \int\dfrac{dy}{y} = \int\dfrac{dx}{x^2-1}$$

$$\Rightarrow \int\dfrac{1}{y}dy = \dfrac{1}{2}\int\left\{\dfrac{1}{x-1} - \dfrac{1}{x+1}\right\}dx$$

$$\Rightarrow \ln y = \tfrac{1}{2}\{\ln(x-1) - \ln(x+1)\} + \ln C$$

$$\Rightarrow 2\ln y = \ln\left(\dfrac{x-1}{x+1}\right) + 2\ln C$$

$$\Rightarrow y^2 = C^2\left(\dfrac{x-1}{x+1}\right) \quad \text{or} \quad y^2 = A\left(\dfrac{x-1}{x+1}\right)$$

$x = 0, \quad y = 1$ satisfies this equation

$$\therefore A = -1$$

Hence the equation is $y^2 + \left(\dfrac{x-1}{x+1}\right) = 0$

- **Exercise 11.8**   See page 239.

11.9 Problems Involving Differential Equations

Example 27   A water tank of uniform cross-sectional area $2\,\mathrm{m}^2$ has a tap at the base of the tank. When the tap is open the rate of flow of the water is proportional to the depth of water in the tank, h. Show that $\dfrac{dh}{dt} = -\lambda h$ where $\lambda > 0$. If the depth of the water is $1\,\mathrm{m}$ when the tap is opened, find the time taken until the depth of water is $50\,\mathrm{cm}$.

Let volume of water $= V =$ area of cross-section $\times$ depth.

$$V = 2h$$

$$\Rightarrow \frac{dV}{dh} = 2 \tag{1}$$

Rate of flow of water $= -\dfrac{dV}{dt}$ ⟵ | Negative sign shows volume is decreasing |

It is given that

$$-\frac{dV}{dt} = kh \qquad k > 0$$

$$\Rightarrow \frac{dV}{dt} = -kh \tag{2}$$

But $\dfrac{dh}{dt} = \dfrac{dh}{dV} \times \dfrac{dV}{dt}$ ⟵ | Chain rule—see 5.7 |

(1) and (2) $\Rightarrow \dfrac{dh}{dt} = \tfrac{1}{2} \times (-kh) = -\dfrac{k}{2}h$

Putting $\lambda = \dfrac{k}{2}$ which is > 0

$$\frac{dh}{dt} = -\lambda h \qquad \text{Q.E.D.}$$

Separating the variables

$$\Rightarrow \frac{dh}{h} = -\lambda\, dt$$

Method 1

$$\int_{1}^{1/2} \frac{dh}{h} = \int_{0}^{T} -\lambda\, dt \qquad \boxed{\begin{array}{l} \text{when } t = 0, \quad h = 1 \\ \text{and when } t = T, \quad h = \tfrac{1}{2} \end{array}}$$

$$\Rightarrow [\ln h]_{1}^{1/2} = [-\lambda t]_{0}^{T}$$

$$\Rightarrow \ln\tfrac{1}{2} = -\lambda T$$

$$\Rightarrow T = \frac{1}{\lambda}\ln 2 \qquad \boxed{\ln\tfrac{1}{2} = -\ln 2 \text{—see 9.1}}$$

Method 2

$$\int \frac{dh}{h} = \int -\lambda\, dt$$

$$\Rightarrow \ln h = -\lambda t + A$$

when $t = 0$, $h = 1$

$$\Rightarrow \ln 1 = 0 + A$$
$$\Rightarrow A = 0$$
$$\therefore \ln h = -\lambda t$$

when $t = T$, $h = \frac{1}{2}$

$$\Rightarrow \ln \tfrac{1}{2} = -\lambda T$$

$$\Rightarrow T = \frac{1}{\lambda} \ln 2$$

- Exercise 11.9   See page 240.

11.10 Applications of Calculus: Mean Values

1 Mean Value of a Function

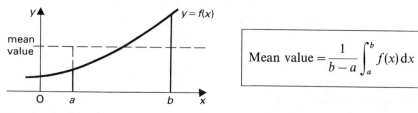

$$\text{Mean value} = \frac{1}{b-a} \int_a^b f(x)\,dx$$

Example 28   If $y = \tan x$ find the mean value of (a) y (b) y^2 over the range 0 to $\pi/4$.

(a)   Mean value of $y = \dfrac{1}{\dfrac{\pi}{4} - 0} \displaystyle\int_0^{\pi/4} \tan x\,dx$ $= \left[\dfrac{4}{\pi}(-\ln \cos x)\right]_0^{\pi/4}$

$$= \frac{4}{\pi}\left[-\ln \frac{1}{\sqrt{2}}\right] + \frac{4}{\pi}[\ln 1] = \frac{4}{\pi}\ln\sqrt{2}$$

$$= \frac{2}{\pi}\ln 2$$

(b)   Mean value of $y^2 = \dfrac{1}{\dfrac{\pi}{4} - 0} \displaystyle\int_0^{\pi/4} \tan^2 x\,dx$ $= \dfrac{4}{\pi}\displaystyle\int_0^{\pi/4}(\sec^2 x - 1)\,dx$

$$= \frac{4}{\pi}[\tan x - x]_0^{\pi/4}$$ $$= \frac{4}{\pi}\left[\left(1 - \frac{\pi}{4}\right) - 0\right]$$

$$= \frac{4}{\pi} - 1$$

2   *Areas between curves and axes*   Revision of 5.4

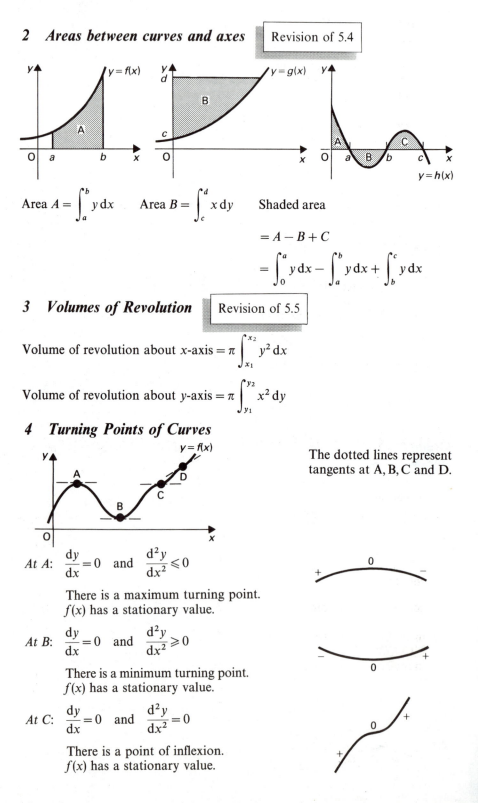

Area $A = \displaystyle\int_a^b y\,dx$ Area $B = \displaystyle\int_c^d x\,dy$ Shaded area

$$= A - B + C$$

$$= \int_0^a y\,dx - \int_a^b y\,dx + \int_b^c y\,dx$$

3   *Volumes of Revolution*   Revision of 5.5

Volume of revolution about x-axis $= \pi \displaystyle\int_{x_1}^{x_2} y^2\,dx$

Volume of revolution about y-axis $= \pi \displaystyle\int_{y_1}^{y_2} x^2\,dy$

4   *Turning Points of Curves*

The dotted lines represent tangents at A, B, C and D.

At A: $\dfrac{dy}{dx} = 0$ and $\dfrac{d^2y}{dx^2} \leqslant 0$

There is a maximum turning point. $f(x)$ has a stationary value.

At B: $\dfrac{dy}{dx} = 0$ and $\dfrac{d^2y}{dx^2} \geqslant 0$

There is a minimum turning point. $f(x)$ has a stationary value.

At C: $\dfrac{dy}{dx} = 0$ and $\dfrac{d^2y}{dx^2} = 0$

There is a point of inflexion. $f(x)$ has a stationary value.

At D: $\dfrac{dy}{dx} \neq 0$ but $\dfrac{d^2y}{dx^2} = 0$

There is a point of inflexion.
$f(x)$ does *not* have a stationary value.

Important Note:

If $\dfrac{d^2y}{dx^2} = 0$ there may exist a maximum, a minimum or a point of inflexion.

If $\dfrac{d^2y}{dx^2} = 0$ you should consider the gradients on either side of the turning point.

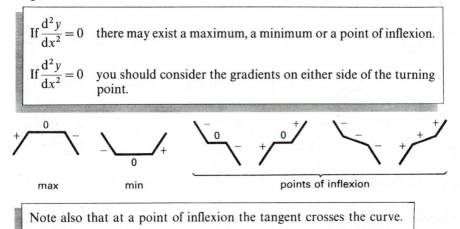

| max | min | points of inflexion |

Note also that at a point of inflexion the tangent crosses the curve.

- Exercise 11.10   See page 241.

11.11 +   Revision of Calculus Units: A Level Questions

- Miscellaneous Exercise 11B   See page 243.

Unit 11 EXERCISES

Exercise 11.1

Differentiate w.r.t. x:

1 $\cos^2 x$ 2 $\tan^3 x$ 3 $\sin 5x$ 4 $\ln(x^2 + 1)$

5 $\ln(\sin x)$ 6 $\sin^2 6x$ 7 e^{3x+1} 8 e^{x^3}

9 $\ln\cos 2x$ 10 $\cos(3x - \pi/4)$ 11 $\sin^{-1} 3x$ 12 $\tan^{-1} x/2$

13 $e^{\sin 3x}$ 14 $x\ln x$ 15 $\dfrac{x}{\ln x}$ 16 $x^2 \tan x$

17 $e^x \sin x$ 18 $\dfrac{x^2 + 1}{\sqrt{x}}$ 19 $x\tan^{-1} 2x$ 20 $\ln(x\sin x)$

Find $\dfrac{dy}{dx}$:

21 $x^2 + 2y^2 = 3$ **22** $xy + y^2 = 3x$

*23 $\dfrac{1}{x} + \dfrac{1}{y} = \ln y$

*24 Find $\dfrac{dy}{dx}$ and $\dfrac{d^2 y}{dx^2}$ if $x = at^2$, $y = 2at$

Evaluate the following integrals:

25 $\displaystyle\int_9^{16} \dfrac{dx}{\sqrt{x}}$

26 $\displaystyle\int_4^9 (1 + \sqrt{x})^2 \, dx$

27 $\displaystyle\int_{\pi 6}^{\pi/4} \sin 2x \, dx$

28 $\displaystyle\int_0^2 (e^x + e^{-x}) \, dx$

29 $\displaystyle\int_2^3 \dfrac{dx}{2x - 1}$

30 $\displaystyle\int_1^3 \left(3x - \dfrac{1}{3x}\right)^2 dx$

31 $\displaystyle\int_0^1 \dfrac{dx}{x^2 + 1}$

32 $\displaystyle\int_0^1 \dfrac{x}{x^2 + 1} \, dx$

33 $\displaystyle\int_0^{\pi/3} \sec x \cdot \tan x \, dx$

34 $\displaystyle\int_0^{\pi/4} \sin x \cdot \cos x \, dx$

35 $\displaystyle\int_1^2 \dfrac{(x^3 + x^2 + x + 1)}{x^3} \, dx$

36 $\displaystyle\int_{-3}^{-1} \dfrac{dx}{e^x}$

37 $\displaystyle\int_0^1 (2x + 1)^4 \, dx$

38 $\displaystyle\int_2^6 \dfrac{1}{\sqrt{e^x}} \, dx$

39 $\displaystyle\int_{1/2}^1 \dfrac{3}{\sqrt{1 - x^2}} \, dx$

40 $\displaystyle\int_0^{\pi/4} \tan x \, dx$

41 $\displaystyle\int \dfrac{3}{4 + 2x^2} \, dx$

42 $\displaystyle\int \dfrac{5}{\sqrt{1 - 3x^2}} \, dx$

Exercise 11.2

Use the given substitutions to evaluate:

1 $\displaystyle\int (5x - 4)^6 \, dx$; let $u = 5x - 4$

2 $\displaystyle\int \dfrac{dx}{(3x - 2)^4}$; let $u = 3x - 2$

3 $\displaystyle\int \dfrac{x \, dx}{\sqrt{x^2 - 3}}$; let $u = x^2 - 3$

4 $\displaystyle\int \sin^4 x \cdot \cos x \, dx$; let $\sin x = s$

5 $\displaystyle\int \dfrac{dx}{3 + 2\sqrt{x}}$; let $u^2 = x$

6 $\displaystyle\int \dfrac{\ln x}{x} \, dx$; let $u = \ln x$

7 $\displaystyle\int \dfrac{x \, dx}{\sqrt{1 - x^2}}$; let $x = \sin \theta$

8 $\displaystyle\int_0^{\pi/4} \tan^5 \theta \cdot \sec^2 \theta \, d\theta$; let $\tan \theta = u$

9 $\int_0^1 \dfrac{3x\,dx}{\sqrt{1+x^2}}$; let $u^2 = 1 + x^2$ **10** $\int_{-1}^2 \dfrac{x^3}{(x+2)^2}\,dx$; let $u = x + 2$

11 $\int e^x\sqrt{1+e^x}\,dx$; let $u = 1 + e^x$ **12** $\int \dfrac{x+2}{x\sqrt{x-1}}\,dx$; let $u^2 = x - 1$

13 $\int x\,e^{x^2+3}\,dx$; let $u = x^2 + 3$ **14** $\int_1^4 \sqrt{x}(1+x^{3/2})\,dx$; let $u = 1 + x^{3/2}$

*15 $\int_0^1 \dfrac{dx}{e^x + e^{-x}}$; let $u = e^x$ ***16** $\int_1^2 \dfrac{dx}{x(x^4+1)}$; let $u = x^4 + 1$

Exercise 11.3

Use integration by substitution to evaluate:

1 $\int \dfrac{dx}{\sqrt{x+1}}$ **2** $\int \dfrac{dx}{(5x+2)^7}$ **3** $\int_0^3 \sqrt{9-x^2}\,dx$

4 $\int_0^{\pi/3} \sin^3 x\cdot dx$ **5** $\int \dfrac{dx}{\sqrt{1-16x^2}}$ **6** $\int_0^2 \dfrac{dx}{(4+x^2)^{3/2}}$

7 $\int \dfrac{x^2}{(x-2)^4}\,dx$ **8** $\int_3^4 \dfrac{x+3}{x\sqrt{x-3}}\,dx$ **9** $\int_4^9 \dfrac{\sqrt{x}-1}{1+\sqrt{x}}\,dx$

10 $\int \cos^3 x\cdot dx$ **11** $\int \dfrac{x}{(3x-1)^5}\,dx$ **12** $\int_3^8 \dfrac{x}{\sqrt{x+1}}\,dx$

13 $\int (x+3)(x-2)^7\,dx$ **14** $\int \sin^3 x\cos^2 x\,dx$ **15** $\int_0^{\pi/6} \cos^5 x\,dx$

16 $\int \dfrac{x}{\sqrt{25-x^2}}\,dx$ ***17** $\int_0^{1/\sqrt{2}} \dfrac{2x}{1-x^4}\,dx$

*Exercise 11.4

Use integration by substitution to evaluate:

1 $\int_0^{\pi/2} \dfrac{1}{1+\sin x}\,dx$ **2** $\int_1^3 \dfrac{dx}{2\sqrt{x}(1+x)}$

3 $\int \sin\theta\sqrt{1-\cos\theta}\,d\theta$ **4** $\int \dfrac{2}{x\sqrt{2x-1}}\,dx$

5 $\int_0^1 \dfrac{x^2}{1+x^6}\,dx$ **6** $\int \dfrac{12x\,dx}{\sqrt{3x^2+1}}$

7 $\displaystyle\int \frac{e^x}{(e^x+2)^{1/2}}\,dx$

8 $\displaystyle\int_{\sqrt{2}}^{2} \frac{x^2}{\sqrt{4-x^2}}\,dx$

9 $\displaystyle\int \frac{x^2}{1+3x^3}\,dx$

10 $\displaystyle\int_{\pi/2}^{\pi} \frac{dx}{1-\cos x}$

11 $\displaystyle\int x\sqrt{9x^2-1}\,dx$

12 $\displaystyle\int_{1}^{3} \frac{x-1}{\sqrt{x+1}}\,dx$

13 $\displaystyle\int_{0}^{1} \frac{2x}{1+x^4}\,dx$

14 $\displaystyle\int_{0}^{2} \frac{dx}{e^x+1}$

15 $\displaystyle\int_{0}^{1/2\sqrt{2}} \sqrt{1-4x^2}\,dx$

*16 $\displaystyle\int_{1/2}^{3/4} \frac{dx}{\sqrt{x}\sqrt{1-x}}$

*17 $\displaystyle\int_{0}^{2} x^2\sqrt{4-x^2}\,dx$

*18 Show that $\displaystyle\int_{0}^{1/2} \frac{dx}{(9+x^2)^{3/2}} = \frac{1}{9\sqrt{37}}$

Exercise 11.5

Integrate by parts:

1 $\displaystyle\int x\cos x\,dx$

2 $\displaystyle\int x\,e^x\,dx$

3 $\displaystyle\int_{0}^{\pi/2} x\sin x\,dx$

4 $\displaystyle\int x\cos 3x\,dx$

5 $\displaystyle\int x^3 \ln x\,dx$

6 $\displaystyle\int_{0}^{1} x\,e^{2x}\,dx$

7 $\displaystyle\int_{1}^{2} \ln 3x\,dx$

8 $\displaystyle\int x^2\cos x\,dx$

9 $\displaystyle\int_{1}^{3} \frac{1}{x^2}\ln x\,dx$

10 $\displaystyle\int \sin^{-1} x\,dx$

11 $\displaystyle\int_{0}^{1} \cos^{-1} x\,dx$

*12 $\displaystyle\int_{1}^{\sqrt{3}} x\tan^{-1} x\,dx$

*13 $\displaystyle\int_{0}^{\pi/2} x\cos^2 x\,dx$

*14 $\displaystyle\int_{0}^{\pi/3} x\sin(\pi/3 - x)\,dx$

*15 $\displaystyle\int e^{-x}\cos x\,dx$

In each of questions 16–20 the products can be integrated either: (a) by inspection, (b) by substitution or (c) by parts. In each question choose the best method and evaluate the integral.

16 $\displaystyle\int x\,e^{x^2}\,dx$

17 $\displaystyle\int (x-1)^2\,e^x\,dx$

18 $\displaystyle\int x(1-x^2)^7\,dx$

19 $\int \sin x \, e^{\cos x} \, dx$ **20** $\int 2x\sqrt{3x-4} \, dx$ **21** $\int x \, e^{3x+1} \, dx$

22 $\int \sin x \cos^5 x \, dx$

Exercise 11.6 [odd number]

[Note that the questions in this exercise of A level questions on integration have **not** been cross-referenced to units in the text. It is up to **you** to decide what techniques to use. The questions are suitable for all syllabuses]

1 (a) Find $\int \dfrac{7}{(3-x)(1+2x)} \, dx$

   (b) By using the substitution $x = 2\sin\theta$, or otherwise, evaluate
   $$\int_0^1 \sqrt{(4-x^2)} \, dx \tag{L}$$

2 Use the substitution $u^2 = x + 1$ to prove that
$$\int_3^8 \frac{\sqrt{(x+1)}}{x+5} \, dx = \pi + 2 - 4\tan^{-1}(\tfrac{3}{2}) \tag{A}$$

3 Write down (or obtain) the derivative of $\tan^{-1} 2x$

   Find (a) $\int \dfrac{x+2}{4x^2+1} \, dx$ (b) $\int \dfrac{4x}{4x^4+1} \, dx$ \hspace{1em} (J)

4 (a) Find $\int (3x+4) e^{2x} \, dx$

   (b) By using the substitution $x = 2\tan\theta$, evaluate
   $$\int_0^2 \frac{1}{(4+x^2)^2} \, dx \tag{L}$$

5 Using the substitution $t = \sin x$, evaluate to two decimal places the integral
$$\int_{\pi/6}^{\pi/2} \frac{4\cos x}{3+\cos^2 x} \, dx \tag{A}$$

6 Evaluate:

   (a) $\int_1^4 \left(\sqrt{x} + \dfrac{1}{\sqrt{x}}\right)^3 \, dx$ •

   (b) $\int_0^{x/2} \cos 2x \sin 4x \, dx$

   (c) $\int_0^{3/4} \dfrac{1-x}{(x+1)(x^2+1)} \, dx$ \hspace{1em} (L)

7 Evaluate (a) $\displaystyle\int_0^{(1/6)\pi} \tan 2\theta \, d\theta$ (b) $\displaystyle\int_0^1 x(1-x)^{1/2} \, dx$ (J)

8 Find (a) $\displaystyle\int x \ln x \, dx$ (b) $\displaystyle\int \frac{x}{\sqrt{(x-2)}} \, dx$ (L)

9 (a) Find $\displaystyle\int \sin 2x (\sin x)^{1/2} \, dx$ (b) Prove that $\displaystyle\int_1^2 \frac{dx}{x^3+x} = \tfrac{1}{2}\log_e(\tfrac{8}{5})$ (A)

10 (a) Evaluate (i) $\displaystyle\int_0^{\pi/4} \sin^2 x \, dx$ (ii) $\displaystyle\int_0^1 x^2\sqrt{(1-x)} \, dx$

(b) Using the substitution $u = e^x - 1$ and leaving your answers in terms of e, evaluate

$$\int_1^2 \frac{e^{2x}}{e^x - 1} \, dx \qquad\qquad \text{(L)}$$

Exercise 11.7

1 If $y = x^3 - 2x$, find the approximate increase in y when x increases from 3 to 3.01

2 Find the approximate increase in the area of a circle if the radius is increased from 5 to 5.2

3 If $y = 3x^2 - 1$ find the approximate decrease in y when x decreases from 4 to 3.9

4 If $T = 2\pi\sqrt{\dfrac{l}{10}}$ find the approximate increase in T if l is increased from 49 to 50

5 If $y = 2x^3$ find the approximate percentage increase in y when x is increased by 0.5%

6 If $y = x^{5/2}$ find $\dfrac{dy}{dx}$ when $x = 4$. Hence find approximate values for

(a) $(4.01)^{5/2}$ (b) $(3.97)^{5/2}$

7 Find approximately the values of

(a) $\sqrt{9.003}$ (b) $\dfrac{1}{4.04}$ (c) $\ln(1.05)$

8 Given that $T = k\sqrt{x}$ where k is a positive constant, find $\dfrac{dT}{dx}$. Calculate the approximate percentage change in the value of T when x increases by 0.2%   (J)

9 Given that $y = e^{-2x} \sin 3x$, write down ln y as a sum of two terms. Hence, or otherwise, find $\dfrac{1}{y} \cdot \dfrac{dy}{dx}$ at $x = \dfrac{\pi}{12}$

Deduce, to 1 significant figure, the change in y as x increases from $\dfrac{\pi}{12}$ to $\dfrac{\pi}{12} + 0.2$

(J)

*10 The volume of a sphere is $V = \frac{4}{3}\pi r^3$ where $r = $ radius. Find a formula connecting V and the diameter of the sphere d.

If a 3% error is made in measuring the diameter of a sphere estimate the percentage error in the volume and surface area of the sphere (surface area $= 4\pi r^2$.)

Exercise 11.8

Find the general solutions of the following differential equations:

1   $x\,dx = y^3\,dy$

2   $\dfrac{dy}{dx} = x^2 + 1$

3   $\dfrac{dy}{dx} = \dfrac{y}{x}$

4   $\dfrac{dy}{dx} = \cos x \sec y$

5   $\dfrac{dy}{dx} = y e^x$

6   $\dfrac{dy}{dx} = y^3$

7   $\dfrac{dy}{dx} = \dfrac{xy}{x^2 - 1}$

8   $(x^2 + 1)\dfrac{dy}{dx} = 1$

9   $2x\dfrac{dy}{dx} = y^2 + 1$

10   $5\dfrac{ds}{dt} = t + 3$

11   $\tan y \dfrac{dy}{dx} = 6$

12   $\dfrac{y^2}{x}\dfrac{dy}{dx} = e^x$

13   $\operatorname{cosec} x \dfrac{dy}{dx} = e^y$

14   $xy\dfrac{dy}{dx} = \ln x$

15   $(\sin\theta + \cos\theta)\dfrac{dr}{d\theta} = \cos\theta - \sin\theta$

16   $\cos^2 x \dfrac{dy}{dx} = 2\cos^2 y$

17 Find the particular solution of $e^x \dfrac{dy}{dx} = \sqrt{y}$ given that $y = 9$ when $x = 0$

18 Find the equation of the curve that satisfies $\dfrac{dy}{dx} = \dfrac{y}{x^2 - 4}$ and passes through $(4, 1)$

19 Find y in terms of x given that

$$x\frac{dy}{dx} = y(y+1)$$

and $y = 4$ when $x = 2$ (L)

20 Given that $e^{2x+y}\frac{dy}{dx} = x$ and that $y = 0$ when $x = 0$, express e^y in terms of x

     (L)

21 Solve the differential equation

$$\frac{dy}{dx} = \frac{(y^2 - 1)}{x}$$

where $y = 2$ when $x = 1$, giving y in terms of x (L)

22 Solve the differential equation

$$xy\frac{dy}{dx} = 1 - x^2 \qquad x > 0$$

given that $y = 2$ when $x = 1$ (L)

23 The gradient of a curve at any point (x, y) on the curve is directly proportional to the product of x and y. The curve passes through the point $(1, 1)$ and at this point the gradient of the curve is 4. Form a differential equation in x and y and solve this equation to express y in terms of x

     (L)

Exercise 11.9

1 Given that $\frac{dp}{dt} = kp$, where k is a positive constant, and that $p = 100$ at time $t = 0$, express t in terms of p and k

The variable p takes time T to increase from 100 to 200. Find the time, in terms of T only, for p to increase from 100 to 150   (A)

2 A container is shaped so that when the depth of the water is x cm the volume of water in the container is $(x^2 + 3x)$ cm³. Water is poured into the container so that, when the depth of water is x cm, the rate of increase of volume is $(x^2 + 4)$ cm³ s^{-1}. Show that $\frac{dx}{dt} = \frac{(x^2 + 4)}{(2x + 3)}$ where t is the time measured in seconds.

Solve the differential equation to obtain t in terms of x, given that initially the container is empty   (A)

3 A water tank has the shape of an open rectangular box of length 1 m, width 0.5 m and height 0.5 m. Water may be drained from the tank through a tap at the bottom of the tank, and it is known that, when the tap is open, water leaves at a rate of $100h$ litres per minute, where h m is the depth of water in the tank. When the tap is open, water is also fed into the tank at a constant rate of 50 litres per minute and no water is fed into the tank when the tap is closed. Show that, t minutes after the tap has been opened, the variable h satisfies the differential equation

$$10\frac{dh}{dt} = 1 - 2h$$

On a particular occasion the tap was opened when $h = 0.25$ and closed when $h = 0.375$. Show that the tap was opened for $5\ln 2$ minutes   (L)

4 A radioactive substance decays so that the rate of decrease of mass at any time is proportional to the mass present at that time. Denoting by x the mass remaining at time t, write down a differential equation satisfied by x and show that $x = x_0\,e^{-kt}$ where x_0 is the initial mass and k is a constant.

The mass is reduced to $\frac{4}{5}$ of its initial value in 30 days. Calculate, correct to the nearest day, the time required for the mass to be reduced to half its initial value. A mass of 625 milligrammes of the substance is prepared. Determine the mass which is present 90 days after the preparation   (J)

*5 A plant grows in a pot which contains a volume V of soil. At time t the mass of the plant is m and the volume of the soil utilised by the roots is αm, where α is a constant. The rate of increase of the mass of the plant is proportional to the mass of the plant times the volume of soil not yet utilised by the roots. Obtain a differential equation for m and verify that it can be written in the form:

$$V\beta\frac{dt}{dm} = \frac{1}{m} + \frac{\alpha}{V - \alpha m} \qquad \text{where } \beta \text{ is a constant}$$

The mass of the plant is initially $V/4\alpha$. Find, in terms of V and β the time taken for the plant to double its mass. Find also the mass of the plant at time t (J)

Exercise 11.10

1 Find the area between $y = 1 + \cos x$, the x-axis and the ordinates $x = 0$, $x = \pi$

2 Find the coordinates of A and B and the area shaded

3 Find the area between the curve $y^2 = 4x$ and the double ordinate $x = 4$

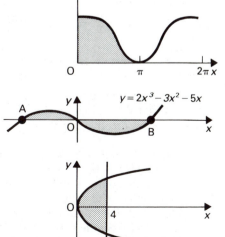

4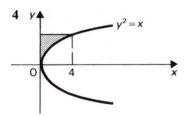

Find the shaded area

5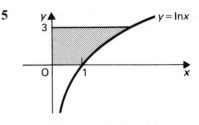

Find the shaded area

6 Find the mean value of e^{-x} in the interval $1 \leqslant x \leqslant 4$

7 Given the curve whose equation is $y = x^{-1/2}$ find

(a) the mean value of $\dfrac{1}{y}$, in the interval $1 \leqslant x \leqslant 4$

(b) the area of the region R bounded by the curve, the x-axis and the lines $x = 1$ and $x = 4$

(c) the volume of the solid generated when the region R is rotated through an angle of 2π radians about the x-axis   (J)

8 Express $y = \dfrac{1}{(x+2)(x+3)}$ in partial fractions and hence find the mean value of y for $x \in [0, 6]$ (A)

9 Find the coordinates of all points of inflexion of the graph of the function

$$y = x - \sin x$$

in the range $0 \leqslant x \leqslant 4\pi$. Sketch the graph in this range. The region bounded by the x-axis, the line $x = 4\pi$ and the curve whose equation is $y = x - \sin x$ is rotated once about the x-axis. Prove that the volume swept out is $\frac{2}{3}\pi^2(32\pi^2 + 15)$

Find the mean value of y^2 w.r.t. x over the range $0 \leqslant x \leqslant 4\pi$ (J)

10 Given that $y = e^{-x}\sin x$ find $\dfrac{d^2 y}{dx^2}$ as a function of x and hence show that

$$\left(\dfrac{d^2 y}{dx^2}\right)^2 = 4(e^{-2x} - y^2)$$

Prove that the curve $y = e^{-x}\sin x$ has infinitely many points of inflexion and that they all lie on the curve $y^2 = e^{-2x}$. Prove that the mean value of $e^{-x}\sin x$ in the interval $0 \leqslant x \leqslant \pi$ is $(e^{-\pi} + 1)/2\pi$ (J)

*11 Find the x coordinates of the stationary points of the curve whose equation is $y = x(x - 3)^4$ and determine whether these points are maxima, minima or points of inflexion. Sketch the curve. Find the area bounded by the curve and that part of the x-axis lying between $x = 0$ and $x = 3$. Show that, for values of x lying between 0 and 3, the mean value w.r.t. x of the perpendicular distance from a point on the curve to the line $y = -1$ is $\frac{91}{10}$ (J)

Miscellaneous Exercise 11 B

1 Show that the area of the finite region bounded by the x-axis, the line $x = e$ and the curve $y = \ln x$ is 1   (L: 5.4, 9.3, 11.5)

2 Express the function $f(x) = \dfrac{x+2}{(x^2+1)(2x-1)}$ as the sum of partial fractions.

Hence find $\displaystyle\int f(x)\,dx$ (J: 10.9)

3 The parametric equations of a curve are
$$x = \cos 2\theta + 2\cos\theta, \quad y = \sin 2\theta - 2\sin\theta$$
Show that
$$\frac{dy}{dx} = \tan\frac{\theta}{2}$$

Find the equation of the normal to the curve at the point where $\theta = \dfrac{\pi}{2}$

       (A: 8.5)

4 Apply the small increment formula
$$f(x + \delta x) - f(x) \approx \delta x f'(x)$$
to $\tan x$ to find an approximate value of
$$\tan\left(\frac{100\pi + 4}{400}\right) - \tan\frac{\pi}{4} \qquad \text{(L: 11.7)}$$

5 Prove that $\dfrac{d}{d\theta}\left(\log_e \tan\dfrac{\theta}{2}\right) = \dfrac{1}{\sin\theta}$

Given that $\dfrac{dy}{dx} = 2\sin^2 x \sin y$ and that $y = \dfrac{\pi}{2}$ when $x = 0$, prove that $y = 2\tan^{-1}(e^x)$ when $x = \pi$ (L: 11.8)

6 Determine the maximum and the minimum values of the function $y = \sqrt{3}\sin 2x + \cos 2x$ and find the values of x in the interval $0 \leqslant x \leqslant \pi$ for which they occur. Find also the points in the interval $0 \leqslant x \leqslant \pi$ at which $y = 0$ and sketch the graph of y in this interval. Calculate, in terms of π, the volume swept out when the region bounded by the curve $y = \sqrt{3}\sin 2x + \cos 2x$, the line $x = \pi/4$ and the coordinate axes is rotated about the x-axis through an angle of 2π radians   (J: 7.2)

7 With the help of a suitable substitution, or otherwise, find $\displaystyle\int \frac{e^x + e^{2x}}{1 + e^{2x}}\,dx$

       (J: 11.4, 8.6)

8 Find the mean value of $[x + (1 - x^2)^{1/2}]$ in the interval $\frac{1}{2} \leqslant x \leqslant 1$, leaving your answer in terms of π (A: 11.10, 11.3)

9 Express $\dfrac{2}{(1+x)(1+3x)}$ in partial fractions.

Hence, or otherwise, solve the differential equation

$$\frac{dy}{dx} = \frac{2(y+2)}{(1+x)(1+3x)}$$

given that $y = -1$ when $x = 0$ (L: 10.6, 11.8)

10 A curve whose equation is $y = (1-x)e^x$ meets the x-axis at A and the y-axis at B. The region bounded by the arc AB of the curve and the line segments OA and OB, where O is the origin, is rotated through a complete revolution about the x-axis. Show that the volume swept out is $\frac{1}{4}\pi(e^2 - 5)$
(J: 11.5, 5.5)

11 Find (a) $\displaystyle\int \frac{e^{2x}}{(1+e^{2x})^2}\,dx$ (b) $\displaystyle\int_0^1 \frac{1}{(4-3x^2)^{1/2}}\,dx$ (J: 11.4, 11.5, 8.6)

12 (a) Evaluate

(i) $\displaystyle\int_0^{1/2} \sqrt{(1-x^2)}\,dx$ by the substitution $x = \sin\theta$

(ii) $\displaystyle\int_0^1 x e^x\,dx$

(b) Find the area of the region in the coordinate plane for which

$$0 \leqslant y \leqslant \frac{1}{2x+3} \qquad 0 \leqslant x \leqslant 12$$

leaving your answer in terms of a natural logarithm
(L: 5.4, 9.4, 11.2, 11.6)

13 Given that $\dfrac{dy}{dx} = \dfrac{2x}{1+x^2} - 2x\,e^{-x^2}$ and that $y = 0$ when $x = 0$, express y in terms of x

Show that, for small values of $|x|$, $y \approx px^6 + qx^8$ finding the values of the constants p and q (A: 11.8, 9.5)

14 Given that $y = \sqrt{3}\cos x + \sin x$, where $0 \leqslant x \leqslant 2\pi$, find the maximum and minimum values of y. Sketch the graph of y. Calculate the value of x for which $\dfrac{dy}{dx} = -1$. Show that the mean value of y^2 w.r.t. x in the interval $0 \leqslant x \leqslant \pi/2$ is $2(\pi + \sqrt{3})/\pi$ (J: 7.2, 11.10)

15 Evaluate the integrals

(a) $\displaystyle\int_0^{3/4} x\sqrt{(1+x^2)}\,dx$ (b) $\displaystyle\int_0^{\pi/4} \tan^2 x\,dx$

(c) $\displaystyle\int_0^\pi x \sin x \, dx$

(d) $\displaystyle\int_0^4 \frac{1}{x^2 - 3x + 2} \, dx$

Express your answer to (d) as a natural logarithm

(L: 1.6, 11.2, 11.5, 10.9)

16 Find the general solution of the differential equation

$$\frac{dy}{dx} = y^2 + 4$$

Given that $y = 2$ when $x = 0$, show that $y = 2 \tan\left(2x + \dfrac{\pi}{4}\right)$

Find the mean value of y in the interval $-\dfrac{\pi}{8} \leqslant x \leqslant 0$ (A: 11.8, 11.10)

17 The locus of a point $P(x, y)$ which is moving in the x-y plane is such that its distance from the y-axis is always equal to its distance from the point $(4, -1)$. Prove that the equation of the locus of P is

$$(y + 1)^2 = 8(x - 2)$$

(a) Find the coordinates of P when the rate of increase of x is twice the rate of increase of y

(b) When x increases by a small amount δx, the change in y is δy. Find an approximate expression for δy in terms of y and δx

(J: 8.4, 11.7)

18 Referred to an origin O and coordinate axes Ox and Oy, a curve is given by

$$x = \sec t + \tan t, \quad y = \operatorname{cosec} t + \cot t, \quad 0 < t < \pi/2$$

where t is a parameter. Prove that $\dfrac{dy}{dx} = -\dfrac{1 - \sin t}{1 - \cos t}$

Show that the normal to the curve at the point S, where $t = \tan^{-1}\left(\frac{3}{4}\right)$, has equation $x - 2y + 4 = 0$. Find an equation of the normal to the curve at the point T, where $t = \tan^{-1}\left(\frac{4}{3}\right)$. These normals meet at the point N. Find the coordinates of N

Hence calculate

(a) the area of the triangle SNT

(b) the tangent of the angle SNT   (A: 8.5, 8.7)

19 A point $P(x, y)$ moves on the curve $y = e^{\sin x} + e^{1/2} \cos^2 x$ in such a way that the x-coordinate of P increases at a constant rate of 4 units per second

(a) Find the rate of change with respect to time of the y-coordinate of P and show that this rate is zero when $x = \dfrac{\pi}{6}$

(b) Find the rate of change with respect to time of the gradient of the curve when $x = \dfrac{\pi}{6}$ (A: 3.2, 11.1)

*20 Prove that the curve C given by the equation $y = xe^{-x}$ has only one stationary point and that it is a maximum. Prove also that C has a point of inflexion at $P(2, 2e^{-2})$. Find the equation of the tangent to the curve C at the point P. Sketch C and L on the same diagram in the interval $0 \leqslant x \leqslant 4$

Obtain $\int xe^{-x}dx$ and calculate the area of the finite region enclosed by C, L and the x-axis, leaving your answer in terms of e

(J: 3.4, 3.5, 3.3, 5.4)

*21 The area of the region enclosed between two concentric circles of radii x and y $(x > y)$ is denoted by A.
Given that: x is increasing at a rate of $2\,\text{m/s}$
 y is increasing at a rate of $3\,\text{m/s}$
and when $r = 0$, $x = 4\,\text{m}$ and $y = 1\,\text{m}$, find

(a) the rate of increase of A when $r = 0$.
(b) the ratio of x to y when A begins to decrease
(c) the time at which A is zero   (J: 5.2)

*22 Prove that the x-axis is a tangent to the curve C given by the equation $y = x \sin x$. Find the equation of the tangent to C at $x = \pi/2$ and show that for $0 < x < \pi/2$, C lies between this line and the x-axis. Find, in terms of π, the volume swept out when the region bounded by C for $0 \leqslant x \leqslant \pi/2$, the ordinate $x = \pi/2$ and the x-axis, is rotated about the x-axis through an angle of 2π rads   (J: 1.7, 3.3, 3.5, 11.5)

*23

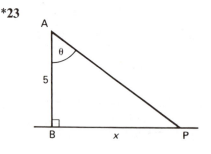

The line AB is perpendicular to a horizontal plane through the point B and $AB = 5\,\text{m}$. A variable point P moves on the plane along a straight line through B, and when $BP = x$ metres the angle PAB is θ radians, as shown in the diagram. Show that, when $BP = (x + \delta x)\text{m}$ and δx is small, the angle PAB is $(\theta + \delta\theta)$ where $\delta\theta \approx \dfrac{\delta x}{5} \cdot \cos^2 \theta$

Given that P is moving towards B with a constant speed of $3\,\text{m s}^{-1}$, find the rate of change of θ w.r.t. time when $x = 5$ (J: 11.7)

24 Differentiate $x \sin^{-1} mx$ where m is a constant. Hence, or otherwise, integrate $\sin^{-1} mx$ (J: 8.6, 11.3 or 11.5)

12.1 Sketching quadratic functions

Met briefly earlier in 1.7

Method One: Graphs of equations of the form $y = (x - a)(x - b)$

Example 1 Sketch the graph of $y = (x + 2)(x - 1)$. State the equation of the line of symmetry and the coordinates of the vertex. Use the graph to solve the inequation $(x + 2)(x - 1) \geqslant 0$.

(a) The coefficient of x^2 is positive
$\Rightarrow$ the curve is $\cup$.
(b) It crosses the x-axis when $y = 0$
$\Rightarrow$ when $x = -2$ and $x = 1$
Thus the graph is
The line of symmetry is $x = -\frac{1}{2}$

When $x = -\frac{1}{2}, y = -\frac{9}{4}$
$\therefore$ the vertex is $\left(-\frac{1}{2}, -\frac{9}{4}\right)$

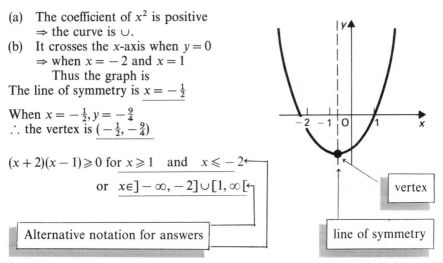

$(x + 2)(x - 1) \geqslant 0$ for $x \geqslant 1$ and $x \leqslant -2$

or $x \in \,] -\infty, -2] \cup [1, \infty [$

vertex

line of symmetry

Alternative notation for answers

Method Two: Using translations and scalings

An equation of the form $y = ax^2 + bx + c$ can be expressed in the form

$$y = \mu(x - \alpha)^2 + \beta$$

Completed square form: see 1.9

or $y - \beta = \mu(x - \alpha)^2$

The graph of this equation may be obtained from the graph of $y = x^2$ by:
a translation of α in the positive x-direction
a translation of β in the positive y-direction
and a y-scaling of factor μ.

Example 2 Express $2x^2 - 6x + 5$ in the form $\mu(x - \alpha)^2 + \beta$. Sketch the graph of $y = 2x^2 - 6x + 5$. What is the range of $2x^2 - 6x + 5$?

$$2x^2 - 6x + 5 = 2[x^2 - 3x + \tfrac{5}{2}]$$
$$= 2[(x - \tfrac{3}{2})^2 - \tfrac{9}{4} + \tfrac{5}{2}]$$
$$= 2[(x - \tfrac{3}{2})^2 + \tfrac{1}{4}]$$
$$= 2(x - \tfrac{3}{2})^2 + \tfrac{1}{2}$$

The first step is always to make the coefficient of x^2 into 1, before completing the square.

So, we translate the graph of $y = x^2$ by $\frac{3}{2} \rightarrow$ and $\frac{1}{2}\uparrow$ and scale it by a factor 2, to give The range is $[\frac{1}{2}, \infty[$

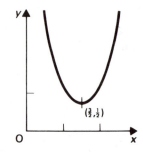

> **Note:** A negative scaling 'flips over' the graph from $\cup$ to $\cap$

Example 3   Show how the graph of $y = 2(x + 1)^2 + 3$ may be obtained from the graph of $y = x^2$ by appropriate translations and scalings. Illustrate your answer with a sequence of sketches.

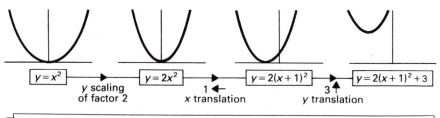

> **Note:**   There are several possible sequences of these transformations. The x-translation can come at any point in the sequence but the y-scaling must come before the y-translation.

● **Exercise 12.1**   See page 263.

12.2 General Curve Sketching Using Translations and Scalings

Let $y = f(x)$ be the basic function whose graph we know.

$\boxed{y = f(x - \alpha)}$ is the graph of $y = f(x)$ given an x-translation of $\alpha \rightarrow$.

$\boxed{y - \beta = f(x)}$ is the graph of $y = f(x)$ given a y-translation of $\beta \uparrow$.

$\boxed{\dfrac{y}{\mu} = f(x)}$ is the graph of $y = f(x)$ given a y-scaling of factor μ.

$\boxed{y = f\left(\dfrac{x}{\lambda}\right)}$ is the graph of $y = f(x)$ given an x-scaling of factor λ.

Example 4   Sketch the graphs of (a) $y = \sin(x - \pi/2)$ (b) $y = 1 + \sin x$

(c) $y = 2\sin x$ (d) $y = \sin\left(\dfrac{x}{2}\right)$ for $x \in [0, 2\pi]$.

The basic graph is $y = \sin x$

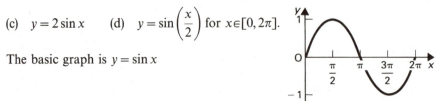

(a) $y = \sin(x - \pi/2)$ is the basic
graph x-translated
through $\pi/2 \rightarrow$

(b) $y = 1 + \sin x$
or $y - 1 = \sin x$
is the basic graph y-translated
through $1\uparrow$

(c) $y = 2 \sin x$

or $\dfrac{y}{2} = \sin x$

is the basic graph y-scaled by
factor 2

(d) $y = \sin\left(\dfrac{x}{2}\right)$ is the basic graph

x-scaled by a factor of 2

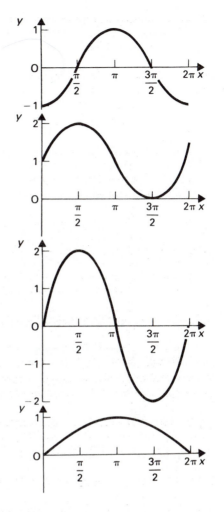

Useful Basic Graphs

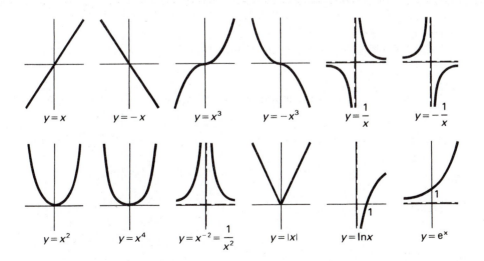

$y = x$ $y = -x$ $y = x^3$ $y = -x^3$ $y = \dfrac{1}{x}$ $y = -\dfrac{1}{x}$

$y = x^2$ $y = x^4$ $y = x^{-2} = \dfrac{1}{x^2}$ $y = |x|$ $y = \ln x$ $y = e^x$

The function $\dfrac{ax+b}{cx+d}$

The graph of $y = \dfrac{ax+b}{cx+d}$ can be

formed from the graph of $y = \dfrac{1}{x}$ using

translations and scalings.

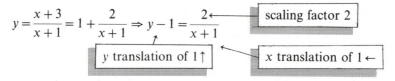

$$y = \frac{1}{x}$$

The curve approaches, but never meets, the lines $x = 0$ and $y = 0$. Such lines are called **asymptotes**. They are usually sketched on graphs as dotted lines.

Example 5   Show that the graph of $y = \dfrac{x+3}{x+1}$ may be obtained from the

graph of $y = \dfrac{1}{x}$ by appropriate translations and scalings.

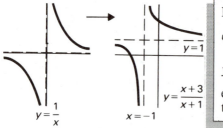

$$y = \frac{x+3}{x+1} = 1 + \frac{2}{x+1} \Rightarrow y - 1 = \frac{2}{x+1}$$

scaling factor 2

y translation of 1 ↑

x translation of 1 ←

Note   In this family of curves it is difficult to distinguish between $y = \dfrac{1}{x}$, $y = \dfrac{2}{x} \cdots$. For these equations it is easier to determine the points where the curve crosses the axes.

Example 6
Sketch the curves (a) $y = \dfrac{2x+3}{x+3}$ (b) $y = \dfrac{1}{2x+1}$

(a) $y = \dfrac{2x+3}{x+3} = 2 - \dfrac{3}{x+3}$

$\Rightarrow y - 2 = \dfrac{-3}{x+3}$

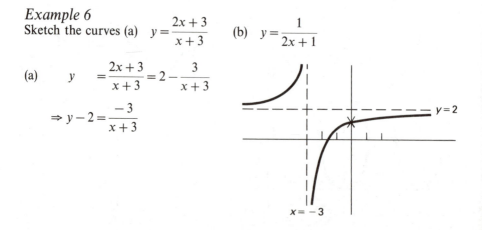

The asymptotes are $x = -3$ and $y = 2$ and the basic graph is $y = -\dfrac{1}{x}$.

When $x = 0$, $y = 1$
When $y = 0$, $x = -\frac{3}{2}$

(b) $y = \dfrac{1}{2x+1} = \dfrac{\frac{1}{2}}{x + \frac{1}{2}}$

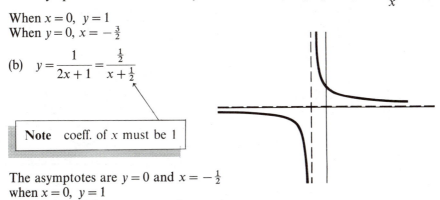

Note   coeff. of x must be 1

The asymptotes are $y = 0$ and $x = -\frac{1}{2}$
when $x = 0$, $y = 1$
but $y \neq 0$ ∴ it does not cross the x-axis.

● **Exercise 12.2**   See page 264.

12.3 Graphs of Related Functions

1   *Graphs of the Modulus of a Function*

> *To obtain the graph of* $y = |f(x)|$ *from that of* $y = f(x)$, *the part of the graph above the x-axis remains the same, and the part below the x-axis is reflected in the x-axis.*

Example 7   Sketch the graph of $y = |x^2 - 2|$.

The graph of $y = x^2 - 2$ is   ∴ the graph of $y = |x^2 - 2|$ is

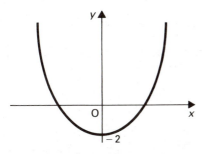
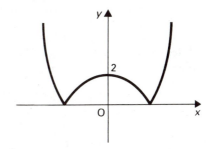

2   *Graphs of Functions of* $|x|$

> *Functions of* $|x|$
>
> If $g(x) = f(|x|)$,
> then, for $x \geqslant 0$, $g(x) = f(x)$
> and for $x < 0$, the graph is a reflection, in the y-axis, of the graph for $x > 0$.

Example 8   Sketch the graph of $f(x) = x^3 + 1$ and $g(x) = |x^3| + 1$

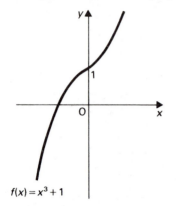

$f(x) = x^3 + 1$

For $x \geqslant 0$ $g(x) = f(x) = x^3 + 1$

For $x < 0$ $g(-2) = g(2)$

               $g(-3) = g(3)$

               $g(-8) = g(8)$

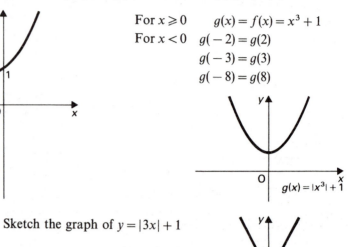

$g(x) = |x^3| + 1$

Example 9   Sketch the graph of $y = |3x| + 1$

Method 1
$y = |3x| + 1$ is the same as $y - 1 = 3|x|$
which is the graph of $y = |x|$ translated
$1\uparrow$ and y-scaled by a factor 3.

Method 2
When $x \geqslant 0$, $y = |3x| + 1$ is the same as $y = 3x + 1$.
When $x < 0$, $y = |3x| + 1$ is the reflection, in the y-axis, of the graph for $x > 0$.

3   Graph of $y = 1/f(x)$

The graph of $y = 1/f(x)$ may be deduced from the graph of $y = f(x)$ by noting the following points:
(a)   As $f(x)$ increases, $1/f(x)$ decreases (and conversely)
(b)   A maximum turning point on $y = f(x)$ becomes a minimum turning point on $y = 1/f(x)$ (and conversely)
(c) $f(x)$ and $1/f(x)$ have the same sign, for a given value of x
(d)   As $f(x) \to \infty$, $1/f(x) \to 0$
(e)   If $f(a) = 0$ then $x = a$ is an asymptote for the graph of $y = 1/f(x)$.

Example 10   Sketch the graphs of $y = x^2 - 1$ and $y = \dfrac{1}{x^2 - 1}$

$y = x^2 - 1$

Hence for $y = \dfrac{1}{x^2 - 1}$

(a) $\dfrac{1}{x^2 - 1}$ decreases for $x > 0$ and

increases for $x < 0$

(b)   there is a maximum t.p. at $(0, -1)$

(c) $\dfrac{1}{x^2-1}$ is positive for $x>1$ and $x<-1$,

$\dfrac{1}{x^2-1}$ is negative for $-1<x<1$

(d) As $x\to\pm\infty$, $x^2-1\to\infty\Rightarrow\dfrac{1}{x^2-1}\to0$

(e) $x=1$ and $x=-1$ are asymptotes

Thus the graph of $y=\dfrac{1}{x^2-1}$ is

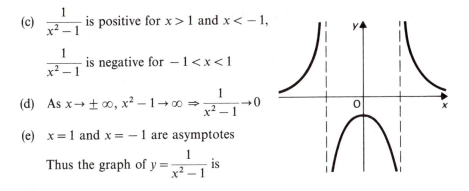

4 Graph of $y^2=f(x)$

The graph of $y^2=f(x)$ may be deduced from the graph of $y=f(x)$ by noting the following points:
(a) the graph of $y^2=f(x)$ is symmetrical about the x-axis
(b) $y^2\geqslant0\Rightarrow$ the graph only exists for those values of x for which $f(x)\geqslant0$
(c) $y^2=f(x)$ and $y=f(x)$ meet the x-axis at the same points
(d) $y=\pm\sqrt{f(x)}\Rightarrow$ when $f(x)>1$ $\sqrt{f(x)}<f(x)$
              when $f(x)<1$ $\sqrt{f(x)}>f(x)$
(e) when $f(x)=1$, $\sqrt{f(x)}=\pm1$

Example 11   Sketch the graphs of $y=x(x-1)$ and $y^2=x(x-1)$.

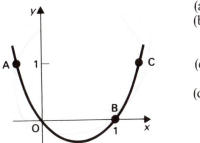

(a) Symmetry about $y=0$
(b) $x(x-1)\leqslant0$ for $0\leqslant x\leqslant1\Rightarrow$ $y^2=x(x-1)$ only exists for $x>1$ and $x<0$
(c) $y^2=x(x-1)$ meets the x-axis at $x=0$ and $x=1$
(d) Between A and O, B and C, $f(x)$ $<1\Rightarrow\sqrt{f(x)}>f(x)$. To the left of A and to the right of C$\sqrt{f(x)}$ $<f(x)$

(e)   A and C are on both graphs.
Thus the graph of $y^2=x(x-1)$ is

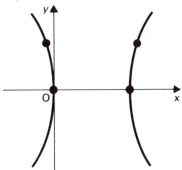

● Exercise 12.3   See page 264.

12.4 Graphical Solutions of Equations and Inequations

Example 12   Sketch the graph of $f(x) = |x-1| + |x+2|$ and hence solve $|x-1| + |x+2| = 3$.

First let us consider the graphs of $y = |x-1|$ and $y = |x+2|$.
The sketch indicates that the behaviour of $f(x)$ is likely to change when $x = -2$ and $x = 1$.

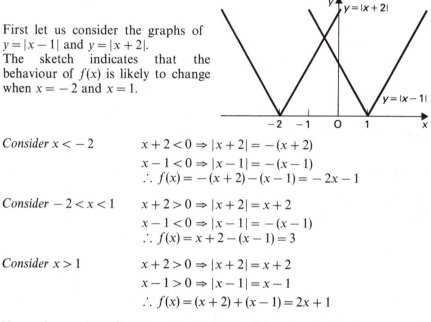

Consider $x < -2$

$$x+2 < 0 \Rightarrow |x+2| = -(x+2)$$
$$x-1 < 0 \Rightarrow |x-1| = -(x-1)$$
$$\therefore f(x) = -(x+2) - (x-1) = -2x - 1$$

Consider $-2 < x < 1$

$$x+2 > 0 \Rightarrow |x+2| = x+2$$
$$x-1 < 0 \Rightarrow |x-1| = -(x-1)$$
$$\therefore f(x) = x+2 - (x-1) = 3$$

Consider $x > 1$

$$x+2 > 0 \Rightarrow |x+2| = x+2$$
$$x-1 > 0 \Rightarrow |x-1| = x-1$$
$$\therefore f(x) = (x+2) + (x-1) = 2x+1$$

Hence the graph of $f(x) = |x-1| + |x+2|$ can now be sketched.
$y = f(x)$ is given by the solid line on the diagram.
From the diagram $|x-1| + |x+2| = 3$ for $x \in [-2, 1]$

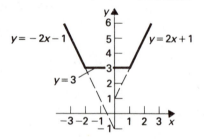

Example 13   Solve $|x| = |(x-2)(x+1)|$ giving solutions to 2 d.p.

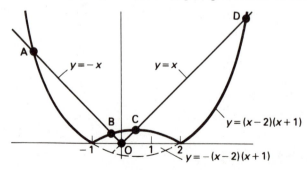

The four solutions of the equations are the x-coordinates of A, B, C, D.

At D $(x-2)(x+1)=x$ (1)   At C $-(x-2)(x+1)=x$ (2)

$\Rightarrow x^2-2x-2=0$ $\Rightarrow$ $x^2=2$

$\Rightarrow$ $x=1\pm\sqrt{3}$ $\Rightarrow$ $x=\pm\sqrt{2}$

Since $x_D>0$ But $x_C>0$

$x_D\approx2.73$ $\therefore x_C\approx1.41$

At B $-(x-2)(x+1)=-x$ At A $(x-2)(x+1)=-x$
which is equivalent to   which is equivalent to
equation (1).   equation (2)

$\therefore x_B=1-\sqrt{3}\approx-0.73$ $\therefore x_A=-\sqrt{2}\approx-1.41$

$\therefore$ solutions are $x=\pm1.41,\ 2.73,\ -0.73$ (to 2.d.p.)

Example 14   Solve $|x|>|(x-2)(x+1)|$.

From the diagram in Example 12, $|x|>|(x-2)(x+1)|$ between A and B and
between C and D.

$$\therefore |x|>|(x-2)(x+1)|$$

for $x\in\ ]-\sqrt{2},1-\sqrt{3}[\cup]\sqrt{2},1+\sqrt{3}[$

Note $x\in\ ]-1.41,-0.73[$ and $x\in\ ]1.41,2.73[$ is not strictly correct. e.g.
$x=2.732$ is less than $1+\sqrt{3}$ and $\therefore 2.732\in[\sqrt{2},1+\sqrt{3}[$
but $2.732\notin\ ]1.41,2.73[$
The answer given is an exact answer. The decimal approximation is not
strictly true.

Example 15   Shade the region defined by $\dfrac{y+1}{x+1}>0$.

Either $y>-1$ and simultaneously $x>-1$ $\left.\vphantom{\dfrac{1}{1}}\right\}$
or $y<-1$ and simultaneously $x<-1$ $\Rightarrow$

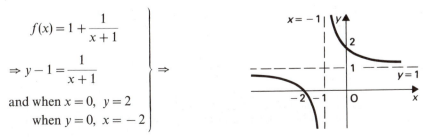

Example 16   Sketch $f(x)=\dfrac{x+2}{x+1}$ and use the graph to solve $\left|\dfrac{x+2}{x+1}\right|<2$.

$f(x)=1+\dfrac{1}{x+1}$

$\Rightarrow y-1=\dfrac{1}{x+1}$ $\Rightarrow$

and when $x=0,\ y=2$
when $y=0,\ x=-2$

$\left|\dfrac{x+2}{x+1}\right| < 2$ can also be written as $-2 < \dfrac{x+2}{x+1} < 2$

Let $\dfrac{x+2}{x+1} = 2 \Rightarrow x+2 = 2x+2 \Rightarrow x = 0$

Let $\dfrac{x+2}{x+1} = -2 \Rightarrow x+2 = -2x-2 \Rightarrow x = -\tfrac{4}{3}$

Hence, from the graph $\left|\dfrac{x+2}{x+1}\right| < 2$ for $x > 0$ and $x < -\tfrac{4}{3}$

●   **Exercise 12.4**   See page 265.

12.5 Composite and Inverse Functions Revisited

Composite Functions

> Met earlier in 5.1, but there the accent was on differentiation.

Example 17   If $f:x \to x+2$, $g:x \to x^2$, $h:x \to \sin x$ find fg, gf, gh, hg, ff, ggg.

$$f(x) = x+2 \qquad g(x) = x^2 \qquad h(x) = \sin x$$

Hence

$$
\begin{aligned}
f(g(x)) &= f(x^2) &&= x^2+2 &&\Rightarrow fg:x \to x^2+2 \\
g(f(x)) &= g(x+2) &&= (x+2)^2 &&\Rightarrow gf:x \to (x+2)^2 \\
g(h(x)) &= g(\sin x) &&= (\sin x)^2 &&\Rightarrow gh:x \to \sin^2 x \\
h(g(x)) &= h(x^2) &&= \sin(x^2) &&\Rightarrow hg:x \to \sin(x^2) \\
f(f(x)) &= f(x+2) &&= x+4 &&\Rightarrow ff:x \to x+4 \\
g(g(g(x))) &= g(g(x^2)) = g((x^2)^2) &&= x^8 &&\Rightarrow ggg:x \to x^8
\end{aligned}
$$

Example 18   If $f:x \to \ln x$, $x > 0$ and $g:x \to x-1$, $x > 1$ find an expression for $f(g(x))$ and sketch its graph. State the domain and range of $f(g(x))$. If the domain of g was changed to $x > 0$ why would the formation of $f(g(x))$ be impossible?

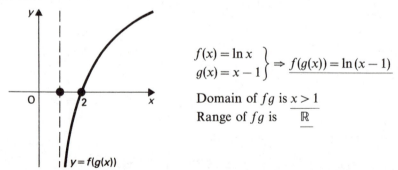

$\left. \begin{aligned} f(x) &= \ln x \\ g(x) &= x-1 \end{aligned} \right\} \Rightarrow f(g(x)) = \ln(x-1)$

Domain of fg is $x > 1$

Range of fg is $\mathbb{R}$

The domain of fg is the domain of g. If the domain of g was $x > 0$ then for $0 < x \leqslant 1$, $x-1$ would be negative and $\ln$ (negative number) does not exist.

Revision of Inverse Functions   Met in 1.4

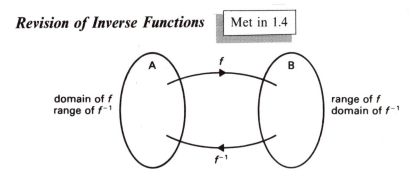

domain of *f*
range of *f*⁻¹

range of *f*
domain of *f*⁻¹

If *f* maps A onto B then the **inverse function** f^{-1} maps B onto A.
The domain of f^{-1} is the range of f.
The range of f^{-1} is the domain of f.

> **Important**   The inverse function f^{-1} only exists if f is one–one for the given domain.

Example 18   Find the inverse functions of (a) $f: x \to x^2, x \geqslant 0$
(b) $g: x \to x^2, x \in \mathbb{R}$, if they exist. If the inverse function does not exist, explain why.

(a) $g: x \to x^2$, $x \geqslant 0$ has   the   graph:
For each value of x there is only one   value   of $y \Rightarrow f$ is   one–one $\Rightarrow f^{-1}$ exists
$\underline{f^{-1}: x \to +\sqrt{x}, x \geqslant 0}$

(b) $g: x \to x^2$, $x \in \mathbb{R}$ has   the   graph:
g is not one–one
e.g. $g(2) = 4 = g(-2)$
$\Rightarrow \underline{g^{-1} \text{ does not exist}}$

Finding More Complex Inverse Functions

Example 19   Find the inverse function of $g(x) = e^{2x} - 3, x \in \mathbb{R}$.

Method 1—using flow charts

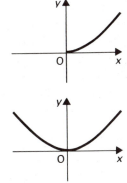

$$x \to \boxed{\times 2} \xrightarrow{2x} \boxed{e^x} \xrightarrow{e^{2x}} \boxed{-3} \to e^{2x-3}$$

Now reversing the flow chart operations

$$\tfrac{1}{2}\ln(x+3) \leftarrow \boxed{\div 2} \xleftarrow{\ln(x+3)} \boxed{\ln x} \xleftarrow{x+3} \boxed{+3} \leftarrow x$$

$\therefore \underline{g^{-1}(x) = \tfrac{1}{2}\ln(x+3)}$

Method 2—using algebraic techniques

Let
$$y = g(x)$$
$$\Rightarrow \quad y = e^{2x} - 3$$
$$\Rightarrow \quad e^{2x} = y + 3$$
$$\Rightarrow \quad 2x = \ln(y + 3)$$
$$\Rightarrow \quad x = \tfrac{1}{2}\ln(y + 3)$$
$$\Rightarrow g^{-1}(x) = \tfrac{1}{2}\ln(x + 3)$$

The notations $g:x \to e^{2x} - 3$ and $g(x) = e^{2x} - 3$ are alternate ways of describing the same function. Either may be used but it is usual to give the answer to a question in the same format as in the question.

Graphs of Inverse Functions

Example 20
On one diagram sketch the graphs of $y = \ln x$, $y = e^x$ and $y = x$. What is the relationship between the graphs? State the domain and range of $\ln x$ and e^x.

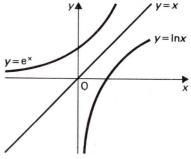

$y = e^x$ is the reflection of $y = \ln x$ in the line $y = x$.

Domain of e^x is $\mathbb{R}$: range of e^x is $]0, \infty[$

Range of $\ln x$ is $\mathbb{R}$: domain of $\ln x$ is $]0, \infty[$

The graph of f^{-1} is the reflection of the graph of f in the line $y = x$.

- Exercise 12.5   See page 266.

12.6 Some Properties of Functions I

1   A function is said to be **even** if $f(x) = f(-x)$ for all values of x. The graphs of all even functions are symmetrical about the y-axis.
For example:

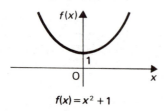

$f(x) = x^2 + 1$

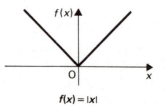

$f(x) = |x|$

2 A function is said to be **odd** if $f(x) = -f(-x)$. The graphs of all odd functions are symmetrical about the origin and the graph is unchanged under a half turn rotation. For example:

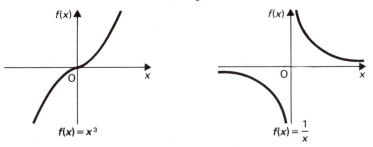

3 A function is said to be **continuous** if its graph can be drawn without removing pencil from paper.

4 A function has a **discontinuity** at $x = a$ if there is a break in the graph at that point. For example: These are examples of **infinite discontinuities.**

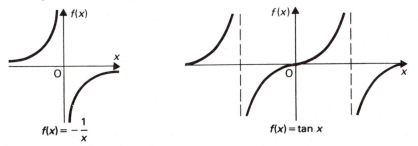

Example 21   Sketch the graph of $f(x) = [x]$ where $[x]$ denotes the greatest integer less than, or equal to, x.

Consider some values: $[2.4] = 2$, $[-0.8] = -1$, $[3] = 3$

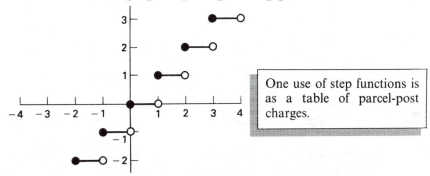

One use of step functions is as a table of parcel-post charges.

This is known as "the step-function" and it has **finite discontinuities.**

5 A function is said to be **not differentiable** at $x = a$ if when $x = a$ there is either a discontinuity or a 'sharp point' on the graph. For all points where the graph is smooth and continuous the function is said to be **differentiable**.

Example 22   For what values of x is f differentiable if (a) $f(x)=|\cos x|$

(b) $f(x)=\dfrac{1}{x}$.

(a)

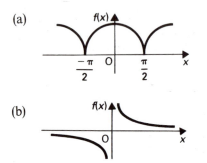

f is differentiable for all values of x

except $\pm\dfrac{\pi}{2},\pm\dfrac{3\pi}{2},\dots$

(b)

f is differentiable for all values of x
except $x=0$.

6 f is said to be **integrable over the interval $[a,b]$** if the area under the curve
can be evaluated for that interval.

Example 23   Which of the following integrals can be evaluated?

(a) $\displaystyle\int_{0}^{1}\dfrac{1}{x}\,dx$ (b) $\displaystyle\int_{1}^{4}[x]\,dx$ (c) $\displaystyle\int_{-1}^{1}|x|\,dx$.

(a)

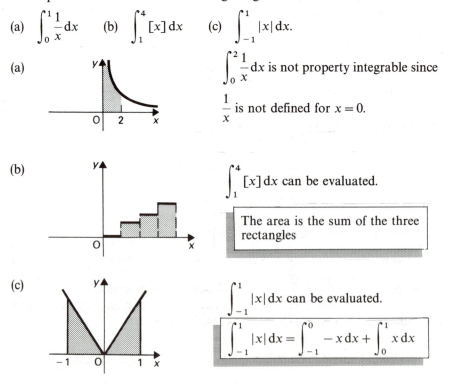

$\displaystyle\int_{0}^{2}\dfrac{1}{x}\,dx$ is not property integrable since

$\dfrac{1}{x}$ is not defined for $x=0$.

(b)

$\displaystyle\int_{1}^{4}[x]\,dx$ can be evaluated.

> The area is the sum of the three
> rectangles

(c)

$\displaystyle\int_{-1}^{1}|x|\,dx$ can be evaluated.

> $\displaystyle\int_{-1}^{1}|x|\,dx=\int_{-1}^{0}-x\,dx+\int_{0}^{1}x\,dx$

7   A function is said to be **periodic** if its graph is a pattern which repeats at
regular intervals. The width of the part that repeats is the **period** of the
function. If $f(x+a)=f(x)$ for all values of x, then a is the period of f.

Example 24   Find the period of tan x.

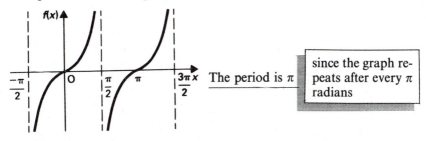

The period is π

since the graph re-peats after every π radians

• Exercise 12.6   See page 267.

12.7 Some Properties of Functions II

Example 25   Without sketching a graph determine whether each of the following functions is odd, even or neither.
(a)   sec x   (b)   x sec x.

(a)   Let $f(x) = \sec x = \dfrac{1}{\cos x}$

then $f(-x) = \dfrac{1}{\cos(-x)} = \dfrac{1}{\cos x} = f(x)$

⇒ sec x is an even function

(b)   Let $f(x) = x \sec x = \dfrac{x}{\cos x}$

then $f(-x) = \dfrac{-x}{\cos(-x)} = -\dfrac{x}{\cos x} = -f(x)$

⇒ x sec x is an odd function

Periods of Trigonometric Functions

sin x has period 2π

sin mx has period $\dfrac{2\pi}{m}$

The graph of sin mx is the graph of sin x with an x-scaling of factor $\dfrac{1}{m}$ – see section 12.2

The other trig. functions can be treated in the same way.

Compound Periodic Functions

Example 26   A function f is periodic with a period of 3. Sketch the graph of the function from -4 to $+6$ given that

$f(x) = x$ for $0 < x \leqslant 2$
$f(x) = 3$ for $2 < x \leqslant 3$

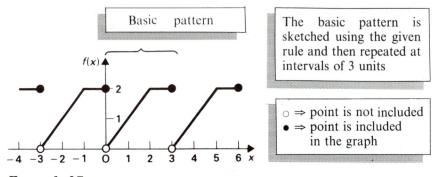

Basic pattern	The basic pattern is sketched using the given rule and then repeated at intervals of 3 units

$\circ \Rightarrow$ point is not included
$\bullet \Rightarrow$ point is included in the graph

Example 27   An even function g is defined by

$$g(x) = x^2 \quad \text{for} \quad 0 < x \leqslant 2$$
$$g(x) = 12 - 4x \quad \text{for} \quad 2 < x \leqslant 3$$

and $g(x + 6) = g(x)$ for all values of x.

Sketch g for the domain $[-3, 9]$. Evaluate $g(20)$.

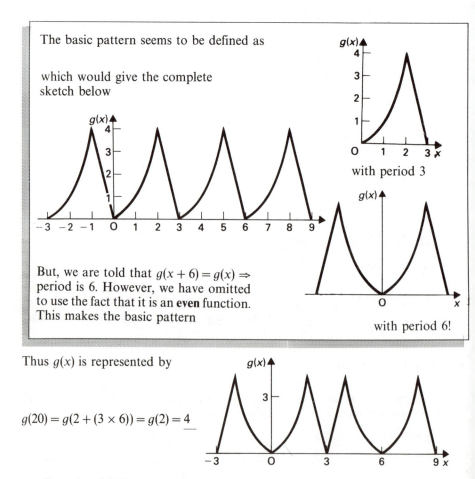

The basic pattern seems to be defined as

which would give the complete sketch below

with period 3

But, we are told that $g(x + 6) = g(x) \Rightarrow$ period is 6. However, we have omitted to use the fact that it is an **even** function. This makes the basic pattern

with period 6!

Thus $g(x)$ is represented by

$$g(20) = g(2 + (3 \times 6)) = g(2) = 4$$

• Exercise 12.7   See page 269.

12.8 + Revision of Unit: A Level Questions

- Miscellaneous Exercise 12B   See page 270

Unit 12 EXERCISES

Exercise 12.1

On separate diagrams, **sketch** each of the curves given by the following equations. Mark on each sketch the coordinates of the vertex and label the axis of symmetry

1 $y = (x-1)(x-3)$ 2 $y = x(x-1)$ 3 $y = (x+1)(x-2)$

4 $y = x^2$ 5 $y = x^2 + 3$ 6 $y = (x-4)^2$

7 $y = x^2 - 4$ 8 $y = (x-3)^2 + \frac{1}{2}$ 9 $y = (x - \frac{1}{2})^2 - \frac{3}{4}$

10 $y = (x-1)^2 + 1$ 11 $y = (x + \frac{3}{2})^2 + \frac{3}{2}$ 12 $y = (x+2)^2 - 3$

Write each of the following quadratic functions in completed square form and sketch the graph of each. Mark and label each vertex

13 $x^2 - 2x + 5$ 14 $x^2 + 4x - 5$ 15 $x^2 - 3x + 4$

16 $x^2 - 2x$ 17 $x^2 - x + \frac{1}{2}$ 18 $x^2 + ax + a^2$

19 Sketch and label each of the following curves:
    (a) $y = x^2$ (b) $y = 2x^2$ (c) $y = \frac{1}{3}x^2$ (d) $y = -x^2$
    (e) $y = -2x^2$ (f) $x = y^2$ (g) $x = 3y^2$ (h) $x = -\frac{1}{2}y^2$

20 Sketch the curves given by each of the functions:
    (a) $2x^2 + 1$ (b) $-3x^2$ (c) $-3x^2 + 2$ (d) $4(x+2)^2 - 3$
    (e) $5 - 2(x-2)^2$

21 Express each of the following functions in completed square form and sketch the graph of each:
    (a) $x^2 - 12x + 25$ (b) $2x^2 + 8x$ (c) $4x - x^2$
    (d) $4x^2 - 12x + 11$

22 Express each of the following functions in completed square form. For each function, sketch the graph and state the range of the function
    (a) $4x^2 - 12$ (b) $1 + x - 3x^2$ (c) $5x^2 + 9x$
    (d) $4x - 6x^2 - 9$

23 Show how the graph of $g = 2 - 3(x-1)^2$ may be obtained from the graph of $y = x^2$ by appropriate translations and scalings. Illustrate your answer with a sequence of sketches

24 Find the range of each of the following functions:
    (a) $x^2 - 3$ (b) $x^2 - 4x$ (c) $2x^2 + 2x + 3$ (d) $7 - 5x - x^2$

*25 Use sketches to solve the following:

(a) $x^2 - 2x < 0$ (b) $4 - x^2 \geqslant 0$ (c) $x^2 + 8 \leqslant 6x + 3$
(d) $4x + 1 < x^2 + 4$ (e) $3x^2 + 2x < 1$ (f) $2x^2 + 7x + 3 \geqslant 0$

*26 Sketch the graph of the function $3x^2 - 12x + 3$. Find the range of values of c for which the function $3x^2 - 12x + c$ is positive for all values of x

Exercise 12.2

Sketch the graphs of the following functions:

1 (a) $y = \dfrac{1}{x^2}$ (b) $y = \dfrac{1}{x^2} + 4$ (c) $y = \dfrac{1}{(x + 4)^2}$

  (d) $y = \dfrac{1}{(x - 1)^2} + 1$

2 (a) $y = |x|$ (b) $y = 2|x|$ (c) $y = |x| - 1$ (d) $y = |x - 1|$

3 (a) $y = \cos x$ (b) $y = 2 + \cos x$ (c) $y = \cos(x - \pi/3)$
  (d) $y = \cos x/4$ where $x \in [0, 2\pi]$

4 (a) $y = \tan x$ (b) $y = \tan(x + \pi/2)$
  (c) $y = 2 + \tan(x - \pi/2)$ where $x \in [0, 2\pi]$

5 (a) $y = \ln x$ (b) $y = 2\ln x$ (c) $y = \ln(x + 3)$

6 (a) $y = e^x$ (b) $y = e^{(x-2)}$ (c) $y = 2e^x$ (d) $y = e^{x/2}$

Sketch the following curves, labelling the asymptotes, and the points where they cross the axes:

7 $y = \dfrac{1}{x - 1}$ **8** $y = \dfrac{2}{x - 4}$ **9** $y = \dfrac{4}{2x - 5}$

10 $y = \dfrac{x}{x - 1}$ **11** $y = \dfrac{x - 1}{x + 1}$ **12** $y = \dfrac{2x + 3}{x - 1}$

13 $y = \dfrac{1 - 3x}{x - 1}$ ***14** $y = \dfrac{2x + 5}{2x - 5}$ ***15** $y = \dfrac{3 - 2x}{2 - 3x}$

Exercise 12.3

Sketch graphs of each of the following:

1 $y = |x|$ **2** $y = 2|x|$ **3** $y = |x| - 1$

4 $y = |x - 1|$ **5** $y = -|x|$ **6** $y = 2 - |x|$

7 $y = 2|x| - 1$ ***8** $y = |2x - 1|$ **9** $y = |1 - x^2|$

10 $y = |\sin x|, \quad x \in [0, 2\pi]$ **11** $y = \sin |x|, \quad x \in [-\pi, \pi]$

12 $y = |(x+1)(x-3)|$

For questions 13–15 sketch the graph of $y = f(x)$ $y = 1/f(x)$ and $y^2 = f(x)$:

13 $f(x) = x^2 - 2x - 3$ **14** $f(x) = x^2 + 1$ **15** $f(x) = \sin x, \quad x \in [0, \pi]$

For questions 16–18 sketch the graphs of $y = f(x)$ and $y = |f(x)|$:

16 $f(x) = 5 - 2x$ **17** $f(x) = x^2 - 3x + 2$ **18** $f(x) = 6x - x^2 - 9$

19 $f(x) = 1/x + 3$

Sketch graphs of the following:

20 $y = \dfrac{x+3}{x+2}$ **21** $y = \left| \dfrac{x+3}{x+2} \right|$ **22** $y = \dfrac{|x|+3}{|x|+2}$

*23 $y = x^2 - |x| - 12$ [Hint: $|x|^2 \equiv x^2$]

24 $y = 2x^2 - 7|x| + 3$ **25** $y = |x^2 + 3x|$ **26** $y = x^2 + 3|x|$

27 $y = \left| \dfrac{2x+1}{x-1} \right|$ **28** $y = \dfrac{2|x|+1}{|x|-1}$

For questions 29–31 sketch graphs of $y = g(x)$, $y = |g(x)|$ and $y = 1/g(x)$.

29 $g(x) = x(x-1)(x-2)$ **30** $g(x) = \ln x$

31 $g(x) = \tan x, \quad x \in \,] -\pi/2, 3\pi/2 [$

Exercise 12.4

1 Sketch the graphs of (a) $y = |x| + |x-2|$ (b) $y = |x-1| + |2x-1|$

Solve the following equations:

2 $|x| = |x-2|$ **3** $|x-1| = |x+3|$

*4 $2 - |x+1| = |4x-3|$ **5** $|x^2-1| = |3x-2| + 1$

Solve the following inequations:

6 $|x| < |x-2|$ **7** $|x-1| > |x+3|$

8 $2 - |x+1| < |4x-3|$ **9** $|x^2-1| > |3x-2| + 1$

10 Sketch the graph of $f(x) = \dfrac{x-2}{x+2}$ and use it to solve the inequality

$$\frac{x-2}{x+2} < 2$$

11 Sketch the graph of $f(x) = \dfrac{x-1}{x+2}$ and use it to solve $\left|\dfrac{x-1}{x+2}\right| > 2$

12 Shade the regions given by:

(a) $\dfrac{x+1}{y-1} > 0$ (b) $\dfrac{y+2}{x-1} < 0$ (c) $\dfrac{x+y}{x-1} < 1$

*13 Find the set of values of x for which $|x-1| - |2x+1| > 0$ (L)

*14 Sketch the graphs of the functions f and g given by

(a) $f(x) = \dfrac{3x-1}{2x-3}$, x real, $x \neq \frac{3}{2}$

(b) $g(x) = \dfrac{3|x|-1}{2|x|-3}$, x real, $|x| \neq \frac{3}{2}$

Two geometrical progressions have common ratios

$\dfrac{3a-1}{2a-3}$ and $\dfrac{3|a|-1}{2|a|-3}$, respectively, where a is real.

Find, in each case, the set of values of a for which the progression has a sum to infinity   (J)

Exercise 12.5

1 If $f:x \to x^3, g:x \to x-3, h:x \to \sin x$, find $fg, gf, hf, hg, fh, gh, fff, gg, g^{-1}$

2 If $f:x \to 2x, x \in [0, 2\pi]$ and $g:x \to \sin x, x \in \mathbb{R}$ find an expression for $g(f(x))$ and sketch the graph of $y = g(f(x))$. What is the domain of gf?

3 If $f(x) = \dfrac{x+2}{x+1}$ show that $f(f(x)) = \dfrac{3x+4}{2x+3}$ and find $f^{-1}(x)$

In questions 4–7 sketch the graph of the function and the graph of its inverse on the same diagram. State the range of the given function.

4 $y = x^3$, $x \in [0, 3]$ **5** $y = -x^3$, $x \in [0, 3]$

6 $y = \sin x$, $x \in [-\pi/2, \pi/2]$ **7** $y = \cos x$, $x \in [0, \pi]$

8 Find the inverses of
(a) $f(x) = x - 3$, $x \in \mathbb{R}$
(b) $g(x) = \ln(x-1)$, $x \in]1, \infty[$

In questions 9–16 sketch and label the graph of the function. State whether the inverse exists. If it does not exist give brief reasons why. If it does exist, sketch the inverse function on the same diagram as the original function. State the domain and range of each inverse function.

9 $f(x) = x^2$, $x \in [0, \infty[$ **10** $g(x) = x^2$, $x \in \mathbb{R}$

11 $h(x) = x^2 + 1, \quad x\in[0, \infty[$ **12** $j(x) = x^2 - 1, \quad x\in[0, \infty[$

13 $k(x) = (x - 1)^2, \quad x\in[0, 3]$ **14** $l(x) = \ln(x) + 1, \quad x\in[0, \infty[$

15 $m(x) = e^x + 2, \quad x\in\mathbb{R}$ **16** $n(x) = \sin x, \quad x\in[0, 2\pi]$

17 The function f is defined by

$$f:x\mapsto\frac{x + 3}{x - 1}, \quad x\in\mathbb{R}, \quad x\neq 1$$

Find

(a) the range of f
(b) $ff(x)$
(c) $f^{-1}(x)$ (L)

18 The functions f and g, each with domain D, where

$$D = \{x: x\in\mathbb{R} \quad \text{and} \quad 0\leqslant x\leqslant\pi\}$$

and defined by

$$f:x\to\cos x \quad \text{and} \quad g:x\to x - \tfrac{1}{2}\pi$$

Write down and simplify an expression for $f[g(x)]$, giving its domain of definition. Sketch the graph of $y = f[g(x)]$ (L)

*19 The function f, with domain $x > -3$, is defined by $f(x) = \ln(x + 3) - 1$. Find $f^{-1}(x)$ and state the domain and range of f^{-1}. Explain how the graph of $y = f^{-1}(x)$ may be constructed from the graphs of $y = f(x)$ and $y = x$. Sketch the graphs of $y = f(x)$ and $y = f^{-1}(x)$ on the same axes. Show that the x-coordinates of the points of intersection of the graphs satisfy the equation $\ln(x + 3) = x + 1$ (J)

Exercise 12.6

For each of the following functions, sketch its graph and state whether it is continuous. If it is not continuous, state whether its discontinuity is finite or infinite

1 e^x **2** $\ln x$ **3** $\ln|x|$ **4** $\pm\sqrt{x}$

5 $\dfrac{2}{x}$ **6** $|x + 1|$ **7** $\cot x$ **8** $\tan^{-1}x$

For each of the following functions:

(a) sketch its graph for $-\pi\leqslant x\leqslant\pi$
(b) state whether it is odd, even, or neither
(c) state whether it is periodic and, if it is, give its period

9 $\sin x$ **10** $\sin 2x$ **11** $\cos x$

12 $\cos x/2$ **13** $|\sin x|$ **14** $|\tan x|$

15 For what values of x is $f(x)$ differentiable if:

(a) $f(x) = |\sin x|$ (b) $f(x) = \tan x$ (c) $f(x) = \ln x$

16 State whether each of the following integrals can be evaluated:

(a) $\displaystyle\int_{-1}^{1} \frac{1}{x^2}\,dx$ (b) $\displaystyle\int_{0}^{\pi} \tan x\,dx$ (c) $\displaystyle\int_{-2}^{0} |x+1|\,dx$

[Do not evaluate any of the integrals]

17 A function g is defined by

$$g(x) = \cos x \quad \text{for } x \geqslant 0$$
$$g(x) = x^2 + k \quad \text{for } x < 0$$

For what value of k is g a continuous function?

18 A periodic function f, of period 2, is defined by

$$f(x) = 0 \quad 0 < x \leqslant 1$$
$$f(x) = 1 \quad 1 < x \leqslant 2$$

Evaluate $f(\frac{1}{2})$, $f(2)$, $f(2\frac{1}{2})$, $f(3)$, $f(-1)$

19 Evaluate the following integrals: (a) $\displaystyle\int_{1}^{2} [x]\,dx$ (b) $\displaystyle\int_{2}^{3} |x-1|\,dx$

For each of the following graphs, state whether it represents a function that is:

(a) continuous
(b) odd, even or neither
(c) periodic

20

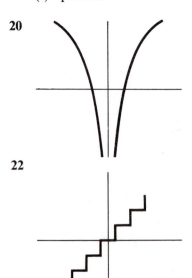

21

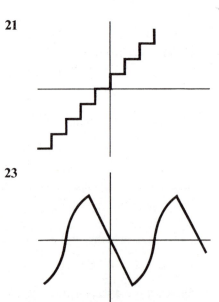

22

23

24

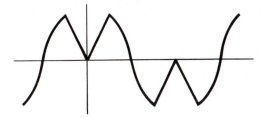

*25 Sketch the graph of $x - [x]$. State whether it is (a) odd   (b) even
(c) periodic

Exercise 12.7

1 State the period of each of the following functions:
(a) $\cos x$ (b) $\tan x$ (c) $\cot x$ (d) $\sin 3x$

(e) $\sec x$ (f) $\tan(x - \alpha)$ (g) $\sin 4x$ (h) $\cos \dfrac{x}{3}$

2 Without sketching a graph, determine whether each of the following
functions is odd, even or neither:
(a) $\tan x$ (b) $\tan|x|$ (c) x^3 (d) $x^3 + 1$
(e) $x \cos x$ (f) $x^2 \cos x$

3 Express $1 + 2\cos^2 x$ in terms of $\cos 2x$. Hence deduce the period and range
of $1 + 2\cos^2 x$

4 A function f is defined by
$$f(x) = 9 - x^2 \qquad \text{for } 0 < x \leqslant 2$$
$$f(x) = 2x + 1 \qquad \text{for } 2 < x \leqslant 5$$
and $f(x + 5) = f(x)$ for all x.
Sketch f for the domain $[-3, 10]$.

Evaluate $f(20)$ and $\displaystyle\int_0^7 f(x)\,dx$

5 The function f is defined by
$$f(x) = \sin x \qquad \text{for } x \geqslant 0$$
$$f(x) = -x \qquad \text{for } x < 0$$
Sketch the graphs of $f(x)$ and $f'(x)$ for $-\pi < x < \pi$.
Is f differentiable at $x = 0$? Are f and f' continuous at $x = 0$?

6 An even function f, of period π, is defined by
$$f(x) = 4x^2 \qquad \text{for } 0 \leqslant x \leqslant \pi/4$$
$$f(x) = \pi^2/4 \qquad \text{for } \pi/4 < x \leqslant \pi/2$$
Sketch the graph of f for $-\pi \leqslant x \leqslant \pi$ (L)

7 A periodic function for which $f(x + 2\pi) \equiv f(x)$ is defined in the interval $-\pi < x \leqslant \pi$ by

$$f(x) = \sin x \qquad -\pi < x < -\pi/2$$
$$f(x) = \cos x \qquad -\pi/2 \leqslant x \leqslant \pi/2$$
$$f(x) = \sin x \qquad \pi/2 < x \leqslant \pi$$

Obtain the values of $f(11\pi/4)$, $f(-9\pi/4)$ and $f(-9\pi/2)$ (L)

8 The function $f(x)$ is periodic with period π and

$$f(x) = \sin x \qquad \text{for } 0 < x \leqslant \pi/2$$
$$f(x) = \frac{2(\pi - x)}{\pi} \qquad \text{for } \pi/2 < x \leqslant \pi$$

Sketch the graph of the function for $-\pi < x < 2\pi$ (L)

9 State the relation between $f(x)$ and $f(-x)$ when (a) f is an even function (b) f is an odd function. In each case describe a geometrical transformation which maps the graph of $y = f(x)$ onto itself.

The function g has domain $\mathbb{R}$; it is even, periodic with period, π and for $0 \leqslant x \leqslant \frac{1}{2}\pi$, $g(x) = \sin x$. Sketch a graph of $y = g(x)$ for $-2\pi \leqslant x \leqslant 2\pi$. State the domain of the derived function g' (J)

10 For each of the functions f and g defined below determine whether it is
(a) odd   (b) even   (c) neither. Give reasons for your answers.

(i) $f(x) = \cos ax + \sin ax$, where x is real and a is a real and non-zero constant.

(ii) $g(x) = \dfrac{e^x}{1 + e^{2x}}$ where x is real   (J)

11 Determine the period of the function f defined by

$$f(x) = \sin^2 \left(\frac{x}{3} \right) \qquad x \in \mathbb{R}$$ (J)

12 The function f is periodic with period 2 and

$$f(x) = 1 - \cos \pi x \qquad \text{for } -1 < x \leqslant 0$$
$$f(x) = \lambda \ln(1 + x) \qquad \text{for } 0 < x \leqslant 1$$

Find the value of λ which makes f continuous for all x (L)

Miscellaneous Exercise 12B

1 Find the ranges of the functions f, g, h whose definitions are as given:
(a) $f(x) = x^2 + 6x + 11$ all real x
(b) $g(x) = x^2 - 8x + 12$ $1 \leqslant x \leqslant 5$
(c) $h(x) = \cos x + \sin x$ all real x (J: 12.1, 12.2, 7.2)

2   Using the same axes sketch the curves

$$y = \frac{1}{x-1}, \quad y = \frac{x}{x+3}$$

giving the equations of the asymptotes. Hence, or otherwise, find the set of values of x for which

$$\frac{1}{x-1} > \frac{x}{x+3} \qquad\qquad \text{(L: 12.2, 12.4)}$$

3   Solve the inequality $x^2 - |x| - 12 < 0$ (J: 12.3, 12.4)

4   Sketch a graph of the function f given by

$$f(x) = 2|x| + |x-1| \qquad x \text{ real}$$

Solve the equation $f(x) = 4$ (J: 12.4)

5   Find the set of real values of x for which $|x+4| < |x+3|$ (L: 12.4)

6   Given that $f(x) \equiv x^2 + 4x + 5$, find the least value of $f(x)$. With the same axes, sketch the curves $y = f(x)$ and $y = \dfrac{1}{f(x)}$ (L: 12.3)

7   Write down, or obtain, the stationary points of the curve whose equation is

$$y = p(x+q)^2 + r$$

where $p\,(\neq 0)$, q and r are real constants. Obtain the condition, in terms of p and r, that the curve does not meet the x-axis.

Given that

$$y = 4x^2 - 6x + 5$$

find the numerical values of p, q and r and sketch the curve given by this equation   (J: 12.1)

8   Find the set of real values of x for which $|x-2| - 2|2x-1| > 0$ (L: 12.4)

9   Sketch the curve $y = 1 + 2\cos x$ for $-\dfrac{2\pi}{3} \leqslant x \leqslant \dfrac{2\pi}{3}$.

The region between the curve and the x-axis is rotated through $360°$ about the x-axis. Find the volume generated, leaving your answer in terms of π (A: 12.1, 5.5)

10   The function g is defined by

$$g: x \to \frac{2x+5}{x-3} \qquad x \in \mathbb{R} \qquad x \neq 3$$

Sketch the graph of the function g. Find an expression for $g^{-1}(x)$, specifying its domain   (L: 12.2, 12.5)

11 Sketch a graph of the function f given by

$$f(x) = \frac{x-2}{x+1} \qquad x \text{ real} \qquad x \neq -1$$

Hence, or otherwise, solve $\left|\dfrac{x-2}{x+1}\right| < 2$

On a separate diagram, sketch a graph of the inverse function f^{-1}

(J: 12.2, 12.4, 12.5)

12 The functions f, g have domain $\mathbb{R}$ and are defined by

$$f: x \to \sin \tfrac{1}{2}x \qquad g: x \to 3x \sin \tfrac{1}{2}x$$

State which of these functions are
(a) odd   (b) even   (c) periodic, giving the period. State the value of $f'(0)$ and sketch, with the same axes, the graph of f for $-\pi \leqslant x \leqslant \pi$ and the graph of f^{-1} for $-1 \leqslant x \leqslant 1$ (L: 12.6)

13 Find the set of real values of x for which $|x-2| > 2|x+1|$ (L: 12.4)

14 Given that $f(x) \equiv x^2 - 6x + 10$, show that $f(x) > 0$ for all real values of x.

Using the same axes sketch the graphs of $y = f(x)$ and $y = 1/f(x)$

(L: 12.3)

15 The functions f and g are defined by

$$f: x \mapsto 2 + x - x^2 \qquad x \in \mathbb{R}$$

$$g: x \mapsto \frac{1}{1 + \tan x} \qquad 0 \leqslant x < \pi/2$$

Determine the range of each function and state, in each case, whether or not an inverse function exists   (L: 12.5)

16 The function f is defined for $x \geqslant 0$ by $f(x) = 1 + \ln(2 + 3x)$. Find

(a) the range of f
(b) an expression for $f^{-1}(x)$
(c) the domain of f^{-1}

On one diagram sketch the graphs of f and f^{-1}. Show that the x-coordinate of the point of intersection of the graphs of f and f^{-1} satisfies the equation

$$\ln(2 + 3x) = x - 1 \qquad \text{(J: 9.3, 12.2, 12.5)}$$

17 (a) Sketch the curve $y^2 = (1-x)^2(x+3)$ which is defined for $x \geqslant -3$

(b) Show that the area of the loop is $2 \displaystyle\int_{-3}^{1} (1-x)\sqrt{(x+3)}\, dx$. By means of the substitution $u^2 = x + 3$, or otherwise, find the area of the loop
(A: 5.4, 11.3, 12.3)

18 Show that there is no real value of x for which

$$|x| + |x - 2| = x^2 - 2x + 4$$

and sketch on the same axes the graphs of

(a) $y = |x| + |x - 2|$, and
(b) $y = x^2 - 2x + 4$

Shade the region within which all the following inequalities are satisfied:

(a) $y > 2$ (b) $y < x^2 - 2x + 4$ (c) $y > 2x - 2$ (d) $y > 2 - 2x$
(J: 12.4)

19 Obtain the three sets of values of x for which

(a) $x > \dfrac{1}{x}$

(b) $\dfrac{1}{x - 1} > \dfrac{x}{2 - x}$

(c) $3|x - 1| > |x + 1|$ (L: 10.5, 12.4)

20 The functions f and g are defined by

$$f(x) = \sin 2x \qquad x \in \mathbb{R}$$
$$g(x) = \cot x \qquad x \in \mathbb{R} \qquad x \neq k\pi \, (k \in \mathbb{Z})$$

State the periods of f and g

Find the period of the function $f \cdot g$

On separate axes sketch the graphs of f, g and $f \cdot g$ for the interval $\{x: -\pi < x < \pi, x \neq 0\}$
Find the range of the function $f \cdot g$ (J: 12.6)

*21 Given that $y = ax^2 + bx + c$, that $y = 8$ when $x = 1$, and that $y = 2$ when $x = -1$, show that $b = 3$ and find a in terms of c

Find the range of values of c for which $y > 0$ for all real values of x
(J: 1.7, 12.1)

*22 Given that

$$f(x) \equiv 2 - \frac{3}{x^2 - 2x + 4}$$

Show that $f(x)$ is always positive

Sketch the graph of the curve $y = f(x)$, stating the equations of any asymptotes and the coordinates of any points of intersection with the coordinate axes (L: 12.1, 12.2, 12.3)

13.1 Introduction

The Modulus of a Vector is its magnitude.
The modulus of **a** is written as $|\mathbf{a}|$ or a.
The modulus of $\overrightarrow{PQ}$ is written as $|\overrightarrow{PQ}|$ or PQ.

Unit Vectors are vectors with unit magnitude.
$\hat{\mathbf{a}}$ is the unit vector in the direction of **a**. Hence $\mathbf{a} = a\hat{\mathbf{a}}$.
i, **j**, **k** are unit vectors parallel to the axes Ox, Oy, Oz.

Position Vectors

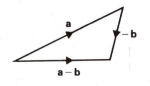

$$\overrightarrow{OP} = a\mathbf{i} + b\mathbf{j} + c\mathbf{k} = \begin{pmatrix} a \\ b \\ c \end{pmatrix}$$

$$|\overrightarrow{OP}| = \sqrt{a^2 + b^2 + c^2}$$

Sum of Two Vectors

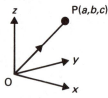

$$\text{and} \quad \begin{pmatrix} a \\ b \\ c \end{pmatrix} + \begin{pmatrix} p \\ q \\ r \end{pmatrix} = \begin{pmatrix} a+p \\ b+q \\ c+r \end{pmatrix}$$

Difference Between Two Vectors

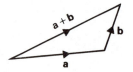

$$\text{and} \quad \begin{pmatrix} a \\ b \\ c \end{pmatrix} - \begin{pmatrix} p \\ q \\ r \end{pmatrix} = \begin{pmatrix} a-p \\ b-q \\ c-r \end{pmatrix}$$

Parallel Vectors $\begin{pmatrix} ka \\ kb \\ kc \end{pmatrix} = k \begin{pmatrix} a \\ b \\ c \end{pmatrix}$

$\begin{pmatrix} ka \\ kb \\ kc \end{pmatrix}$ is parallel to $\begin{pmatrix} a \\ b \\ c \end{pmatrix}$ and k times its length.

Example 1 $\mathbf{r}_1 = 3\mathbf{i} + 2\mathbf{j} - \mathbf{k}$, $\mathbf{r}_2 = 2\mathbf{i} - \mathbf{j}$. Find the unit vector in the direction of $2\mathbf{r}_1 - \mathbf{r}_2$.

$$2\mathbf{r}_1 - \mathbf{r}_2 = 2\begin{pmatrix} 3 \\ 2 \\ -1 \end{pmatrix} - \begin{pmatrix} 2 \\ -1 \\ 0 \end{pmatrix} = \begin{pmatrix} 4 \\ 5 \\ -2 \end{pmatrix}$$

The length of $2\mathbf{r}_1 - \mathbf{r}_2 = |2\mathbf{r}_1 - \mathbf{r}_2| = \sqrt{16 + 25 + 4} = \sqrt{45} = 3\sqrt{5}$

$\therefore$ unit vector $= \dfrac{1}{3\sqrt{5}}(4\mathbf{i} + 5\mathbf{j} - 2\mathbf{k})$

Example 2   If $a\mathbf{i} + 5\mathbf{j} - c\mathbf{k}$ is parallel to $4\mathbf{i} - 10\mathbf{j} + \mathbf{k}$, find a and c.

$$a\mathbf{i} + 5\mathbf{j} - c\mathbf{k} = \mu(4\mathbf{i} - 10\mathbf{j} + \mathbf{k})$$

Equating $\mathbf{j}$ components: $5 = -10\mu \Rightarrow \mu = -\tfrac{1}{2}$
Equating $\mathbf{i}$ components: $a = 4\mu = -2$
Equating $\mathbf{k}$ components: $-c = \mu$

$\left.\begin{array}{c}\\ \\ \\\end{array}\right\}$ $\therefore$ $a = -2$ $c = \tfrac{1}{2}$

Example 3
If $\mathbf{X}$ is the midpoint of OP and $\overrightarrow{OX} = \mathbf{a}$, $\overrightarrow{OQ} = \mathbf{b}$ express $\overrightarrow{QP}$ in terms of $\mathbf{a}$ and $\mathbf{b}$.

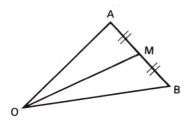

$$\overrightarrow{QP} = \overrightarrow{QO} + \overrightarrow{OP} = -\overrightarrow{OQ} + 2\overrightarrow{OX} = 2\mathbf{a} - \mathbf{b}$$

Example 4

If M is the midpoint of AB prove that $\overrightarrow{OM} = \tfrac{1}{2}(\overrightarrow{OA} + \overrightarrow{OB})$.

$$\overrightarrow{OM} = \overrightarrow{OA} + \overrightarrow{AM}$$
$$= \overrightarrow{OA} + \tfrac{1}{2}\overrightarrow{AB}$$
$$= \overrightarrow{OA} + \tfrac{1}{2}(\overrightarrow{AO} + \overrightarrow{OB})$$
$$= \overrightarrow{OA} + \tfrac{1}{2}(-\overrightarrow{OA} + \overrightarrow{OB})$$
$$\therefore \overrightarrow{OM} = \tfrac{1}{2}(\overrightarrow{OA} + \overrightarrow{OB}) \qquad \text{Q.E.D.}$$

Note   This result may be quoted and used in questions but you should know the technique for finding it.

- Exercise 13.1   See page 288.

13.2 The Scalar Product (Dot Product)

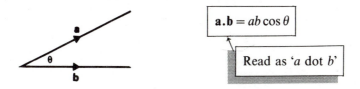

$$\boxed{\mathbf{a}.\mathbf{b} = ab\cos\theta}$$

Read as '*a* dot *b*'

Properties of the Scalar Product

1 $\mathbf{a}.\mathbf{b} = ab \cos \theta = ba \cos \theta = \mathbf{b}.\mathbf{a}$.

2 $\mathbf{a}.(\mathbf{b} + \mathbf{c}) = \mathbf{a}.\mathbf{b} + \mathbf{a}.\mathbf{c}$.

3   **a** is perpendicular to $\mathbf{b} \Rightarrow \mathbf{a}.\mathbf{b} = 0$. In particular, $\mathbf{i}.\mathbf{j} = \mathbf{j}.\mathbf{k} = \mathbf{k}.\mathbf{i} = 0$.

4   **a** is parallel to $\mathbf{b} \Rightarrow \mathbf{a}.\mathbf{b} = ab$.

5 $\mathbf{a}.\mathbf{a} = a^2$. In particular $\mathbf{i}.\mathbf{i} = \mathbf{j}.\mathbf{j} = \mathbf{k}.\mathbf{k} = 1$.

Calculation of the Scalar Product

$$\begin{pmatrix} a_1 \\ b_1 \\ c_1 \end{pmatrix} \cdot \begin{pmatrix} a_2 \\ b_2 \\ c_2 \end{pmatrix} = a_1 a_2 + b_1 b_2 + c_1 c_2$$

To Find the Resolved Part (Component) of a Vector in a Given Direction

The component of **F** in the direction of **d** is $F \cos \theta$. But, $\mathbf{F}.\mathbf{d} = (F \cos \theta)d$.

$$\therefore \quad \text{component of } \mathbf{F} \text{ in direction of } \mathbf{d} = \frac{\mathbf{F}.\mathbf{d}}{d}$$

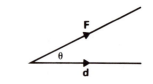

Example 5   Find the resolved part of $\mathbf{i} + 2\mathbf{j} + \mathbf{k}$ in the direction of $\mathbf{i} - \mathbf{j}$.

$$\text{Resolved part} = \frac{\begin{pmatrix} 1 \\ 2 \\ 1 \end{pmatrix} \cdot \begin{pmatrix} 1 \\ -1 \\ 0 \end{pmatrix}}{\sqrt{1+1+0}} = \frac{1}{\sqrt{2}}[1 - 2 + 0] = -\frac{1}{\sqrt{2}}$$

To Find the Angle Between Two Vectors

$$\mathbf{a}.\mathbf{b} = ab \cos \theta \Rightarrow \boxed{\cos \theta = \frac{\mathbf{a}.\mathbf{b}}{ab}}$$

Example 6   Find the cosine of the angle between $2\mathbf{i} - 3\mathbf{j} + 4\mathbf{k}$ and $2\mathbf{i} + 2\mathbf{j} + \mathbf{k}$.

$$\cos \theta = \frac{\begin{pmatrix} 2 \\ -3 \\ 4 \end{pmatrix} \cdot \begin{pmatrix} 2 \\ 2 \\ +1 \end{pmatrix}}{\sqrt{4+9+16}\sqrt{4+4+1}} = \frac{4 - 6 + 4}{\sqrt{29}\sqrt{9}}$$

$$\therefore \cos \theta = \frac{2}{3\sqrt{29}}$$

Example 7   Prove that $\cos\theta\mathbf{i} + \sin\theta\mathbf{j} + 3\mathbf{k}$ is perpendicular to $\sin\theta\mathbf{i} - \cos\theta\mathbf{j}$.

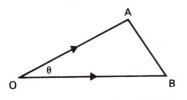

$$\begin{pmatrix} \cos\theta \\ \sin\theta \\ 3 \end{pmatrix} \cdot \begin{pmatrix} \sin\theta \\ -\cos\theta \\ 0 \end{pmatrix} = \sin\theta.\cos\theta - \sin\theta.\cos\theta + 3.0 = 0$$

$\therefore$ since the scalar product equals zero, the vectors are perpendicular

Example 8   If A is $(1, 2, 3)$, B is $(2, 2, 0)$ and O is the origin, find the scalar product of $\overrightarrow{OA}$ and $\overrightarrow{OB}$ and hence find the area of $\triangle$ OAB.

$$\overrightarrow{OA}.\overrightarrow{OB} = \begin{pmatrix} 1 \\ 2 \\ 3 \end{pmatrix} \cdot \begin{pmatrix} 2 \\ 2 \\ 0 \end{pmatrix} = 2 + 4 = \underline{6}$$

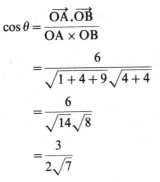

Also $\cos\theta = \dfrac{\overrightarrow{OA}.\overrightarrow{OB}}{OA \times OB}$

| Note   To find $\cos\theta$ your must consider $\overrightarrow{OA}.\overrightarrow{OB}$ or $\overrightarrow{AO}.\overrightarrow{BO}$. Taking $\overrightarrow{AO}.\overrightarrow{OB}$ gives the cosine of $(\pi - \theta)$. |

$$= \dfrac{6}{\sqrt{1+4+9}\sqrt{4+4}}$$

$$= \dfrac{6}{\sqrt{14}\sqrt{8}}$$

$$= \dfrac{3}{2\sqrt{7}}$$

Area of $\triangle$ OAB is

$$\tfrac{1}{2}OA \times OB \sin\theta$$

$$= \tfrac{1}{2}OA \times OB \times \sqrt{1 - \cos^2\theta}$$

$$= \tfrac{1}{2} \times \sqrt{14} \times \sqrt{8} \times \sqrt{1 - \tfrac{9}{28}}$$

$$= \tfrac{1}{2} \times \sqrt{14} \times \sqrt{8} \times \dfrac{\sqrt{19}}{\sqrt{28}} = \sqrt{19}$$

$\therefore$ Area of $\triangle$ OAB $= \sqrt{19}$

● **Exercise 13.2**   See page 289.

13.3 Vector Equation of a Line

Case 1: Given a Point on the Line and a Direction Vector

If P is a variable point on the line then

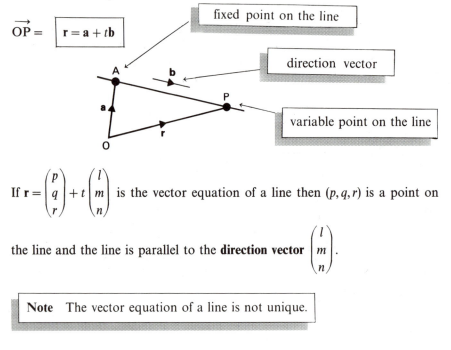

$$\overrightarrow{OP} = \boxed{\mathbf{r} = \mathbf{a} + t\mathbf{b}}$$

If $\mathbf{r} = \begin{pmatrix} p \\ q \\ r \end{pmatrix} + t \begin{pmatrix} l \\ m \\ n \end{pmatrix}$ is the vector equation of a line then (p, q, r) is a point on

the line and the line is parallel to the **direction vector** $\begin{pmatrix} l \\ m \\ n \end{pmatrix}$.

Note   The vector equation of a line is not unique.

Case 2: Given Two Points on the Line

$$\mathbf{r} = \mathbf{a} + t(\overrightarrow{AB})$$

$$\Rightarrow \boxed{\mathbf{r} = \mathbf{a} + t(\mathbf{b} - \mathbf{a})}$$

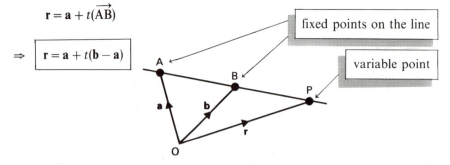

Example 9   Find, in their simplest forms, the vector equations of the following lines:

(a)   the line passing through the point $(3, 1, 2)$ parallel to $2\mathbf{i} - 8\mathbf{j} + 4\mathbf{k}$
(b)   the line passing through the points $(1, -3, 2)$ and $(5, 2, -1)$
(c)   the line passing through the origin and the point $(0, \frac{1}{2}, \frac{3}{2})$

(a) $\mathbf{r} = \begin{pmatrix} 3 \\ 1 \\ 2 \end{pmatrix} + t' \begin{pmatrix} 2 \\ -8 \\ 4 \end{pmatrix}$ which simplifies to $\mathbf{r} = \begin{pmatrix} 3 \\ 1 \\ 2 \end{pmatrix} + t \begin{pmatrix} 1 \\ -4 \\ 2 \end{pmatrix}$

(b) $\mathbf{r} = \begin{pmatrix} 1 \\ -3 \\ 2 \end{pmatrix} + t \left[\begin{pmatrix} 5 \\ 2 \\ -1 \end{pmatrix} - \begin{pmatrix} 1 \\ -3 \\ 2 \end{pmatrix} \right] \Rightarrow \qquad \mathbf{r} = \begin{pmatrix} 1 \\ -3 \\ 2 \end{pmatrix} + t \begin{pmatrix} 4 \\ 5 \\ -3 \end{pmatrix}$

(c) $\mathbf{r} = \begin{pmatrix} 0 \\ 0 \\ 0 \end{pmatrix} + t' \left[\begin{pmatrix} 0 \\ \frac{1}{2} \\ \frac{3}{2} \end{pmatrix} - \begin{pmatrix} 0 \\ 0 \\ 0 \end{pmatrix} \right]$ which simplifies to $\quad \mathbf{r} = t \begin{pmatrix} 0 \\ 1 \\ 3 \end{pmatrix}$

Pairs of Lines

Given a pair of lines there are three possible geometric situations:

1 The lines are parallel.

2 The lines are not parallel and they intersect.

3 The lines are not parallel and do not intersect: they are **skew**.

Example 10 Find whether the following pairs of lines are parallel, intersecting or skew. If they intersect find the point of intersection.

(a) $\mathbf{r}_1 = \mathbf{i} - 3\mathbf{i} + 5\mathbf{k} + s(3\mathbf{i} - \mathbf{j} + 2\mathbf{k})$
$\mathbf{r}_2 = 4\mathbf{i} + \mathbf{k} + t(12\mathbf{i} - 4\mathbf{j} + 8\mathbf{k})$

Look at the direction vectors: $12\mathbf{i} - 4\mathbf{j} + 8\mathbf{k} = 4(3\mathbf{i} - \mathbf{j} + 2\mathbf{k})$
$\therefore$ the lines are parallel

(b) $\mathbf{r}_1 = \mathbf{i} - \mathbf{j} + 3\mathbf{k} + \lambda(\mathbf{i} - \mathbf{j} + \mathbf{k})$ $\hspace{2cm}$ (1)
$\mathbf{r}_2 = 2\mathbf{i} + 4\mathbf{j} + 6\mathbf{k} + \mu(2\mathbf{i} + \mathbf{j} + 3\mathbf{k})$

$$2\mathbf{i} + \mathbf{j} + 3\mathbf{k} \neq k(\mathbf{i} - \mathbf{j} + \mathbf{k}) \therefore \text{ lines are not parallel}$$

Assume they intersect. Let $\mathbf{r}_1 = \mathbf{r}_2$

$$\Rightarrow \begin{pmatrix} 1 \\ -1 \\ 3 \end{pmatrix} + \lambda \begin{pmatrix} 1 \\ -1 \\ 1 \end{pmatrix} = \begin{pmatrix} 2 \\ 4 \\ 6 \end{pmatrix} + \mu \begin{pmatrix} 2 \\ 1 \\ 3 \end{pmatrix}$$

Equating coeff. of $\mathbf{i}$: $\quad 1 + \lambda = 2 + 2\mu$
Equating coeff. of $\mathbf{j}$: $-1 - \lambda = 4 + \mu$ $\left.\right\} \Rightarrow \mu = -2 \quad$ and $\quad \lambda = -3$

Equating coeff. of $\mathbf{k}$: $\quad 3 + \lambda = 6 + 3\mu$

If the lines intersect then $\mu = -2$ and $\lambda = -3$ will satisfy the third equation. When $\mu = -2$, $\lambda = -3$, LHS $= 0$; RHS $= 0 =$ LHS.
$\therefore$ lines intersect

Putting $\lambda = -3$ into (1) gives the point of intersection

$$\mathbf{r}_1 = \begin{pmatrix} 1 \\ -1 \\ 3 \end{pmatrix} - 3 \begin{pmatrix} 1 \\ -1 \\ 1 \end{pmatrix} = \begin{pmatrix} -2 \\ 2 \\ 0 \end{pmatrix}$$

$\therefore$ point of intersection is $(-2, 2, 0)$

(c) $r_1 = i + k + p(i + 3j + 4k)$
      $r_2 = 2i + 3j + q(4i - j + k)$

$$4i - j + k \neq k(i + 3j + 4k) \therefore \text{ lines are not parallel}$$

Assume they intersect. Let $r_1 = r_2$

Equating coeff. of i : $1 + p = 2 + 4q$ ⎱
Equating coeff. of j : $3p = 3 - q$ ⎰ $\Rightarrow p = 1, \quad q = 0$

Equating coeff. of k: $1 + 4p = q$

When $p = 1$, $q = 0$ LHS $= 5$, RHS $= 0 \neq$ LHS

$\therefore$ these lines are not parallel and they do not intersect.
$\therefore$ they are skew

Note   It is not sufficient to prove that they do not intersect. They are only skew if they do not intersect **and** are not parallel.

Angle Between Two Lines

The angle between two lines (parallel, intersecting or skew) is the angle between their two direction vectors. If the two lines are given by

$$r = a_1 + sb_1 \quad \text{and} \quad r = a_2 + tb_2$$

then the angle between the two lines θ, is given by

$$\cos \theta = \frac{b_1 . b_2}{b_1 b_2}$$

Example 11   Find the angle between the lines given by

$$r = \begin{pmatrix} 1 \\ 2 \\ 3 \end{pmatrix} + t \begin{pmatrix} 2 \\ 4 \\ 5 \end{pmatrix} \text{ and } r = \begin{pmatrix} 3 \\ 0 \\ -1 \end{pmatrix} + s \begin{pmatrix} 4 \\ -7 \\ 4 \end{pmatrix}.$$

$$\cos \theta = \frac{\begin{pmatrix} 2 \\ 4 \\ 5 \end{pmatrix} . \begin{pmatrix} 4 \\ -7 \\ 4 \end{pmatrix}}{\sqrt{4 + 16 + 25}\sqrt{16 + 49 + 16}} = \frac{8 - 28 + 20}{\sqrt{45}\sqrt{81}} = 0$$

$\Rightarrow$ angle between the lines is $\pi/2$

● **Exercise 13.3**   See page 291.

13.4 Scalar Product Form for the Vector Equation of a Plane

Case 1: Given a Normal to the Plane and the Perpendicular Distance From the Origin

The equation of a plane can be written as

$$\boxed{\mathbf{r}.\hat{\mathbf{n}} = d}$$

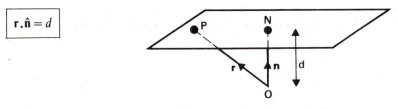

where $\hat{\mathbf{n}}$ is a unit vector perpendicular to the plane and d is the perpendicular distance from the origin to the plane. But it is often used in the form

$$\boxed{\mathbf{r}.\mathbf{n} = D}$$ where $\mathbf{n}$ is normal to the plane.

Case 2: Given a Normal to the Plane and a Point in the Plane

$$\boxed{\mathbf{r}.\mathbf{n} = \mathbf{a}.\mathbf{n}}$$

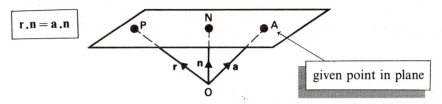

given point in plane

Example 12 Find the equation of the plane normal to $2\mathbf{i} - \mathbf{j} - 3\mathbf{k}$ which passes through the point $(3, 4, 1)$.

Equation of plane is

$$\mathbf{r}.\begin{pmatrix} 2 \\ -1 \\ -3 \end{pmatrix} = \begin{pmatrix} 3 \\ 4 \\ 1 \end{pmatrix}.\begin{pmatrix} 2 \\ -1 \\ -3 \end{pmatrix} = 6 - 4 - 3 = -1$$

i.e.

$$\mathbf{r}.\begin{pmatrix} 2 \\ -1 \\ -3 \end{pmatrix} = -1 \quad \text{or} \quad \mathbf{r}.\begin{pmatrix} -2 \\ 1 \\ 3 \end{pmatrix} = 1$$

Example 13 Show that the point $(2, -1, 2)$ lies in the plane $\mathbf{r}.\begin{pmatrix} 1 \\ 3 \\ 5 \end{pmatrix} = 9$.

If $(2, -1, 2)$ lies in the plane then $\mathbf{r} = \begin{pmatrix} 2 \\ -1 \\ 2 \end{pmatrix}$ should satisfy the equation of the plane.

Let $\mathbf{r} = \begin{pmatrix} 2 \\ -1 \\ 2 \end{pmatrix}$ then $\mathbf{r} \cdot \begin{pmatrix} 1 \\ 3 \\ 5 \end{pmatrix} = \begin{pmatrix} 2 \\ -1 \\ 2 \end{pmatrix} \cdot \begin{pmatrix} 1 \\ 3 \\ 5 \end{pmatrix} = 2 - 3 + 10 = 9.$

$\therefore (2, -1, 2)$ lies in the plane

Example 14 Find the point of intersection of the line given by

$\mathbf{r} = \begin{pmatrix} 1 \\ 2 \\ 3 \end{pmatrix} + \lambda \begin{pmatrix} 3 \\ 5 \\ 0 \end{pmatrix}$ and the plane $\mathbf{r} \cdot \begin{pmatrix} 2 \\ 4 \\ 1 \end{pmatrix} = 65.$

For each value of λ, $\begin{pmatrix} 1 \\ 2 \\ 3 \end{pmatrix} + \lambda \begin{pmatrix} 3 \\ 5 \\ 0 \end{pmatrix}$ is a point on the line, so we need to find the value of λ at the point where the line meets the plane.

We put $\mathbf{r} = \begin{pmatrix} 1 \\ 2 \\ 3 \end{pmatrix} + \lambda \begin{pmatrix} 3 \\ 5 \\ 0 \end{pmatrix}$ into the equation of the plane.

$$\Rightarrow \left[\begin{pmatrix} 1 \\ 2 \\ 3 \end{pmatrix} + \lambda \begin{pmatrix} 3 \\ 5 \\ 0 \end{pmatrix} \right] \cdot \begin{pmatrix} 2 \\ 4 \\ 1 \end{pmatrix} = 65$$

$$\Rightarrow \begin{pmatrix} 1 \\ 2 \\ 3 \end{pmatrix} \cdot \begin{pmatrix} 2 \\ 4 \\ 1 \end{pmatrix} + \lambda \begin{pmatrix} 3 \\ 5 \\ 0 \end{pmatrix} \cdot \begin{pmatrix} 2 \\ 4 \\ 1 \end{pmatrix} = 65$$

$$\Rightarrow (2 + 8 + 3) + \lambda(6 + 20) = 65$$

$$\therefore 26\lambda = 52 \qquad \lambda = 2$$

$$\therefore \mathbf{r_T} = \begin{pmatrix} 1 \\ 2 \\ 3 \end{pmatrix} + 2 \begin{pmatrix} 3 \\ 5 \\ 0 \end{pmatrix} = \begin{pmatrix} 7 \\ 12 \\ 3 \end{pmatrix}$$

$\therefore$ point of intersection is $(7, 12, 3)$

Note   If a line meets the plane for all values of λ then the line is contained in the plane.

Angle Between A Line and A Plane

If θ is the angle between the line and the plane then $\dfrac{\pi}{2} - \theta$ is the angle between the line and the normal.

$$\cos\left(\frac{\pi}{2} - \theta\right) = \frac{\mathbf{n.b}}{nb}$$

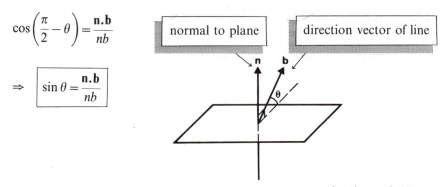

normal to plane	direction vector of line

$$\Rightarrow \quad \boxed{\sin\theta = \frac{\mathbf{n.b}}{nb}}$$

Example 15   Show that the line given by $\mathbf{r} = \begin{pmatrix} 1 \\ 2 \\ -1 \end{pmatrix} + t\begin{pmatrix} 12 \\ -3 \\ 4 \end{pmatrix}$ is

parallel to the plane $\mathbf{r}.\begin{pmatrix} -2 \\ 4 \\ 9 \end{pmatrix} = 12.$

$$\mathbf{n} \text{ is } \begin{pmatrix} -2 \\ 4 \\ 9 \end{pmatrix} \text{ and } \mathbf{b} \text{ is } \begin{pmatrix} 12 \\ -3 \\ 4 \end{pmatrix}.$$

$$\Rightarrow \sin\theta = \frac{\mathbf{n.b}}{nb} = \frac{\begin{pmatrix} -2 \\ 4 \\ 9 \end{pmatrix}.\begin{pmatrix} 12 \\ -3 \\ 4 \end{pmatrix}}{\sqrt{101}\sqrt{169}} = \frac{-24 - 12 + 36}{13\sqrt{101}} = 0$$

$$\Rightarrow \theta = 0$$

$\Rightarrow$ the line is parallel to the plane

Angle Between Two Planes

If θ is the angle between the planes then θ is also the angle between the normals.

$$\Rightarrow \quad \boxed{\cos\theta = \frac{\mathbf{n_1.n_2}}{n_1 n_2}}$$

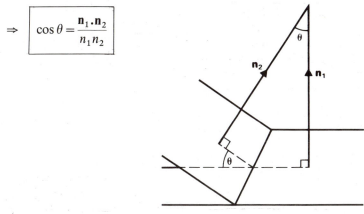

● Exercise 13.4   See page 291.

13.5 Alternative Ways of Describing Lines and Planes.

The Cartesian Equations of a Line

The vector equation of a line is of the form

$$\mathbf{r} = \begin{pmatrix} p \\ q \\ r \end{pmatrix} + t \begin{pmatrix} a \\ b \\ c \end{pmatrix}$$

where $\mathbf{r}$ is the position vector of any point (x, y, z) on the line.

Let

$$\mathbf{r} = \begin{pmatrix} x \\ y \\ z \end{pmatrix}$$

Hence

$$x = p + ta$$
$$y = q + tb$$

and

$$z = r + tc$$

Solving each of these equations for t gives the Cartesian equations of the line:

$$\boxed{\dfrac{x-p}{a} = \dfrac{y-q}{b} = \dfrac{z-r}{c}} \qquad (= t)$$

where $\begin{pmatrix} a \\ b \\ c \end{pmatrix}$ is the direction vector of the line

The Cartesian Equation of a Plane

The scalar product form for the vector equation of a plane is

$$\mathbf{r} . \begin{pmatrix} a \\ b \\ c \end{pmatrix} = D$$

where $\mathbf{r}$ is the position vector of any point (x, y, z) in the plane

Let

$$\mathbf{r} = \begin{pmatrix} x \\ y \\ z \end{pmatrix}$$

Hence the Cartesian equation of the plane is

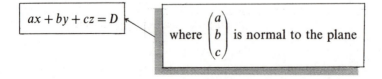

$$\boxed{ax + by + cz = D}$$

where $\begin{pmatrix} a \\ b \\ c \end{pmatrix}$ is normal to the plane

The Parametric Form for the Vector Equation of a Plane

$$\mathbf{r} = \mathbf{a} + s\mathbf{b} + t\mathbf{c}$$

represents a plane parallel to the vectors **b** and **c** and containing the point with position vector **a**.

Example 16   A plane contains the points $A(0, 1, 1)$, $B(2, 1, 0)$, $C(-2, 0, 3)$. Find the vectors $\vec{AB}$ and $\vec{AC}$. Hence find the vector equation of the plane
(a)   in scalar product form   (b)   in parametric form.

$$\vec{AB} = \vec{AO} + \vec{OB} = \vec{OB} - \vec{OA} = \begin{pmatrix} 2 \\ 1 \\ 0 \end{pmatrix} - \begin{pmatrix} 0 \\ 1 \\ 1 \end{pmatrix} = \begin{pmatrix} 2 \\ 0 \\ -1 \end{pmatrix}$$

$$\vec{AC} = \vec{AO} + \vec{OC} = \vec{OC} - \vec{OA} = \begin{pmatrix} -2 \\ 0 \\ 3 \end{pmatrix} - \begin{pmatrix} 0 \\ 1 \\ 1 \end{pmatrix} = \begin{pmatrix} -2 \\ -1 \\ 2 \end{pmatrix}$$

$$\therefore \vec{AB} = 2\mathbf{i} - \mathbf{k} \quad \text{and} \quad \vec{AC} = -2\mathbf{i} - \mathbf{j} + 2\mathbf{t}$$

(a) | Any vector which is normal to the plane will be normal to all vectors in the plane.

Let $\begin{pmatrix} a \\ b \\ c \end{pmatrix}$ be perpendicular to $\vec{AB}$ and $\vec{AC}$.

Hence $\quad \begin{pmatrix} a \\ b \\ c \end{pmatrix} \cdot \begin{pmatrix} 2 \\ 0 \\ -1 \end{pmatrix} = 0 \Rightarrow \quad 2a - c = 0 \Rightarrow 2a = c$ $\qquad$ (1)

Also $\quad \begin{pmatrix} a \\ b \\ c \end{pmatrix} \cdot \begin{pmatrix} -2 \\ -1 \\ 2 \end{pmatrix} = 0 \Rightarrow -2a - b + 2c = 0 \Rightarrow \quad b = 2c - 2a$

and (1) $\Rightarrow \quad b = c$

$\therefore$ any vector of the form $\begin{pmatrix} \frac{1}{2}c \\ c \\ c \end{pmatrix}$ is normal to the plane.

So let $\mathbf{n} = \begin{pmatrix} 1 \\ 2 \\ 2 \end{pmatrix}$. $\Rightarrow$ Scalar product form of the vector equation of the plane is

$$\mathbf{r} \cdot \begin{pmatrix} 1 \\ 2 \\ 2 \end{pmatrix} = \begin{pmatrix} 0 \\ 1 \\ 1 \end{pmatrix} \cdot \begin{pmatrix} 1 \\ 2 \\ 2 \end{pmatrix} \longleftarrow \begin{pmatrix} 1 \\ 2 \\ 2 \end{pmatrix} \text{ is the position vector of A}$$

$$\Rightarrow \mathbf{r} \cdot \begin{pmatrix} 1 \\ 2 \\ 2 \end{pmatrix} = 4$$

(b) Parametric form of the vector equation of the plane is

$$\mathbf{r} = \begin{pmatrix} 0 \\ 1 \\ 1 \end{pmatrix} + s \begin{pmatrix} 2 \\ 0 \\ -1 \end{pmatrix} + t \begin{pmatrix} -2 \\ -1 \\ 2 \end{pmatrix}$$

| position vector of A | $\overrightarrow{AB}$ and $\overrightarrow{AC}$ |

Alternative Method for Finding a Vector Perpendicular to Two Given Vectors

Since any vector that is perpendicular to the given vectors will do, it is easier to try a vector of the form $\begin{pmatrix} 1 \\ p \\ q \end{pmatrix}$ rather than $\begin{pmatrix} a \\ b \\ c \end{pmatrix}$.

Example 17   In each of the following cases, find a vector which is perpendicular to the two given lines:

(a) $\mathbf{r} = \begin{pmatrix} 2 \\ 3 \\ 1 \end{pmatrix} + s \begin{pmatrix} 1 \\ -4 \\ 1 \end{pmatrix}$ and $\mathbf{r} = t \begin{pmatrix} 2 \\ -2 \\ -1 \end{pmatrix}$

(b) $\mathbf{r} = s(2\mathbf{i} + \mathbf{j} + 3\mathbf{k})$ and $\mathbf{r} = (2 + t)\mathbf{i} + \mathbf{j}$

(a)   Let $\begin{pmatrix} 1 \\ p \\ q \end{pmatrix}$ be perpendicular to the two direction vectors $\begin{pmatrix} 1 \\ -4 \\ 1 \end{pmatrix}$ and $\begin{pmatrix} 2 \\ -2 \\ -1 \end{pmatrix}$

Hence $\begin{pmatrix} 1 \\ -4 \\ 1 \end{pmatrix} \cdot \begin{pmatrix} 1 \\ p \\ q \end{pmatrix} = 0 \Rightarrow 1 - 4p + q = 0$

And $\begin{pmatrix} 2 \\ -2 \\ -1 \end{pmatrix} \cdot \begin{pmatrix} 1 \\ p \\ q \end{pmatrix} = 0 \Rightarrow 2 - 2p - q = 0$

Solving these equations simultaneously gives $p = \frac{1}{2}, q = 1$.

Therefore $\begin{pmatrix} 1 \\ \frac{1}{2} \\ 1 \end{pmatrix}$ is perpendicular to the two lines

Note Any multiple of $\begin{pmatrix} 1 \\ \frac{1}{2} \\ 1 \end{pmatrix}$ would do for the answer. In practice, it is usual to give the multiple with the smallest integral elements $-\begin{pmatrix} 2 \\ 1 \\ 2 \end{pmatrix}$

(b)   In this case, the direction vectors are $\begin{pmatrix} 2 \\ 1 \\ 3 \end{pmatrix}$ and $\begin{pmatrix} 1 \\ 0 \\ 0 \end{pmatrix}$.

Let $\begin{pmatrix} 1 \\ p \\ q \end{pmatrix}$ be perpendicular to these vectors.

But $\begin{pmatrix} 1 \\ p \\ q \end{pmatrix} \cdot \begin{pmatrix} 1 \\ 0 \\ 0 \end{pmatrix} = 0$ gives $1 = 0$ which is not possible.

So let $\begin{pmatrix} p \\ q \\ 1 \end{pmatrix}$ be perpendicular to the vectors.

This gives $\begin{pmatrix} 2 \\ 1 \\ 3 \end{pmatrix} \cdot \begin{pmatrix} p \\ q \\ 1 \end{pmatrix} = 0$ and $\begin{pmatrix} 1 \\ 0 \\ 0 \end{pmatrix} \cdot \begin{pmatrix} p \\ q \\ 1 \end{pmatrix} = 0 \Rightarrow p = 0$ and $q = -3$

∴ the required vector is $-3\mathbf{j} + \mathbf{k}$

Note   Whilst calculations are done using column vectors (which is easier) answers are given in the same format as the question.

●   Exercise 13.5   See page 293.

13.6+   Perpendicular from a Point to a Line: Revision of Unit: A Level Questions

Example 18   L is given by $\mathbf{r} = 2\mathbf{i} + \mathbf{j} + \mathbf{k} + t(3\mathbf{i} - \mathbf{k})$. Find the coordinates of the point N where the perpendicular from the origin meets L. Hence find the coordinates of the reflection of the origin in L.

We need to find the value of t that makes $\mathbf{r}$ perpendicular to L.

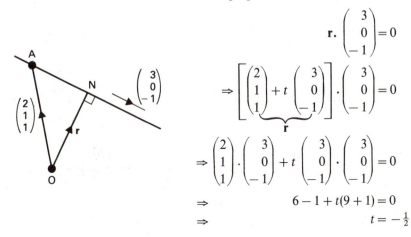

$$\mathbf{r}. \begin{pmatrix} 3 \\ 0 \\ -1 \end{pmatrix} = 0$$

$$\Rightarrow \left[\begin{pmatrix} 2 \\ 1 \\ 1 \end{pmatrix} + t \begin{pmatrix} 3 \\ 0 \\ -1 \end{pmatrix} \right] \cdot \begin{pmatrix} 3 \\ 0 \\ -1 \end{pmatrix} = 0$$

$$\Rightarrow \begin{pmatrix} 2 \\ 1 \\ 1 \end{pmatrix} \cdot \begin{pmatrix} 3 \\ 0 \\ -1 \end{pmatrix} + t \begin{pmatrix} 3 \\ 0 \\ -1 \end{pmatrix} \cdot \begin{pmatrix} 3 \\ 0 \\ -1 \end{pmatrix} = 0$$

$$\Rightarrow 6 - 1 + t(9 + 1) = 0$$

$$\Rightarrow t = -\tfrac{1}{2}$$

$$\therefore \mathbf{r}_N = \begin{pmatrix} 2 \\ 1 \\ 1 \end{pmatrix} - \frac{1}{2}\begin{pmatrix} 3 \\ 0 \\ -1 \end{pmatrix} = \begin{pmatrix} \frac{1}{2} \\ 1 \\ \frac{3}{2} \end{pmatrix}$$

$$\therefore \text{N is } (\tfrac{1}{2}, 1, \tfrac{3}{2})$$

Reflection is (1, 2, 3)

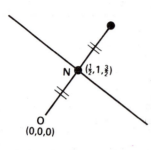

The coordinates of the reflected point were calculated by considering the increase in each coordinate.

x-coordinate: $0 \to \frac{1}{2} \to 1$
    increase is $\frac{1}{2}$ each time
y-coordinate: $0 \to 1 \to 2$
    increase is 1 each time
z-coordinate: $0 \to \frac{3}{2} \to 3$
    increase is $\frac{3}{2}$ each time

Note   To find the coordinates of the foot of the perpendicular from a point c to the line $\mathbf{r} = \mathbf{a} + t\mathbf{b}$, let $(\mathbf{r} - \mathbf{c}).\mathbf{b} = 0$

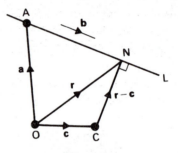

● Miscellaneous Exercise 13B   See page 294.

Unit 13   EXERCISES

Exercise 13.1

1   Find the sum of the vectors $3\mathbf{i} + \mathbf{j} - 2\mathbf{k}$, $\mathbf{i} - 3\mathbf{j} + 5\mathbf{k}$ and $2\mathbf{j} + \mathbf{k}$

2   Evaluate $\begin{pmatrix} 2 \\ 1 \\ 0 \end{pmatrix} + 3\begin{pmatrix} 1 \\ 0 \\ 3 \end{pmatrix} - 2\begin{pmatrix} 1 \\ -1 \\ -2 \end{pmatrix}$

3   Find the modulus of the vector $2\mathbf{i} - \mathbf{j} + 3\mathbf{k}$

4   If $(a + 2)\mathbf{i} + (b - 1)\mathbf{j}$ and $(b - 1)\mathbf{i} - a\mathbf{j}$ are equal vectors find the values of a and b

5   If $a\mathbf{i} - 4\mathbf{j}$ is parallel to $2\mathbf{i} - 6\mathbf{j}$ find the value of a.

6   Find the unit vector in the direction of $2\mathbf{i} + 2\mathbf{j} - \mathbf{k}$

7 Find the vector with modulus four which is parallel to the vector $\begin{pmatrix} 6 \\ -3 \\ 2 \end{pmatrix}$

8 If A is $(2, 3, 1)$ and B is $(4, 5, 2)$ express as column vectors:
   (a) $\overrightarrow{OA}$ (b) $\overrightarrow{OB}$ (c) $\overrightarrow{AB}$ (d) $\overrightarrow{BA}$, where O is the origin

9 If $\overrightarrow{OA} = 4\mathbf{i} + 14\mathbf{j} - 5\mathbf{k}$, $\overrightarrow{OB} = \mathbf{i} + 2\mathbf{j} + 7\mathbf{k}$, $\overrightarrow{OC} = 2\mathbf{i} + 6\mathbf{j} + 3\mathbf{k}$ show that the vectors $\overrightarrow{BC}$, $\overrightarrow{CA}$ are parallel. Hence deduce that the points A, B, C are collinear

10 If $\overrightarrow{OP} = \begin{pmatrix} 2 \\ 3 \\ 5 \end{pmatrix}$ and $\overrightarrow{OQ} = \begin{pmatrix} 4 \\ -1 \\ 1 \end{pmatrix}$ find the vector $\overrightarrow{OR}$ where R is the midpoint of PQ

11 If $\overrightarrow{OA} = 3\mathbf{i} - 6\mathbf{j} + 9\mathbf{k}$ and $\overrightarrow{OB} = 12\mathbf{j} - 3\mathbf{k}$ find $\overrightarrow{OC}$ where C lies on AB and $BC = 2 \times AC$

12 A, B, C, D lie in a straight line with $AB = BC = \frac{1}{2}CD$. If the position vectors of A and B are **a** and **b** respectively find the position vectors of C and D, and the mid point of CD, in terms of **a** and **b**

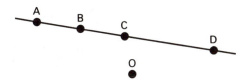

*13 Find, in component form, the vector **r** if **r** $+ 2\mathbf{j} - \mathbf{k}$ is parallel to the axis $\overrightarrow{OX}$ and **r** $- 2\mathbf{i}$ is parallel to $2\mathbf{k} - 4\mathbf{i}$

*14 $\overrightarrow{OP} = \mathbf{p}$, $\overrightarrow{OR} = 3\mathbf{p}$, $\overrightarrow{OQ} = \mathbf{q}$
   M is the mid point of QR.
   (a) Express $\overrightarrow{QP}$ and $\overrightarrow{RQ}$ in terms of **p** and **q**
   (b) Express $\overrightarrow{MQ}$ in terms of **p** and **q**
   (c) If S lies on $\overrightarrow{QP}$ produced so that $\overrightarrow{QS} = k\overrightarrow{QP}$, express $\overrightarrow{MS}$ in terms of **p, q** and k
   (d) Find the value of k if $\overrightarrow{MS}$ is parallel to QO

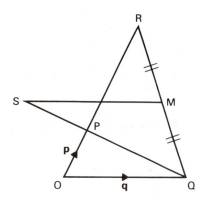

Exercise 13.2

1 Calculate **a.b** if
   (a) $\mathbf{a} = 2\mathbf{i} - 4\mathbf{j} + 3\mathbf{k}$, $\mathbf{b} = 3\mathbf{i} + 2\mathbf{j} + \mathbf{k}$

(b) $\mathbf{a} = \begin{pmatrix} 2 \\ -1 \\ 3 \end{pmatrix}, \quad \mathbf{b} = \begin{pmatrix} 0 \\ 1 \\ 1 \end{pmatrix}$

2   Prove that $3\mathbf{i} + 7\mathbf{j} + 2\mathbf{k}$ is perpendicular to $5\mathbf{i} - \mathbf{j} - 4\mathbf{k}$

3   Find $\mathbf{p.q}$ and the cosine of the angle between $\mathbf{p}$ and $\mathbf{q}$ if:
   (a) $\mathbf{p} = 4\mathbf{i} + 2\mathbf{j} + \mathbf{k}, \quad \mathbf{q} = \mathbf{i} + \mathbf{j} + \mathbf{k}$

   (b) $\mathbf{p} = \begin{pmatrix} -1 \\ 3 \\ -2 \end{pmatrix}, \quad \mathbf{q} = \begin{pmatrix} 1 \\ 1 \\ -6 \end{pmatrix}$

4   Find the resolved part of:
   (a) $2\mathbf{i} - \mathbf{j} + 2\mathbf{k}$ in the direction of $\mathbf{i} + \mathbf{j} - \mathbf{k}$

   (b) $\begin{pmatrix} 1 \\ 2 \\ 3 \end{pmatrix}$ in the direction of $\begin{pmatrix} -1 \\ -2 \\ 2 \end{pmatrix}$

5   In a $\triangle ABC$, $\overrightarrow{AB} = \mathbf{i} - 2\mathbf{j} + 3\mathbf{k}$ and $\overrightarrow{BC} = 4\mathbf{j} + 3\mathbf{k}$. Find the cosine of angle $A\hat{B}C$. Find the vector $\overrightarrow{AC}$ and use it to calculate the angle $B\hat{A}C$

*6   Find the component of $-\mathbf{i} + \mathbf{j} + 2\mathbf{k}$ in the direction of $\mathbf{j} + \mathbf{k}$. Hence find the component of $-\mathbf{i} + \mathbf{j} + 2\mathbf{k}$ perpendicular to $\mathbf{j} + \mathbf{k}$

*7   The angle between two vectors $\mathbf{p}$ and $\mathbf{q}$ is $\arccos \frac{4}{21}$. If $\mathbf{p} = \begin{pmatrix} -2 \\ \lambda \\ -4 \end{pmatrix}$ and

   $\mathbf{q} = \begin{pmatrix} 6 \\ 3 \\ -2 \end{pmatrix}$, find the positive value of λ

*8   In $\triangle OAB$, $\overrightarrow{OA} = \mathbf{a}$, $\overrightarrow{OB} = \mathbf{b}$ and $\overrightarrow{OC} = \mathbf{c}$ where C is a point on AB such that $(\mathbf{b} - \mathbf{a}).\mathbf{c} = 0$. Interpret this equation geometrically and find the area of $\triangle OAB$ in terms of $\mathbf{a}$, $\mathbf{b}$ and $\mathbf{c}$

9   The points A, B and C have coordinates $(2, 1, -1), (1, -7, 3)$ and $(-2, 5, 1)$ respectively, Calculate the cosine of the angle BAC and show that the area of the triangle ABC is $\sqrt{629}$ (J)

*10   Show that if $|\mathbf{a}| = |\mathbf{b}|$ the vectors $\mathbf{a} + \mathbf{b}$ and $\mathbf{a} - \mathbf{b}$ are perpendicular

   O is the centre of the circumscribed circle of the $\triangle ABC$, and the position vectors of A, B and C with respect to O are $\mathbf{a}$, $\mathbf{b}$ and $\mathbf{c}$ respectively. H is a point such that $\overrightarrow{OH} = \mathbf{a} + \mathbf{b} + \mathbf{c}$. Find the vectors $\overrightarrow{BA}$ and $\overrightarrow{CH}$ in ferms of $\mathbf{a}$ and $\mathbf{b}$ and show that BA is perpendicular to CH. Deduce that the perpendiculars from the vertices of the $\triangle ABC$ to the opposite sides intersect at H   (J)

Exercise 13.3

1 Find, in their simplest forms, the vector equations of the lines AD, BC, AB, OB where O is the origin, A is the point $(2, 3, 7)$, B is the point $(9, 3, 6)$, $\overrightarrow{BC} = 3\mathbf{i} - 6\mathbf{j}$ and $\overrightarrow{AD} = 2\mathbf{i} + 5\mathbf{j} - \mathbf{k}$

2 On the line given by the vector equation

$$\mathbf{r} = 3\mathbf{j} + 4\mathbf{k} + t(3\mathbf{i} - 2\mathbf{j} + 5\mathbf{k})$$

find the points represented by the parameters $t = 1, -2$ and 0. For what value of t does the point $(-4\frac{1}{2}, 6, -3\frac{1}{2})$ lie on the line? Show that $(6, -2, 14)$ does not lie on the line

3 Find the angles between the following pairs of lines:
(a) $\mathbf{r}_1 = 2\mathbf{i} - \mathbf{j} + 7\mathbf{k} + \lambda(3\mathbf{i} - \mathbf{j} + 4\mathbf{k})$, $\mathbf{r}_2 = 3\mathbf{j} - \mathbf{k} + \mu(-12\mathbf{i} + 4\mathbf{j} - 16\mathbf{k})$
(b) $\mathbf{r}_1 = 3\mathbf{i} + 2\mathbf{k} + s(2\mathbf{i} - \mathbf{j} + 2\mathbf{k})$, $\mathbf{r}_2 = 3\mathbf{k} + t(4\mathbf{i} - 3\mathbf{j})$

(c) $\mathbf{r}_1 = \begin{pmatrix} 2 \\ 5 \\ 1 \end{pmatrix} + p\begin{pmatrix} 3 \\ -4 \\ 1 \end{pmatrix}$, $\mathbf{r}_2 = \begin{pmatrix} 0 \\ 0 \\ 1 \end{pmatrix} + q\begin{pmatrix} 2 \\ 3 \\ -1 \end{pmatrix}$

(d) $\mathbf{r}_1 = \begin{pmatrix} 4 \\ 6 \\ -11 \end{pmatrix} + s\begin{pmatrix} 4 \\ 0 \\ -1 \end{pmatrix}$, $\mathbf{r}_2 = t\begin{pmatrix} 3 \\ 5 \\ 12 \end{pmatrix}$

4 Show that the following pair of lines intersect and find the position vector of the point of intersection:

$$\mathbf{r}_1 = \begin{pmatrix} 4 \\ 8 \\ 3 \end{pmatrix} + \lambda\begin{pmatrix} 1 \\ 2 \\ 1 \end{pmatrix} \quad \mathbf{r}_2 = \begin{pmatrix} 7 \\ 6 \\ 5 \end{pmatrix} + \mu\begin{pmatrix} 6 \\ 4 \\ 5 \end{pmatrix}$$

5 Prove that $\mathbf{r}_1 = \mathbf{i} + 3\mathbf{k} + \lambda(2\mathbf{i} - \mathbf{j} + \mathbf{k})$, $\mathbf{r}_2 = 2\mathbf{i} - \mathbf{j} + \mu(\mathbf{i} - 2\mathbf{j} - \mathbf{k})$ represent a pair of skew lines

6 L_1 is given by $\mathbf{r} = 2\mathbf{i} + 9\mathbf{j} + 13\mathbf{k} + \lambda(\mathbf{i} + 2\mathbf{j} + 3\mathbf{k})$ L_2 is given by $\mathbf{r} = -3\mathbf{i} + 7\mathbf{j} + p\mathbf{k} + \mu(-\mathbf{i} + 2\mathbf{j} - 3\mathbf{k})$. If L_1 and L_2 intersect, find the value of p and the point of intersection

*7 A parallelogram ABCD has three of its vertices $A(4, 5, 10)$, $B(2, 3, 4)$, $C(1, 2, -1)$. Find column vectors representing $\overrightarrow{AB}$ and $\overrightarrow{BC}$. Hence find vector equations for AD and CD and find the coordinates of the fourth vertex D

Exercise 13.4

1 Find the equations of the planes through the given points which are normal to the given vectors:
(a) $(0, 0, 0)$, $2\mathbf{i} + \mathbf{j} - 3\mathbf{k}$
(b) $(1, -2, 3)$, $-2\mathbf{i} - 2\mathbf{j} + \mathbf{k}$
(c) (a, b, c), $2\mathbf{j} - \mathbf{k}$

2 Find the perpendicular distance of the plane $\mathbf{r}.\begin{pmatrix} 2 \\ 3 \\ 4 \end{pmatrix} = 20$ from the origin

3 Show that the line given by $\mathbf{r} = \begin{pmatrix} 0 \\ 0 \\ 1 \end{pmatrix} + \lambda \begin{pmatrix} 2 \\ 6 \\ 8 \end{pmatrix}$ is perpendicular to the plane

$\mathbf{r}.\begin{pmatrix} 5 \\ 15 \\ 20 \end{pmatrix} = 7$

4 Find the equation of the plane which contains $(-2, 0, 3)$ and is perpendicular to the line $\mathbf{r} = (2 + 3t)\mathbf{i} + (4 - t)\mathbf{j} + (1 + 2t)\mathbf{k}$

5 L_1 is the line given by $\mathbf{r} = 2\mathbf{i} + \mathbf{j} + 3\mathbf{k} + t(\mathbf{i} - 2\mathbf{k})$
L_2 is the line given by $\mathbf{r} = (1 + 2s)\mathbf{i} + s\mathbf{j} + 5\mathbf{k}$

Π_1 is the plane given by $\mathbf{r}.\begin{pmatrix} 2 \\ 2 \\ -1 \end{pmatrix} = 5$

Π_2 is the plane given by $\mathbf{r}.\begin{pmatrix} 3 \\ -1 \\ 0 \end{pmatrix} = 2$

Find the angle between:
(a) L_1 and L_2 (b) L_1 and Π_1 (c) Π_1 and Π_2

6 Show that the point $(2, 3, 1)$ lies in the plane $\mathbf{r}.\begin{pmatrix} 2 \\ -1 \\ 0 \end{pmatrix} = 1$ and

show that L is contained in the same plane where L is given by

$\mathbf{r} = \begin{pmatrix} 2 \\ 3 \\ 4 \end{pmatrix} + t\begin{pmatrix} 1 \\ 2 \\ 5 \end{pmatrix}$

7 Find the point where the line given by $\mathbf{r} = \begin{pmatrix} 0 \\ 0 \\ 1 \end{pmatrix} + t\begin{pmatrix} 2 \\ 1 \\ 1 \end{pmatrix}$ cuts the

plane $\mathbf{r}.\begin{pmatrix} 1 \\ 0 \\ 0 \end{pmatrix} = 4$

*8 The coordinates of the points A and B are $(0, 2, 5)$ and $(-1, 3, 1)$ respectively
and the line L has the vector equation

$$\mathbf{r} = \begin{pmatrix} 3 \\ 2 \\ 2 \end{pmatrix} + t\begin{pmatrix} 2 \\ -2 \\ -1 \end{pmatrix}$$

(a) Find a vector equation for the plane which contains A and is perpendicular to *L*. Verify that B lies in it

(b) Show that the point in which *L* meets is C $(1, 4, 3)$ and find the angle between CA and CB

(c) Given that $D = (5, 0, 1)$ show that D lies on *L* and calculate the volume of the tetrahedron ABCD (J)

Exercise 13.5

1 Find the Cartesian equation of each of the following lines:

(a) $\mathbf{r} = 2\mathbf{i} - \mathbf{k} + t(3\mathbf{i} - \mathbf{j} + 2\mathbf{k})$

(b) $\mathbf{r} = (1 - t)\mathbf{i} + 3t\mathbf{j} - 4t\mathbf{k}$

2 Find the Cartesian equations of the following planes:

(a) $\mathbf{r} . \begin{pmatrix} 2 \\ 1 \\ 3 \end{pmatrix} = 5$

(b) passing through the point $(4, 1, 3)$ and perpendicular to $\mathbf{j}$

3 Find a parametric form of the vector equation of the plane which passes through the points $A(1, 0, 0)$, $B(2, 1, 0)$, $C(3, 1, 1)$

4 A plane passes through the point with position vector $\mathbf{i} - \mathbf{j}$ and is parallel to the two vectors $\mathbf{i} + \mathbf{j} + \mathbf{k}$ and $\mathbf{i} - 2\mathbf{j} + 3\mathbf{k}$, find

(a) the vector equation of the plane in parametric form

(b) the vector equation of the plane in scalar product form

(c) the Cartesian equation of the plane

*5 Find the Cartesian equations of each of the following lines:

(a) $\mathbf{r} = \begin{pmatrix} 4 \\ 1 \\ 2 \end{pmatrix} + t \begin{pmatrix} 3 \\ 2 \\ 0 \end{pmatrix}$ (b) $\mathbf{r} = (1 - t)\mathbf{i} + 2\mathbf{j}$

6 The two lines with vector equations

$$\mathbf{r} = \mathbf{k} + s(\mathbf{i} + \mathbf{j}) \quad \text{and} \quad \mathbf{r} = \mathbf{k} + t(-\mathbf{i} + \mathbf{k})$$

intersect at the point A. Write down the position vector of A. Find a vector perpendicular to both of the lines and hence, or otherwise, obtain a vector equation of the plane containing the two lines (L)

7 The point A has position vector $\mathbf{i} + 4\mathbf{j} - 3\mathbf{k}$ referred to the origin O. The line *L* has vector equation $\mathbf{r} = t\mathbf{i}$. The plane Π contains the line *L* and the point A. Find

(a) a vector which is normal to the plane Π

(b) a vector equation for the plane Π

(c) the cosine of the acute angle between OA and the line *L* (L)

8   With respect to the origin O the points A, B, C have position vectors

$$a(5\mathbf{i} - \mathbf{j} - 3\mathbf{k}), \quad a(-4\mathbf{i} + 4\mathbf{j} - \mathbf{k}), \quad a(5\mathbf{i} - 2\mathbf{j} + 11\mathbf{k})$$

respectively, where a is a non-zero constant. Find

(a)   a vector equation for the line BC,
(b)   a vector equation for the plane OAB,
(c)   the cosine of the acute angle between the lines OA and OB

Obtain, in the form $\mathbf{r.n} = p$, a vector equation for Π, the plane which passes through A and is perpendicular to BC

Find cartesian equations for
(d)   the plane Π
(e)   the line BC   (L)

Miscellaneous Exercise 13B

1   Given that $\mathbf{a} = -\mathbf{i} + \mathbf{j} + 2\mathbf{k}$ and $b = \mathbf{j} + \mathbf{k}$, find $|\mathbf{a}|, |\mathbf{b}|$ and the angle between $\mathbf{a}$ and $\mathbf{b}$ (J: 13.2)

2   Find the resolved part of the vector $\begin{pmatrix} 2 \\ 1 \\ 3 \end{pmatrix}$ in the direction of the

vector $\begin{pmatrix} 6 \\ 3 \\ 2 \end{pmatrix}$ (J: 13.2)

3   Referred to O as origin, A is the point $(6, 3, -2)$ and B is the point $(8, 1, -4)$. Write down the lengths of OA and OB and determine (as a fraction) the cosine of the angle AOB. Find also the area of the triangle AOB, leaving your answer in surd form   (J: 13.2, 4.6)

4   Find a unit vector which is in the opposite direction to the sum of the vectors $(3\mathbf{i} + 2\mathbf{j} + \mathbf{k})$ and $(-5\mathbf{i} - 3\mathbf{j} + 6\mathbf{k})$

Prove that this unit vector is perpendicular to the vector $(9\mathbf{i} - 4\mathbf{j} + 2\mathbf{k})$
                                                                (L: 13.1, 13.2)

5   Two straight lines are given by the equations

$$\mathbf{r} = 17\mathbf{i} - 9\mathbf{j} + 9\mathbf{k} + \lambda(3\mathbf{i} + \mathbf{j} + 5\mathbf{k})$$
$$\mathbf{r} = 15\mathbf{i} - 8\mathbf{j} - \mathbf{k} + \mu(4\mathbf{i} + 3\mathbf{j})$$

where λ and μ are scalar parameters. Show that these lines intersect and find the position vector of their point of intersection

Find also the cosine of the acute angle contained by the lines
                                                                (L: 13.3)

6 The resolved part of the vector $\begin{pmatrix} 2c \\ -2c \\ c \end{pmatrix}$ in the direction of

the vector $\begin{pmatrix} 1 \\ -2 \\ -2 \end{pmatrix}$ is 6. Find the constant c.   (J: 13.2)

7 The lines L_1 and L_2 are given by the vector equations

$$\mathbf{r} = \begin{pmatrix} 0 \\ 1 \\ -1 \end{pmatrix} + t \begin{pmatrix} 2 \\ -1 \\ 2 \end{pmatrix} \quad \text{and} \quad \mathbf{r} = \begin{pmatrix} p \\ 3 \\ 0 \end{pmatrix} + s \begin{pmatrix} 2 \\ 2 \\ -1 \end{pmatrix} \text{ respectively}$$

where t and s are parameters. Given that L_1 and L_2 intersect find the value of the constant p

Find the distance from the point with coordinates $(p, 3, 0)$ to the point of intersection of L_1 and L_2 (J: 13.3)

8 Find a vector equation for the line AB, where A is the point $(6, 4, 2)$ and B is the point $(8, 10, 10)$

The perpendicular from the origin O to AB meets it at N. Using the scalar product of $\overrightarrow{AB}$ and $\overrightarrow{ON}$ or otherwise, calculate the coordinates of N. Deduce the coordinates of the point C (distinct from B) on AB such that $OC = OB$ (J: 13.3, 13.6)

9 Relative to an origin O, the point A has position vector $\mathbf{a}$ where $\mathbf{a} = 4\mathbf{i} - 6\mathbf{j} + 14\mathbf{k}$. Given that P is the point on the line OA, such that $OP:PA = 2:1$, find a vector equation for the line through P parallel to the vector $\mathbf{b}$, where $\mathbf{b} = \mathbf{i} - 2\mathbf{j} + 2\mathbf{k}$

Evaluate $(\mathbf{a} - \mathbf{b}).\mathbf{b}$

Hence determine the cosine of the angle between $(\mathbf{a} - \mathbf{b})$ and $\mathbf{b}$
    (A: 13.2, 13.3)

10 The lines L_1 and L_2 have equations given respectively by

$$\mathbf{r}_1 = \begin{pmatrix} 2 \\ 2 \\ 2 \end{pmatrix} + t \begin{pmatrix} 2 \\ 1 \\ -1 \end{pmatrix} \quad \text{and} \quad \mathbf{r}_2 = \begin{pmatrix} 0 \\ 3 \\ -1 \end{pmatrix} + s \begin{pmatrix} -2 \\ 1 \\ 1 \end{pmatrix}$$

where t and $\mathbf{s}$ are real parameters. Show that L_1 and L_2 do not intersect. The point P on L_1 has parameter p and the point Q on L_2 has parameter q. Write $\overrightarrow{PQ}$ as a column vector in terms of p and q. Given that L_1 and L_2 are both perpendicular to PQ, find p and q. Find the length of PQ, giving your answer in surd form   (J: 13.2, 13.3)

11 Relative to an origin O, the points P and Q have position vectors

$$\mathbf{p} = 4\mathbf{i} - 3\mathbf{j} + 4\mathbf{k}, \quad \mathbf{q} = 6\mathbf{i} + \mathbf{j} - 2\mathbf{k} \text{ respectively}$$

(a) Prove that triangle OPQ is isosceles

(b)   Hence, or otherwise, find a unit vector in the plane OPQ which is perpendicular to the line PQ

(c)   Evaluate $(\mathbf{q} - \mathbf{p}).\mathbf{q}$ and deduce the value of angle PQO, to the nearest $0.1°$ (A: 13.1, 13.2)

12   Given the points $A(4, 2, 6)$ and $B(7, 8, 9)$ find a vector equation for the line AB in terms of a parameter t. The perpendicular to the line AB from the point $C(1, 8, 3)$ meets the line at N. Find the coordinates of N. Obtain a vector equation for the line which is the reflection of the line AC in the line AB   (J: 13.3, 13.6)

13   Relative to the origin O, the position vectors of the points A, B and C are $\mathbf{j} - 4\mathbf{k}$, $6\mathbf{i} - 5\mathbf{j} - \mathbf{k}$ and $4\mathbf{i} + 7\mathbf{j} - 9\mathbf{k}$ respectively, the unit of length being the metre.

(a)   Show that, for all values of the scalar parameter t, the point P with position vector $2t\mathbf{i} + (1 - 2t)\mathbf{j} + (t - 4)\mathbf{k}$ lies on the straight line passing through A and B

(b)   Use the scalar product $\overrightarrow{AB}\cdot\overrightarrow{CP}$ to determine the value of t for which CP is perpendicular to AB

(c)   Hence find the shortest distance from C to AB   (A: 13.3, 13.6)

14   The points A and B have position vectors $\mathbf{a}$ and $\mathbf{b}$, respectively, relative to the origin O, which is not collinear with A and B. The midpoints of OA and OB are C and D respectively. Find vector equations for the lines AD and BC and hence find the position vector of the point G at which the two lines intersect. Find a vector equation for the line OG and hence show that the medians of a triangle are concurrent   (J: 13.1, 13.3)

15   The coordinates of the points A, B, C are $(1, 2, 1)$, $(2, -1, 0)$, $(3, 1, 2)$ respectively. The line L and the plane Π have the vector equations

$$\mathbf{r} = \begin{pmatrix} 0 \\ 3 \\ 5 \end{pmatrix} + t \begin{pmatrix} 2 \\ 1 \\ -2 \end{pmatrix} \quad \text{and} \quad \mathbf{r}.\begin{pmatrix} 2 \\ -2 \\ 1 \end{pmatrix} = -1 \quad \text{respectively}$$

(a)   Show that the plane Π contains both A and L

(b)   Show that BC is parallel to Π and perpendicular to L

(c)   Show that BC is equal to the perpendicular distance from A to L (J: 13.4, 13.6)

16   The point A is $(1, 3, 5)$ and has position vector $\mathbf{a}$ relative to the origin O; B is $(2, 1, 6)$ and has position vector $\mathbf{b}$. The line l is given by $\mathbf{r} = \mathbf{a} + t(\mathbf{i} + 6\mathbf{j} + 5\mathbf{k})$; the line m is given by $\mathbf{r} = \mathbf{b} + s(\mathbf{i} + 2\mathbf{j} + 3\mathbf{k})$. Show that the lines l and m intersect and find the coordinates of the common point. Prove that AB is perpendicular to m and hence write down the coordinates of the reflection of A in m (J: 13.3, 13.6)

17   The points A and B have position vectors $4\mathbf{i} + \mathbf{j} - 7\mathbf{k}$ and $2\mathbf{i} + 6\mathbf{j} + 2\mathbf{k}$ respectively relative to the origin O. Show that the angle AOB is a right angle

Find a vector equation for the median AM of the triangle OAB

Find also, in the form $\mathbf{r}.\mathbf{n} = p$, a vector equation of the plane OAB

(L: 13.2, 13.3, 13.5)

18 The position vectors, with respect to a fixed origin, of the points L, M and N are given by $\mathbf{l}, \mathbf{m}$ and $\mathbf{n}$ respectively, where

$$\mathbf{l} = a(\mathbf{i} + \mathbf{j} + \mathbf{k}), \quad \mathbf{m} = a(2\mathbf{i} + \mathbf{j}), \quad \mathbf{n} = a(\mathbf{j} + 4\mathbf{k})$$

and a is a non-zero constant. Show that the unit vector $\mathbf{j}$ is perpendicular to the plane of the triangle LMN

Find a vector perpendicular to both $\mathbf{j}$ and $(\mathbf{m} - \mathbf{n})$, and hence, or otherwise, obtain a vector equation of that perpendicular bisector of MN which lies in the plane LMN

Verify that the point K with position vector $a(5\mathbf{i} + \mathbf{j} + 4\mathbf{k})$ lies on this bisector and show that K is equidistant from L, M and N

(L: 13.1, 13.5)

19 Let $\mathbf{a} = \mathbf{i} - 2\mathbf{j} + \mathbf{k}$, $\mathbf{b} = 2\mathbf{i} + \mathbf{j} - \mathbf{k}$. Given that $\mathbf{c} = \lambda\mathbf{a} + \mu\mathbf{b}$ and that $\mathbf{c}$ is perpendicular to $\mathbf{a}$, find the ratio of λ to μ

Let A, B be the points with position vectors $\mathbf{a}, \mathbf{b}$ respectively with respect to the origin O. Write down, in terms of $\mathbf{a}$, and $\mathbf{b}$, a vector equation of the line l through A, in the plane of O, A and B, which is perpendicular to OA

Find the position vector of P, the point of intersection of l and OB

(J: 13.2, 13.3)

20 The lines L_1 and L_2 are given by the equations

$$L_1: \quad \mathbf{r} = \begin{pmatrix} 1 \\ 6 \\ 3 \end{pmatrix} + t \begin{pmatrix} 2 \\ -1 \\ 1 \end{pmatrix}, \quad L_2: \quad \mathbf{r} = \begin{pmatrix} 3 \\ 3 \\ 8 \end{pmatrix} + s \begin{pmatrix} 1 \\ 0 \\ 1 \end{pmatrix}$$

(a) Calculate the acute angle between the direction of L_1 and L_2
(b) Show that the lines do not intersect

(c) Verify that the vector $\mathbf{a} = \begin{pmatrix} 1 \\ 1 \\ -1 \end{pmatrix}$ is perpendicular to each

of the lines

The point P on L_1 is given by $t = p$; the point Q on L_2 is given by $s = q$. Write down column vector representing $\overrightarrow{PQ}$. Hence calculate p and q so that the vectors $\overrightarrow{PQ}$ and $\mathbf{a}$ are parallel. (J: 13.1, 13.2, 13.3)

*21 The vertices of a tetrahedron are O, A, B and C. The position vectors of A, B and C with respect to O are $\mathbf{a}, \mathbf{b}$ and $\mathbf{c}$ respectively; P, Q, R, L, M, N are the midpoints of OA, OB, OC, BC, CA, AB respectively. Write down the position vectors of L, M and N (see diagram on next page)

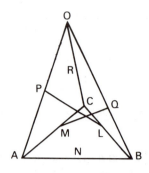

Determine the vector equations of the lines PL and QM, and find the position vector of the point at which they intersect. Deduce, or show otherwise, that PL, QM and RN are concurrent. If, in addition, OABC is regular (i.e. all six edges have the same length) prove that

(a) $\mathbf{a.b} = \mathbf{b.c} = \mathbf{c.a}$
(b)   PL and BC are perpendicular
(J: 13.1, 13.2, 13.3)

*22 The vertices A, B and C of triangle have position vectors $\mathbf{a}, \mathbf{b}$ and $\mathbf{c}$ respectively to an origin O, and H is a point with position vector h such that $(\mathbf{a} - \mathbf{h}).(\mathbf{b} - \mathbf{c}) = 0$ and $(\mathbf{b} - \mathbf{h}).(\mathbf{c} - \mathbf{a}) = 0$.   Deduce   that $(\mathbf{c} - \mathbf{h}).(\mathbf{a} - \mathbf{b}) = 0$. Interpret this result geometrically.

(a)   In the case when A, B and C have coordinates $(1, 1)$, $(5, 5)$ and $(3, 4)$ respectively when referred to two-dimensional Cartesian axes, determine the coordinates of H

(b)   Given that A, B and C have coordinates $(1, 1, 1)$, $(7, 4, 6)$ and $(9, 6, 8)$, respectively when referred to three dimensional Cartesian axes, show that the line with equation

$$\begin{pmatrix} x \\ y \\ z \end{pmatrix} = \begin{pmatrix} 1 \\ 1 \\ 1 \end{pmatrix} + t \begin{pmatrix} 4 \\ -5 \\ 1 \end{pmatrix}$$

is perpendicular to BC and verify that the equation of AH may be written in this way. Find the coordinates of the point of intersection of AH and BC
(J: 13.1, 13.2, 13.3)

COMPLEX NUMBERS

14.1 Definitions, Notation and Properties

Imaginary numbers are numbers of the form ni where n is a real number and $i = \sqrt{-1}$.

Example 1   Simplify (a) $2i + 3i$ (b) i^7 (c) i^{-1}

(a) $2i + 3i = \underline{5i}$

(b) $i^7 = i(i^2)^3$
$= i(-1)^3$
$= \underline{-i}$

(c) $i^{-1} = \dfrac{1}{i}$
$= \dfrac{1}{i} \times \dfrac{i}{i}$
$= \dfrac{i}{-1}$
$= \underline{-i}$

Complex numbers are numbers of the form $a + bi$ where a and b are real $i = \sqrt{-1}$.

Special cases: (a)   when $b = 0$, it becomes a real number
(b)   when $a = 0$, it becomes an imaginary number

	Literally:
Notation : $\text{Re}(a + ib) = a \longleftarrow$	The real part of $a + ib$ is a.
$\text{Im}(a + ib) = b \longleftarrow$	The imaginary part of $a + ib$ is b.

Complex Roots of Quadratic Equations

Example 2   Solve $z^2 + 4z + 5 = 0$

$$z = \frac{-4 \pm \sqrt{16 - 4 \cdot 1 \cdot 5}}{2 \cdot 1} = \frac{-4 \pm \sqrt{-4}}{2} = \frac{-4 \pm 2i}{2} = -2 \pm i$$

$\therefore$ roots are $\underline{-2 + i \text{ and } -2 - i}$

Conjugate numbers

$a + bi$ and $a - bi$ are called conjugate numbers.
If $a + bi = z$ then $a - bi$ is denoted by z^*.

Note: Complex roots of quadratic equations occur as conjugate pairs.

Addition and Subtraction of Complex Numbers

Example 3   Simplify (a) $(2 + 3i) + (4 - 5i)$ (b) $-3i - (2 + 3i)$.

(a) $(2 + 3i) + (4 - 5i) = (2 + 4) + (3 - 5)i = \underline{6 - 2i}$

(b) $-3i - (2 + 3i) = (0 - 2) + (-3 - 3)i = \underline{-2 - 6i}$

Multiplication and Division

Example 4   Simplify (a) $(2 + 3i)(1 + i)$ (b) $(2 + 3i)(2 - 3i)$

(c) $\dfrac{2 + 3i}{1 + i}$.

(a) $(2 + 3i)(1 + i) = 2 + 2i + 3i + 3(-1) = \underline{-1 + 5i}$

(b) $(2 + 3i)(2 - 3i) = 2^2 - (3i)^2 = 4 + 9 = \underline{13}$

(c) $\dfrac{2 + 3i}{1 + i} = \dfrac{(2 + 3i)}{(1 + i)} \times \dfrac{(1 - i)}{(1 - i)} = \dfrac{2 - 2i + 3i - 3(-1)}{1 + 1} = \underline{\tfrac{5}{2} + \tfrac{1}{2}i}$

Example 5   Rationalise $\dfrac{3}{2 + 3i}$.

$$\dfrac{3}{2 + 3i} = \dfrac{3}{(2 + 3i)} \times \dfrac{(2 - 3i)}{(2 - 3i)} = \dfrac{6 - 9i}{4 + 9} = \underline{\tfrac{1}{13}(6 - 9i)}$$

The Zero Complex Number

$$x + yi = 0 \quad \Leftrightarrow \quad x = 0 \quad \text{and} \quad y = 0$$

Equality

Example 6   Solve (a) $(x - 1) + (y - 2)i = 0$ (b) $x + (x + y)i = 2 + 5i$.

(a) $(x - 1) + (y - 2)i = 0$
Equating real parts: $\underline{x = 1}$

Equating imaginary parts: $\underline{y = 2}$

(b) $x + (x + y)i = 2 + 5i$
Equating real parts: $\underline{x = 2}$

Equating imaginary parts: $x + y = 5 \Rightarrow \underline{y = 3}$

The Square Root of a Complex Number

Example 7 Find the square roots of $3 + 4i$.

Let $a + bi = \sqrt{3 + 4i}$ where a and b are real.

Then
$$(a + bi)^2 = 3 + 4i$$
$$\Rightarrow a^2 - b^2 + 2abi = 3 + 4i$$

Equating real parts   : $a^2 - b^2 = 3$
Equating imaginary parts: $2ab = 4$ $\left.\begin{array}{c}\\\\\end{array}\right\} \Rightarrow a^2 - \dfrac{4}{a^2} = 3$

$$\Rightarrow a^4 - 3a^2 - 4 = 0$$
$$\Rightarrow (a^2 + 1)(a^2 - 4) = 0$$

Since a is real, $a = \pm 2$ and $b = \pm 1$.
$\therefore$ square roots of $3 + 4i$ are $2 + i$ and $-2 - i$.

● Exercise 14.1 See page 308.

14.2 The Argand Diagram: Complex Numbers as Vectors

The complex number $z = a + bi$ can be represented geometrically by the point P with coordinates (a, b). P corresponds uniquely to z and is often referred to as the point z.

Example 8 Represent the complex numbers $-i$, $-5 + 4i$, $1 + i$ on the Argand diagram by the points A, B, C. Prove that triangle ABC is right angled.

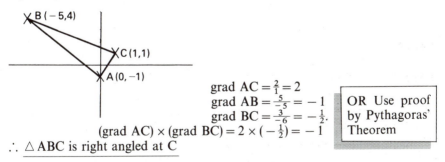

grad AC $= \frac{2}{1} = 2$
grad AB $= \frac{5}{-5} = -1$
grad BC $= \frac{3}{-6} = -\frac{1}{2}$.

OR Use proof by Pythagoras' Theorem

(grad AC) $\times$ (grad BC) $= 2 \times (-\frac{1}{2}) = -1$
$\therefore \triangle ABC$ is right angled at C

Example 9 The complex number $\lambda(3 + i)$ is represented on the Argand diagram. Sketch the locus of $\lambda(3 + i)$ as λ varies.

$x + yi = \lambda(3 + i)$
$\Rightarrow x = 3\lambda$ and $y = \lambda$
$\Rightarrow x = 3y$
$\Rightarrow y = \frac{1}{3}x$

Complex Numbers as Vectors

Addition and subtraction of complex numbers obeys the same rules as addition and subtraction of vectors. Therefore complex numbers can be represented as vectors on the Argand diagram.

Example 10   If $z_1 = 2 + 3i$ and $z_2 = 1 - i$ represent the following as vectors on Argand diagrams:
(a) z_1 (b) z_2 (c) $z_1 + z_2$ (d) $-z_2$ (e) $z_1 - z_2$ (f) $2z_2 - z_1$.

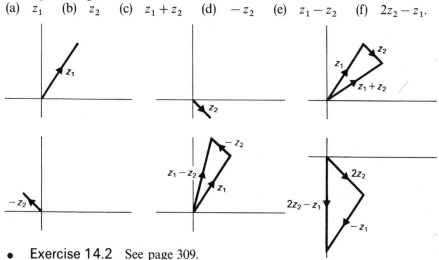

● Exercise 14.2   See page 309.

14.3 The Modulus-Argument Form of a Complex Number

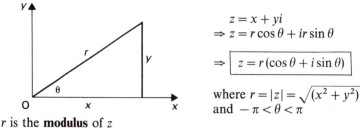

$$z = x + yi$$
$$\Rightarrow z = r\cos\theta + ir\sin\theta$$

$$\Rightarrow \boxed{z = r(\cos\theta + i\sin\theta)}$$

where $r = |z| = \sqrt{(x^2 + y^2)}$
and $-\pi < \theta < \pi$

r is the **modulus** of z
θ is the **argument** of z

Example 11   Find the modulus and argument of (a) $3 + 4i$ (b) $3 - 4i$
(c) $-3 + 4i$ (d) $-3 - 4i$

In all four cases $|z| = \sqrt{(3^2 + 4^2)} = 5$

(a)

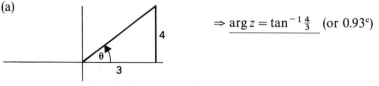

$$\Rightarrow \arg z = \tan^{-1}\tfrac{4}{3} \quad \text{(or } 0.93^c)$$

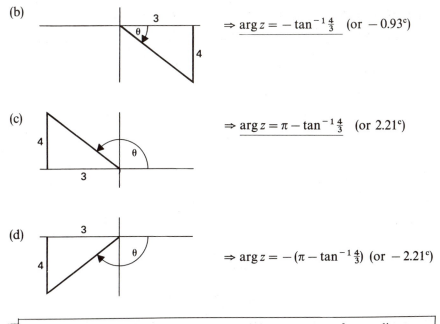

(b)

$$\Rightarrow \arg z = -\tan^{-1}\tfrac{4}{3} \quad (\text{or} -0.93^c)$$

(c)

$$\Rightarrow \arg z = \pi - \tan^{-1}\tfrac{4}{3} \quad (\text{or } 2.21^c)$$

(d)

$$\Rightarrow \arg z = -(\pi - \tan^{-1}\tfrac{4}{3}) \quad (\text{or} -2.21^c)$$

When converting from $a + bi$ to modulus-argument form, diagrams are essential to prevent errors in calculating arguments.

Example 12   Express   (a) $3i$ (b) $\dfrac{2+i}{3-i}$ in   modulus-argument form.

(a)
$\times$ (0,3)

$$\Rightarrow r = 3 \text{ and } \theta = \frac{\pi}{2}$$

$$\Rightarrow z = 3\left(\cos\frac{\pi}{2} + i\sin\frac{\pi}{2}\right)$$

(b)

First, you must express $\dfrac{2+i}{3-i}$ in the form $a + bi$

$$\rightarrow \frac{2+i}{3-i} = \frac{2+i}{3-i} \times \left(\frac{3+i}{3+i}\right) = \frac{5+5i}{10} = \tfrac{1}{2} + \tfrac{1}{2}i$$

$$\Rightarrow r^2 = \left(\frac{1}{2}\right)^2 + \left(\frac{1}{2}\right)^2 = \frac{1}{2} \Rightarrow r = \frac{1}{\sqrt{2}}$$

$$\text{and } \tan\theta = 1 \Rightarrow \theta = \frac{\pi}{4}$$

$$\Rightarrow z = \frac{1}{\sqrt{2}}\left(\cos\frac{\pi}{4} + i\sin\frac{\pi}{4}\right)$$

Example 13   Express in the form $x + yi$ the complex numbers whose moduli and arguments are given by: (a) $(2, \pi/6)$ (b) $(\sqrt{2}, -3\pi/4)$

(a) $z = 2\left(\cos\dfrac{\pi}{6} + i\sin\dfrac{\pi}{6} \right) = 2\left(\dfrac{\sqrt{3}}{2} + i\cdot\dfrac{1}{2} \right) = \underline{\sqrt{3} + i}$

(b) $z = \sqrt{2}\left(\cos\left(-\dfrac{3\pi}{4} \right) + i\sin\left(-\dfrac{3\pi}{4} \right) \right) = \sqrt{2}\left(-\dfrac{1}{\sqrt{2}} + i\left(-\dfrac{1}{\sqrt{2}} \right) \right)$

$= \underline{-1 - i}$

Products and Quotients

If $z_1 = r_1(\cos\theta_1 + i\sin\theta_1)$ and $z_2 = r_2(\cos\theta_2 + i\sin\theta_2)$
then

$$\boxed{\begin{aligned} z_1 z_2 &= r_1 r_2[\cos(\theta_1 + \theta_2) + i\sin(\theta_1 + \theta_2)] \\[4pt] \frac{z_1}{z_2} &= \frac{r_1}{r_2}[\cos(\theta_1 - \theta_2) + i\sin(\theta_1 - \theta_2)] \end{aligned}}$$

Proof

(i) $z_1 z_2 = r_1(\cos\theta_1 + i\sin\theta_1) \times r_2(\cos\theta_2 + i\sin\theta_2)$

$= r_1 r_2[\cos\theta_1 \cos\theta_2 - \sin\theta_1 \sin\theta_2 + i(\sin\theta_1 \cos\theta_2 + \sin\theta_2 \cos\theta_1)]$

$= r_1 r_2[\cos(\theta_1 + \theta_2) + i\sin(\theta_1 + \theta_2)]$

(ii) $z_1/z_2 = \dfrac{r_1(\cos\theta_1 + i\sin\theta_1)}{r_2(\cos\theta_2 + i\sin\theta_2)}$

$= \dfrac{r_1}{r_2} \cdot \dfrac{(\cos\theta_1 + i\sin\theta_1)}{(\cos\theta_2 + i\sin\theta_2)} \times \dfrac{(\cos\theta_2 - i\sin\theta_2)}{(\cos\theta_2 - i\sin\theta_2)}$

$= \dfrac{r_1}{r_2} \cdot \dfrac{[\cos\theta_1 \cos\theta_2 + \sin\theta_1 \sin\theta_2 + i(\sin\theta_1 \cos\theta_2 - \sin\theta_2 \cos\theta_1)]}{(\cos^2\theta_2 + \sin^2\theta_2)}$

$= \dfrac{r_1}{r_2}[\cos(\theta_1 - \theta_2) + i\sin(\theta_1 - \theta_2)]$

Example 14

Find the modulus and argument of (a) $1 + i$ (b) $2\sqrt{3} + 2i$. Hence find
the modulus and argument of (c) $(1 + i)(2\sqrt{3} + 2i)$ (d) $\dfrac{2\sqrt{3} + 2i}{1 + i}$

(a)

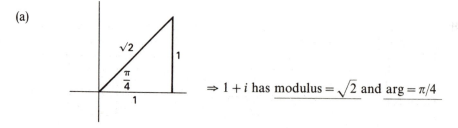

$\Rightarrow \underline{1 + i}$ has modulus $= \underline{\sqrt{2}}$ and $\underline{\arg = \pi/4}$

(b)

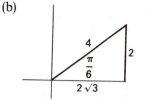

$\Rightarrow 2\sqrt{3} + 2i$ has modulus $= 4$ and arg $= \pi/6$

(c) $(1 + i)(2\sqrt{3} + 2i)$ has modulus $= 4 \times \sqrt{2} = 4\sqrt{2}$

and arg $= \dfrac{\pi}{4} + \dfrac{\pi}{6} = \dfrac{5\pi}{12} \longleftarrow$

$\boxed{\begin{array}{l} |z_1 z_2| = |z_1||z_2| \text{ and} \\ \arg(z_1 z_2) = \arg z_1 + \arg z_2 \end{array}}$

(d) $\dfrac{2\sqrt{3} + 2i}{1 + i}$ has modulus $= \dfrac{4}{\sqrt{2}} = 2\sqrt{2}$

$\boxed{\begin{array}{l} \left|\dfrac{z_1}{z_2}\right| = \dfrac{|z_1|}{|z_2|} \text{ and} \\ \arg\left(\dfrac{z_1}{z_2}\right) = \arg z_1 - \arg z_2 \end{array}}$

and arg $= \left(\dfrac{\pi}{6} - \dfrac{\pi}{4}\right) = -\dfrac{\pi}{12}$

Example 15   Find the modulus and argument of $(\sqrt{3} + i)^4$.

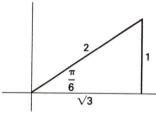

$\Rightarrow \sqrt{3} + i$ has modulus 2 and argument $\dfrac{\pi}{6}$

$\Rightarrow (\sqrt{3} + i)^4$ has modulus $= 2^4 = 16$

and argument $= \dfrac{\pi}{6} \times 4 = \dfrac{2\pi}{3}$

•   Exercise 14.3   See page 310.

14.4 Geometrical Illustrations: Loci

1   $\boxed{\begin{array}{l} |z| = a \text{ represents a circle, centre} \\ (0, 0), \text{ radius } a. \end{array}}$

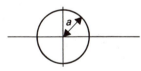

2   $\boxed{\begin{array}{l} |z - z_1| = a \text{ represents a circle, cen-} \\ \text{tre } z_1, \text{ radius } a. \end{array}}$

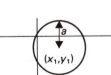

Example 16
Sketch the locus of $P(x, y)$ which represents the complex number $z = x + yi$, given that $|z - 2 + 3i| = 5$.

$|z - 2 + 3i| = 5$ can also be written as
$|z - (2 - 3i)| = 5$
which is a circle, centre $(2, -3)$, radius 5.

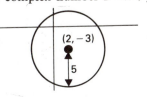

Example 17   Shade the areas represented on Argand diagrams by
(a) $|z+1| < 2$ (b) $|z-i| \geqslant 1$.

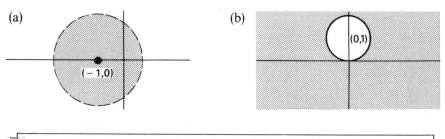

(a)

(b)

$(-1,0)$

$(0,1)$

Note: Solid lines are included in the shaded region. Dotted lines are not.

Example 18   Sketch the locus of z if $|z-4-i| = |z+4+i|$

This represents the set of points which are equidistant from $(4, 1)$ and $(-4, -1)$.
i.e. the perpendicular bisector of the line joining $(4, 1)$ and $(-4, -1)$.

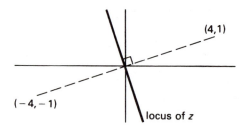

$(4,1)$

$(-4,-1)$

locus of z

3 | $|z-z_1| = |z-z_2|$ represents the perpendicular bisector of the line joining z_1 and z_2.

4 | $\arg(z-z_1) = \theta$ represents a half-line radiating from the point z_1 at an angle θ to the positive direction of the x-axis.

z_1 θ

5 | $\arg(z_1-z) = \phi$ represents a half-line leading to the point z_1 at an angle ϕ to the positive direction of the x-axis.

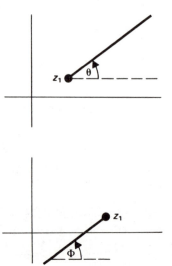

z_1

Φ

Example 19 Sketch the loci of z if (a) $\arg(z-1) = \pi/6$
(b) $\arg(1-z) = \pi/6$.
(a)   (b)

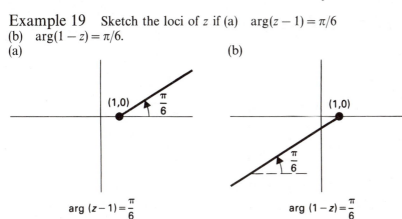

$\arg(z-1) = \dfrac{\pi}{6}$ $\arg(1-z) = \dfrac{\pi}{6}$

Example 20 Shade the area represented by $0 < \arg(z+1+2i) \leqslant \pi/4$.

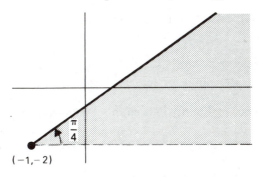

$(-1,-2)$

Example 21 Sketch the locus of z if $\arg(z-2) - \arg(z+2) = \pi/3$

If A us $(2,0)$, B is $(-2,0)$ let $\arg(z-2) = \alpha$ and $\arg(z+2) = \beta$.

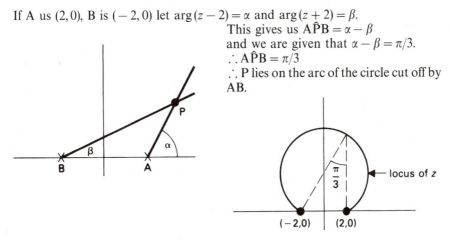

This gives us $\text{A}\hat{\text{P}}\text{B} = \alpha - \beta$
and we are given that $\alpha - \beta = \pi/3$.
$\therefore \text{A}\hat{\text{P}}\text{B} = \pi/3$
$\therefore$ P lies on the arc of the circle cut off by AB.

- Exercise 14.4 See page 311.

14.5+ Revision of Unit: A Level Questions

- Misceiianeous Exercise 14B See page 312.

Unit 14 EXERCISES

Exercise 14.1

1 Simplify:

(a) $3i - i$ (b) i^2 (c) i^3 (d) i^4 (e) i^{13} (f) $\dfrac{1}{i^2}$

(g) $\dfrac{1}{3i^3}$

2 Write down the conjugates of:

(a) $3 + 4i$ (b) $2 - i$ (c) $3i$ (d) $1 - i\sqrt{3}$ (e) $2 + 3i + \sqrt{3}$

3 Solve the following quadratic equations:

(a) $z^2 + 1 = 0$ (b) $z^2 + 25 = 0$
(c) $z^2 + z + 1 = 0$ (d) $z^2 - 8z + 25 = 0$
(e) $2z^2 + 3z + 2 = 0$

4 If $z = 2 - 3i$, write down:

(a) $\text{Re}(z)$ (b) $\text{Im}(z)$ (c) z^* (d) $(z^*)^*$

5 Find the values of x and y in each of the following equations:

(a) $x + yi = 0$ (b) $x - 3 + (y - 1)i = 0$
(c) $x + y + (y - 1)i = 0$ (d) $x + yi = 3 + 4i$
(e) $x + yi = i\sqrt{2}$ (f) $x + yi = 2 - i$
(g) $(x + 1) + (y - 1)i = 3 + 5i$

6 Express in the form $a + bi$:

(a) $(1 + 3i)(1 - i)$ (b) $(1 - 4i)(1 + 4i)$ (c) $(3 - 4i)^2$

(d) $\dfrac{1}{2 - i}$ (e) $\dfrac{3i}{3 + 2i}$ (f) $\dfrac{1 - i}{3 + i}$

(g) $\dfrac{1 + 2i}{4 - i}$ (h) $(1 + i)^2$ (i) $(1 - i)^3$

7 Rationalise $\dfrac{2 + i}{2 - i}$

8 Find the square roots of $21 + 20i$

9 If $z = 1 + i$ find the values of $z + z^*$ and zz^*

10 If $z = 2p + (p - 1)i$ is a real number, evaluate z

11 If $z = a + i$ is such that z^2 is purely imaginary find all the possible values of a

12 If $z = x + yi$ solve:

(a) $z + 2z^* = 6 + i$ (b) $3z - 2z^* = 3 + 10i$

13 If z_1, z_2 are the roots of $z^2 + z + 1 = 0$ write down

(a) $z_1 + z_2$ (b) $z_1 z_2$

14 Verify that $z = 1 - 2i$ is a root of the equation $z^3 - 3z^2 + 7z - 5 = 0$

15 If $1 + i$ is a root of the equation $z^3 + az + b = 0$ find a and b

Exercise 14.2

1 Represent the following numbers on an Argand diagram

(a) $2 + 3i$ (b) $2 - 3i$ (c) $3i$ (d) -4 (e) $-2i$

by the points P, Q, R, S, T respectively

2 Mark $z = 3 - 4i$ on an Argand diagram. On the same diagram mark z^* and $-z$

3 If X and Y represent the complex numbers $1 + i$, $3 - 3i$ in the Argand diagram, what complex number is represented by the midpoint of XY?

4 If $2 \pm i, i$ are represented on an Argand diagram, they form three corners of a square. Find the complex numbers represented by the fourth corner and the centre of the square

5 Show that the points A, B, C, representing the complex numbers $2 + 2i$, $4 + 4i, 6i$ in the Argand diagram, are the vertices of an isosceles triangle

6 If $z_1 = 4 - i$ and $z_2 = -1 + 3i$, represent the following complex numbers as vectors on Argand diagrams:

(a) z_1 (b) z_2 (c) $z_1 + z_2$ (d) $2z_1 - z_2$ (e) z_1^*

7 Find the roots, z_1 and z_2, of the equation

$$z^2 - 5 + 12i = 0$$

in the form $a + bi$, where a and b are real, and give the value of $z_1 z_2$

Draw, on graph paper, an Argand diagram to illustrate the points representing (a) $z_1 z_2$,   (b) $1/z_1 z_2$,   (c) $z_1^* z_2^*$ (L)

8 The non-real cube roots of unity are $-\frac{1}{2} \pm i\sqrt{3}/2$. These roots are represented in an Argand diagram by the points A, B and the number $z = -2$ is represented by the point C. Show that the area of the sector of the circle with centre C through A and B, which is bounded by CA, CB and the minor arc AB is $\frac{1}{2}\pi$ (J)

9 The non-real cube roots of unity are $w_1 = -\frac{1}{2} + i(\sqrt{3}/2)$, $w_2 = -\frac{1}{2} - i\sqrt{3}$. A regular hexagon is drawn in an Argand diagram such that two adjacent vertices represent w_1 and w_2 respectively, and the centre of the circumscribing circle of the hexagon is the point $(1, 0)$. Determine in the form $a + ib$, the complex numbers represented by the other four vertices of the hexagon and find the product of these four complex numbers   (J)

10   In an Argand diagram, P represents the complex number $\lambda(2 - i)$ where λ is real. Sketch the locus of P as λ varies. The point Q represents the complex number $1/(\lambda(2 - i))$ when $\lambda \neq 0$ and R is the midpoint of PQ. Given that $x + yi$ represents R, find x and y in terms of λ (J)

*11   The complex numbers z_1 and z_2 satisfy the relation

$$z_2^2 - z_1 z_2 + z_1^2 = 0$$

Find the ratio z_2/z_1 given that its imaginary part is positive. If $z_1 = a + ib$, where a and b are real, show that $z_2 = \frac{1}{2}(a - b\sqrt{3}) + \frac{i}{2}(b + a\sqrt{3})$

In an Argand diagram, the points P and Q represent z_1 and z_2 respectively and O is the origin. Show that $\triangle OPQ$ is equilateral   (J)

Exercise 14.3

1   Find the modulus and argument of each of the following:
   (a) $4i$ (b) $1 - i$ (c) $-1 - i$ (d) $-1 + i$
   (e) $1 - i\sqrt{3}$ (f) $2 + i\sqrt{5}$ (g) $-2 - i\sqrt{5}$

2   Express in the form $r(\cos\theta + i\sin\theta)$:
   (a) -3 (b) $-2i$ (c) $a + ai$
   (d) $a - ai$ (e) 4 (f) $-3 + i\sqrt{7}$

3   Express in the form $x + yi$ the complex numbers whose modulus and argument are given by:
   (a) $(1, \pi/3)$ (b) $(4, \pi/6)$ (c) $(6, 2\pi/3)$ (d) $(\sqrt{2}, 3\pi/4)$
   (e) $(2, \pi/2)$ (f) $(3, \pi)$ (g) $(4, -\pi/3)$

4   Find the modulus and argument of:
   (a) $1 - i$ (b) $1 + 2i$ (c) $(1 + 2i)^2$ (d) $\dfrac{1 + 2i}{1 - i}$
   (e) $(1 + 2i)(1 - i)$

   giving the arguments correct to 2 d.p.

5   Given that $z = 4(\cos\pi/3 + i\sin\pi/3)$ and $w = 2(\cos\pi/6 - i\sin\pi/6)$ write down the modulus and the argument of each of the following:
   (a) z (b) w (c) z^3 (d) z^3/w (J)

6   Given that $z_1 = 1 + i\sqrt{3}$ and $z_2 = 4 + 3i$, calculate the moduli and arguments of $z_1 z_2$ and z_2/z_1, giving the arguments in degrees to one decimal place.

   Illustrate in one Argand diagram
   (a) $z_1 - z_2$ (b) $z_1 z_2$ (c) z_2/z_1 (L)

7   Given that $z = \sqrt{3} + i$, find the value of $\arg(z^7)$ which lies between $-\pi$ and $+\pi$ (L)

8 Given that $z_1 = 2 + i$ and $z_2 = 3 + 4i$, find the modulus and the tangent of the argument of each of

(a) $z_1 z_2^*$ (b) z_1/z_2 (L)

Exercise 14.4

Sketch the locus of z on an Argand diagram in each of the following cases:

1 $|z| = 3$ **2** $|z - 2| = 4$ **3** $|z + i| = 2$

4 $|z + 3| = 1$ **5** $|z - 1 - i| = 3$ **6** $|z + 2 + 4i| = 4$

7 $\arg z = \pi/4$ **8** $\arg(z - 3) = -\pi/2$ **9** $\arg(z - 1 + i) = \pi/3$

10 $\arg(z - 2i) = \pi$ **11** $\arg(i - z) = \pi/4$ **12** $\arg(1 + 2i - z) = \pi/3$

13 $|z| = |z + 4|$ **14** $|z + 3 - 4i| = |z - 3 + 4i|$

15 $\arg(z - 4) - \arg z = \pi/4$ **16** $\arg(z - 3) - \arg(z + 1) = \pi/2$

Express in complex form the equations of the following:

17 a circle, centre $(2, 3)$, radius 6

18 a circle, centre $(-4, 3)$, radius 1

19 a circle, centre $(0, 3)$, radius 3

20   **21**   **22**

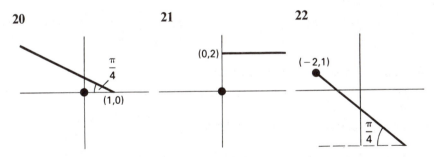

Shade the area represented in each of the following cases, on an Argand diagram:

23 $|z - 2| < 2$ **24** $|z + 2i| \geqslant 3$ **25** $1 \leqslant |z - 1 - i| \leqslant 2$

26 Find the equation in complex form of the perpendicular bisector of the line joining the points P and Q where

(a) P is $(3, 0)$, Q is $(-6, 0)$ (b) P is $(3, -5)$, Q is $(-2, 1)$

27 Find the Cartesian equations of the loci given by:

(a) $|z - i| = |z + 2i|$ (b) $|z - 2 + i| = |z + 2 - i|$

In each of the following questions, shade on an Argand diagram the region defined by:

28 $|z| < 4$ and $-\pi/4 \leqslant \arg z \leqslant \pi/4$

29 $|z + 2i| = 3$ and $0 \leqslant \arg(z + i) \leqslant \pi/2$

30 $1 < |z| < 5$ and $\arg z = 5\pi/6$

31   Show in the separate diagrams the regions of the z-plane in which each of the following inequalities is satisfied:

(a) $|z - 2| \leqslant |z - 2i|$
(b) $0 < \arg(z - 2) \leqslant \pi/3$

Indicate clearly in each case which part of the boundary of the region is to be included in the region. Give the Cartesian equations of the boundaries   (J)

32   Shade on a diagram the region of the z-plane in which one, or the other, but not both, of the following inequalities, is satisfied:

(a) $|z| \leqslant 1$ (b) $|z - 1 - i| \leqslant 2$

Your diagram should show clearly which parts of the boundary are included   (J)

Miscellaneous Exercise 14B

1   The complex number $z_1 = 1 + ic$ where c is a positive real number, is such that z_1^3 is real. Prove that the only possible value for c is $\sqrt{3}$
                                                                                        (J: 14.1)

2   One root of the equation

$$z^2 + az + b = 0$$

where a and b are real constants, is $2 + 3i$. Find the values of a and b
                                                                                (L: 10.4, 14.1)

3   Given that $z = \sqrt{3} + i$. Find the modulus and argument of

(a) z^2 (b) $\dfrac{1}{z}$

Show in an Argand diagram the points representing the complex numbers z, z^2 and $1/z$ (L: 14.2, 14.3)

4   Find, in terms of π, the argument of the complex numbers
(a) $(1 + i)^2$

(b) $\dfrac{3 + i}{1 + 2i}$

(c) $\left(\cos \dfrac{\pi}{3} + i \sin \dfrac{\pi}{3} \right)\left(\cos \dfrac{\pi}{4} - i \sin \dfrac{\pi}{4} \right)$ (A: 14.3)

5 Find the two possible pairs of values of real numbers x and y such that $(x + yi)^2 = -3 + 4i$. Mark in an Argand diagram the points representing your solutions and also the point representing $-3 + 4i$. Prove that the triangle formed by these points is right angled (J: 14.1, 14.2)

6 The roots of the equation

$$(1 + i)z^2 - 2iz + 3 + i = 0$$

are denoted by α and β. Find $\alpha + \beta$ and $\alpha\beta$ in the form $p + iq$ where p and q are real.

Find, in a form not involving α and β, the quadratic equation whose roots are $\alpha + 3\beta$ and $3\alpha + \beta$ (J: 10.4, 14.1)

7 Mark in an Argand diagram the points P_1 and P_2 which represent the two complex numbers z_1 and z_2, where $z_1 = 1 - i$ and $z_2 = 1 + i\sqrt{3}$

On the same diagram, mark the points P_3 and P_4 which represent $(z_1 + z_2)$ and $(z_1 - z_2)$ respectively

Find the modulus and argument of

(a) z_1 (b) z_2 (c) $z_1 z_2$ (d) z_1/z_2 (L: 14.2, 14.3)

8 Determine all pairs of values of the real numbers p, q for which $1 + i$ is a root of the equation

$$z^3 + pz^2 + qz - pq = 0 \qquad \text{(J: 14.1)}$$

9 Expand $z = (1 + ic)^6$ in powers of c and find the five real finite values of c for which z is real (J: 14.1)

10 The complex numbers z_1, z_2 are the roots of the quadratic equation $z^2 + z + 1 = 0$ and $\operatorname{Im} z_1 > \operatorname{Im} z_2$. Express z_1 and z_2 in modulus-argument form

Evaluate

(a) $|z_2 - z_1|$ and $\arg(z_2 - z_1)$
(b) $\operatorname{Re}(z_1 + z_2)$ and $\operatorname{Im}(z_1 + z_2)$
Show that $z_1^3 = z_2^3 = 1$

Mark on an Argand diagram the points representing z_1, z_2 and 1 (L: 14.1, 14.2)

11 Express the complex number $z_1 = \dfrac{11 + 2i}{3 - 4i}$ in the form $x + yi$ where x and y are real. Given that $z_2 = 2 - 5i$, find the distance between the points in the Argand diagram which represent z_1 and z_2

Determine the real numbers α, β such that

$$\alpha z_1 + \beta z_2 = -4 + i \qquad \text{(J: 14.1, 14.2)}$$

12 Find the exact values of the modulus and argument of the complex number

$$z = \frac{1+i}{\sqrt{3}-i}$$

Find the smallest positive integer n such that z^n is a real number. For this value of n, find the value of z^n (J. 14.3)

13 (a) Find the square roots of $5 + 12i$
   (b) Express $(1 + i)^2/(1 - i)$ in the form $r(\cos\theta + i\sin\theta)$, where r is positive and $-\pi < \theta \leqslant \pi$
   (c) z_1 and z_2 are non-zero complex numbers and

$$-\frac{\pi}{2} < \arg z_1 < \frac{\pi}{2}, \quad -\frac{\pi}{2} < \arg z_2 < \frac{\pi}{2}$$

Show that

$$\arg(z_1 z_2) = \arg z_1 + \arg z_2 \qquad \text{(L: 14.1, 14.3)}$$

14 In the quadratic equation

$$x^2 + (p + iq)x + 3i = 0$$

p and q are real. Given that the sum of the squares of the roots is 8, find all possible pairs of values of p and q (J: 10.4, 14.1)

15 The complex numbers $z_1 = 1 + ai$, $z_2 = a + bi$ where a and b are real, are such that $z_1 - z_2 = 3i$. Find a and b and show that $\dfrac{1}{z_1} + \dfrac{1}{z_2} = \dfrac{7-i}{10}$

Hence, or otherwise, find $\dfrac{z_1^2 - z_2^2}{z_1 z_2}$ in the form $x + yi$ where x and y are real

(J: 14.1)

16 Find the modulus and argument of each of the complex numbers z_1 and z_2, where

$$z_1 = \frac{1+i}{1-i} \quad z_2 = \frac{\sqrt{2}}{1-i}$$

Plot the points representing z_1, z_2 and $z_1 + z_2$ on an Argand diagram. Deduce from your diagram that $\tan(3\pi/8) = 1 + \sqrt{2}$

(L: 14.1, 14.2, 14.3)

17 Given that $z = 1 + i\sqrt{2}$, express in the form $a + ib$ each of the complex numbers $p = z + \dfrac{1}{z}$ and $q = z - \dfrac{1}{z}$

In an Argand diagram, P and Q are the points which represent p and q respectively, O is the origin, M is the midpoint of PQ and G is a point on OM such that $OG = \frac{2}{3}OM$. Prove that this angle PGQ is a right angle

(J: 14.1, 14.2)

*18 Given that $z = 2 + 2i$, express z in the form $r(\cos\theta + i\sin\theta)$, where r is a positive real number and $-\pi < \theta \leqslant \pi$

On the same Argand diagram, display and label clearly the numbers z, z^2 and $4/z$

Find the values of $|z + z^2|$ and $\arg(z + 4/z)$ (A: 14.2, 14.3)

*19 (a) Express in modulus-argument form the complex numbers

(i) $-1 + i\sqrt{3}$ (ii) $\dfrac{(1 + i)}{(1 - i)}$

(b) Find the pairs of values of the real constants a and b such that $7 + 24i = -(a + ib)^2$

(c) Three complex numbers α, β and γ are represented in the Argand diagram by the three points A, B and C respectively. Find the complex number represented by D when ABCD forms a parallelogram having BD as a diagonal (L: 14.1, 14.2, 14.3)

*20 (a) Given that $z = \cos\theta + i\sin\theta$, where $z \neq -1$, show that

$$\frac{2}{1 + z} = 1 - i\tan\tfrac{1}{2}\theta$$

(b) One root of the equation

$$z^3 + z^2 + 4z + \lambda = 0$$

where λ is a real number, is $1 - 3i$. Find the other roots and the value of λ (L: 14.1, 14.3)

*21 The sets A, B, C of points z in the complex plane are given by

$$A = \{z : |z - 2| = 2\}$$
$$B = \{z : |z - 2| = |z|\}$$
$$C = \{z : \arg(z - 2) = \arg z\}$$

Determine the locus of z in each case, and illustrate the sets of points on one diagram (J: 14.4)

15 UNIT CURVE SKETCHING II: COORDINATE GEOMETRY

15.1 Systematic Curve Sketching

If we need to sketch a curve that does not fall into the categories studied so far, then there are other techniques that we can employ, many of which we already know and use. Certain features of a curve can be observed, or easily calculated, from the equation of the curve.

Systematic Curve Sketching

1 Look for $\boxed{\text{symmetry}}$ If there are only even powers of x then there is symmetry about the y-axis. If there are only even powers of y then the curve is symmetrical about the x-axis.

2 Look for $\boxed{\text{asymptotes}}$ Consider what happens to the curve as $x \to \pm \infty$ and as $y \to \pm \infty$.

3 Look for $\boxed{\text{intercepts}}$ Find where the curve crosses the axes.

4 Look for and categorise any $\boxed{\text{turning points}}$ The absence of turning points is just as important as their presence. Also, find the points where the gradient is infinite.

5 Look for $\boxed{\text{limitations on } x \text{ and } y}$ Find any regions of the plane where no part of the curve lies.

Example 1 Sketch the graph of $x^2 + y^2 = 9$.

1 There is symmetry about both x and y axes.

2 No asymptotes.

3 When $x = 0$, $y = \pm 3$.

   When $y = 0$, $x = \pm 3$.

4 $2x + 2y \cdot \dfrac{dy}{dx} = 0$ $\therefore \dfrac{dy}{dx} = -\dfrac{x}{y}$

   $\dfrac{dy}{dx} = 0$ when $x = 0, y = \pm 3$

   $\dfrac{dy}{dx} = \infty$ when $y = 0, x = \pm 3$

5 x^2 must be $\geqslant 0$ $\therefore y^2 \leqslant 9$
   $\Rightarrow -3 \leqslant y \leqslant 3$.
   Similarly $-3 \leqslant x \leqslant 3$

This information is sufficient to draw

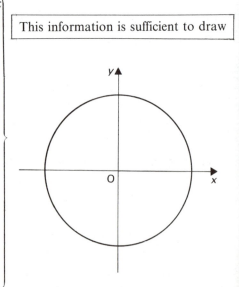

Example 2 Sketch the graph of $y = \dfrac{x(x+1)}{x-1}$.

1 No symmetry.

2 $y \to \infty$ as $x \to 1$ $\therefore x = 1$ is an asymptote.

$$y = \frac{x^2 + x}{x-1} = x + 2 + \frac{2}{x-1}$$

> dividing denominator into numerator as far as possible.

Here, as $x \to \infty$, $\dfrac{2}{x-1} \to 0$ and $y \to x + 2$. $\therefore y = x + 2$ is an asymptote.

3 When $y = 0$, $x = 0$ and $x = -1$ $\therefore (0, 0)$ and $(-1, 0)$ are intercepts.

4 $\dfrac{dy}{dx} = \dfrac{x^2 - 2x - 1}{(x-1)^2} \Rightarrow \dfrac{dy}{dx} = 0$ when $x = 1 \pm \sqrt{2}$

and $y = 3 \pm \sqrt{8}$

$$\frac{d^2y}{dx^2} = \frac{4}{(x-1)^3}$$

When $x = 1 + \sqrt{2}$, $\dfrac{d^2y}{dx^2} = \dfrac{4}{(\sqrt{2})^3} > 0$ $\therefore$ Min at $(1 + \sqrt{2}, 3 + \sqrt{8})$

When $x = 1 - \sqrt{2}$, $\dfrac{d^2y}{dx^2} = \dfrac{4}{(-\sqrt{2})^3} < 0$ $\therefore$ Max at $(1 - \sqrt{2}, 3 - \sqrt{8})$.

This information is sufficient to sketch the following graph without considering limitations on x and y.

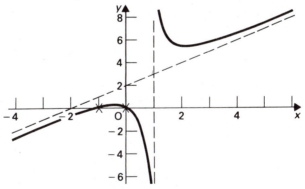

• Exercise 15.1

15.2 Miscellaneous Curve Sketching

1 Whenever possible, **use the techniques developed in Unit 12.**

2 When this is not possible, use the techniques for systematic curve sketching.

3 When a curve is given in **parametric form**, it is better to find its Cartesian equation before sketching it.

Example 3   Sketch the graph of $(x-1)^2 + (y+1)^2 = 1$.

This represents the circle $x^2 + y^2 = 1$
translated $1 \rightarrow$ and $1 \downarrow$

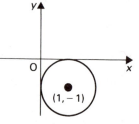

Example 4   Sketch the graph of the curve given by
$x = 1 + \cos\theta$, $y = \sin\theta$.

The Cartesian equation is
$(x-1)^2 + y^2 = 1$
which is the circle
$x^2 + y^2 = 1$
translated $1 \rightarrow$

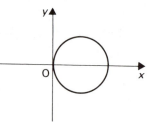

● Exercise 15.2   See page 324.

15.3 Revision of Coordinate Geometry

1   The equation of any straight line
can be expressed in the form
$y = mx + c$ where m is the gradient
of the line and c is the intercept on
the y-axis

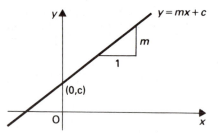

2   The equation of a straight line passing through the point (h, k) with gradient
m is found by using $\dfrac{y-k}{x-h} = m$

3   The equation of a straight line passing through two points (x_1, y_1) and
(x_2, y_2) is found by using

$$\frac{y - y_1}{x - x_1} = \frac{y_2 - y_1}{x_2 - x_1}$$

4   The distance, d, between two points (x_1, y_1) and (x_2, y_2) is given by

$$d^2 = (x_2 - x_1)^2 + (y_2 - y_1)^2$$

5   The midpoint of a line joining (x_1, y_1) and (x_2, y_2) has coordinates

$$\left(\frac{x_1 + x_2}{2}, \frac{y_1 + y_2}{2} \right)$$

6   The angle, θ, between two straight lines with gradients m_1 and m_2 is given by

$$\tan\theta = \frac{m_1 \sim m_2}{1 + m_1 m_2}$$

7   Two lines are parallel if $m_1 = m_2$.

8   Two lines are perpendicular if $m_1 m_2 = -1$.

9   The length of a perpendicular from the point (h, k) to the line $ax + by + c = 0$ is

$$\left| \frac{ah + bk + c}{\sqrt{a^2 + b^2}} \right|$$

10   If a line is included at an angle α to the x-axis then the gradient of the line is given by $\tan\alpha$.

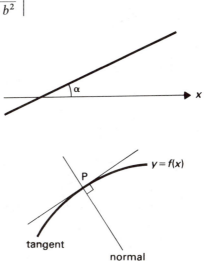

11   If a curve represents the function $y = f(x)$ then the gradient of the tangent at a point P is the value of dy/dx (or $f'(x)$) at that point Since the normal is perpendicular to the tangent, its gradient is equal

to $\dfrac{-1}{\text{gradient of the tangent}}$

12   To find the intersection of a line and a curve, their two equations are solved simultaneously. Often this results in a quadratic expression in x which will either have:
(a)   two real distinct roots ($b^2 > 4ac$) $\Rightarrow$ two distinct points of intersection
(b)   two real equal roots ($b^2 = 4ac$) $\Rightarrow$ the line is a tangent to the curve
(c)   two imaginary roots ($b^2 < 4ac$) $\Rightarrow$ the line and curve do not intersect.

Example 5   Find the equation of the tangent to the curve given by $x = 4\cos\theta$, $y = 3\sin\theta$ at the point where $\theta = \pi/4$.

$$\frac{dx}{d\theta} = -4\sin\theta, \quad \frac{dy}{d\theta} = 3\cos\theta$$

$$\frac{dy}{dx} = \frac{dy}{d\theta} \times \frac{d\theta}{dx} = \frac{3\cos\theta}{-4\sin\theta} = -\tfrac{3}{4}\cot\theta$$

When $\theta = \pi/4$, $x = \dfrac{4}{\sqrt{2}} = 2\sqrt{2}$, $y = \dfrac{3}{\sqrt{2}} = \dfrac{3}{2}\sqrt{2}$ and gradient $= -\tfrac{3}{4}$.

Equation of tangent is given by

$$\frac{y - \frac{3}{2}\sqrt{2}}{x - 2\sqrt{2}} = \frac{-3}{4}$$

$$\Rightarrow 4y - 6\sqrt{2} = -3x + 6\sqrt{2}$$

$$4y + 3x = 12\sqrt{2}$$

Example 6   If A is $(-2, 7)$ and B is $(8, 2)$ find the coordinates of the point P on AB such that AP:PB $= 2:3$.

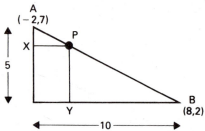

X must divide AZ in the ratio 2:3
$\therefore$ AX $= 2$
$\Rightarrow$ y-coodinate of P $= 7 - 2 = 5$
Y must divide ZB in the ratio 2:3
$\therefore$ ZY $= 4$
$\Rightarrow$ x-coordinate of P $= -2 + 4 = 2$.
$\therefore$ P is $(2, 5)$

Example 7

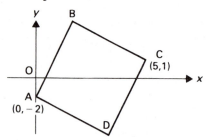

In the diagram, ABCD is a square. A is $(0, -2)$, C is $(5, 1)$ and AC is a diagonal. Find the coordinates of B.

Gradient of AC $= \dfrac{1 - (-2)}{5 - 0} = \dfrac{3}{5}$.

AB and AD make angles of 45° with AC. Let m be the gradient of a line at 45° to AC

$$\tan 45° = \frac{m - \frac{3}{5}}{1 + \frac{3}{5}m}$$

$$\Rightarrow 1 = \frac{5m - 3}{5 + 3m}$$

$$\Rightarrow m = 4 \quad = \text{grad AB}$$

> This must be the gradient of AB or AD and since gradient of AB is positive and gradient of AD is negative....

Since AD is perpendicular to AB, its gradient is $-\frac{1}{4}$. Since BC is parallel to AD, gradient of BC $= -\frac{1}{4}$.

Equation of AB : $\dfrac{y + 2}{x} = 4 \Rightarrow y - 4x + 2 = 0$

Equation of BC : $\dfrac{y - 1}{x - 5} = -\dfrac{1}{4} \Rightarrow 4y + x - 9 = 0$

Solving these equations simultaneously $\Rightarrow$ B is $(1, 2)$

● **Exercise 15.3**   See page 325.

15.4 Circles

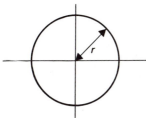

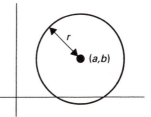

A circle, centre $(0, 0)$ radius r, is given by

$$x^2 + y^2 = r^2$$

A circle, centre (a, b), radius r, is given by

$$(x - a)^2 + (y - b)^2 = r^2$$

The general equation of a circle

$$x^2 + y^2 + 2gx + 2fy + c = 0$$
is a circle, centre $(-g, -f)$, radius $= \sqrt{g^2 + f^2 - c}$.

Note If an equation is a polynomial of order 2, in x and y, such that:
(a) the coefficients of x^2 and y^2 are equal
(b) there is no term in xy
then the equation represents a circle.

Example 8 Find the centre and the radius of the circle given by $4x^2 + 4y^2 + 2x - y - 8 = 0$.

$$4x^2 + 4y^2 + 2x - y - 8 = 0$$

$$(\div 4) \Rightarrow \quad x^2 + y^2 + \frac{x}{2} - \frac{y}{4} - 2 = 0$$

Method 1
Coefficient of $x = 2g = \frac{1}{2} \quad \Rightarrow g = \frac{1}{4}$
Coefficient of $y = 2f = -\frac{1}{4} \Rightarrow f = -\frac{1}{8}$
$\therefore$ Centre of circle $= (-g, -f) = (-\frac{1}{4}, \frac{1}{8})$

Radius $= \sqrt{g^2 + f^2 - c} = \sqrt{\frac{1}{16} + \frac{1}{64} + 2} = \frac{\sqrt{137}}{8}$

Method 2

$$x^2 + \frac{x}{2} + y^2 - \frac{y}{4} = 2$$

'Completing the squares' gives

$$(x + \tfrac{1}{4})^2 + (y - \tfrac{1}{8})^2 - \tfrac{1}{16} - \tfrac{1}{64} = 2$$

$$\Rightarrow (x + \tfrac{1}{4})^2 + (y - \tfrac{1}{8})^2 = \tfrac{1}{16} + \tfrac{1}{64} + 2 = \tfrac{137}{64}$$

$$\Rightarrow \text{centre } (-\tfrac{1}{4}, \tfrac{1}{8}) \text{ radius } \frac{\sqrt{137}}{8}$$

Example 9   Find   the   equation   of   the   tangent   to   the   circle
$x^2 + y^2 - 6x + 4y + 3 = 0$ at the point $(0, -1)$.

Differentiating $2x + 2y \cdot \dfrac{dy}{dx} - 6 + 4 \cdot \dfrac{dy}{dx} = 0$

At $(0, -1)$ $0 - 2\dfrac{dy}{dx} - 6 + 4\dfrac{dy}{dx} = 0$

$\therefore \dfrac{dy}{dx} = 3 = \text{gradient of tangent.}$

Equation of tangent: $\dfrac{y+1}{x} = 3 \Rightarrow y = 3x - 1$

Example 10   Find the equation of the circle that passes through the
points $(1, 1)$, $(1, 0)$, $(3, 2)$.

Let equation be $x^2 + y^2 + 2gx + 2fy + c = 0$

$(1, 1)$ lies on circle: $1 + 1 + 2g + 2f + c = 0$ (1)

$(1, 0)$ lies on circle: $1 + 0 + 2g \quad\quad + c = 0$ (2)

$(3, 2)$ lies on circle: $9 + 4 + 6g + 4f + c = 0$ (3)

Solving (1), (2) and (3) simultaneously $\Rightarrow g = -2\frac{1}{2}, f = -\frac{1}{2}, c = 4.$
$\therefore$ equation is $x^2 + y^2 - 5x - y + 8 = 0$

Systems of Circles

Example 11   Show   that $x^2 + y^2 - 6x - y + 3 + \lambda(x^2 + y^2 + 2x + 4y - 1) = 0$
represents a system of circles passing through the points of intersection of the
circles $x^2 + y^2 - 6x - y + 3 = 0$ and $x^2 + y^2 + 2x + 4y - 1 = 0$.

Consider the circles $x^2 + y^2 - 6x - y + 3 = 0$ (1)

and $x^2 + y^2 + 2x + 4y - 1 = 0$ (2)

and the equation

$$x^2 + y^2 - 6x - y + 3 + \lambda(x^2 + y^2 + 2x + 4y - 1) = 0 \quad\quad (3)$$

In this equation (i)   coefficient of $x^2 = 1 + \lambda = $ coefficient of y^2
             (ii)   there is no term in xy.
$\therefore$ this equation represents a circle for each value of λ.
Also, any pair of values that satisfy both (1) and (2) also satisfy (3)
$\therefore$ the points of intersection of (1) and (2) lie on (3).
$\therefore$ this equation represents a circle passing through the points of intersection

of the first two circles   Q.E.D

• Exercise 15.4   See page 326.

15.5 Conics and Other Curves

Conics

Name	Diagram	Equation	Parametric Equations
Ellipse		$\dfrac{x^2}{a^2} + \dfrac{y^2}{b^2} = 1$	$x = a\cos\theta$ $y = b\sin\theta$
Hyperbola		$\dfrac{x^2}{a^2} - \dfrac{y^2}{b^2} = 1$	—
Rectangular Hyperbola		$xy = c^2$	$x = ct$ $y = c/t$
Parabola		$y^2 = 4ax$	$x = at^2$ $y = 2at$

Other Curves

Name	Diagram	Equation	Parametric Equations
Semi-cubical parabola		$y^2 = x^3$	$x = t^2$ $y = t^3$
Astroid		$x^{2/3} + y^{2/3} = a^{2/3}$ (rarely used)	$x = a\cos^3\theta$ $\underline{y = a\sin^3\theta}$ usual equations for astroid

● **Exercise 15.5** See page 327.

15.6 + A Level Questions

● **Miscellaneous Exercise 15B**   See page 328.

Unit 15   EXERCISES

Exercise 15.1

Sketch the following curves:

1 $y = x^2(x - 1)$ **2** $y = \dfrac{x^2}{x + 1}$ **3** $y = x(2x - 1)(x + 1)$

4 $y^2 = x(x + 1)$ **5** $y^2 = x^3$ **6** $x^2 - y^2 = 9$

7 $y = x - \dfrac{1}{x}$ ***8** $y = |x| - \dfrac{1}{x}$

Exercise 15.2

Sketch the following curves:

1 $y^2 = x$ **2** $y^2 = 4ax$ **3** $(x - a)^2 + (y - b)^2 = 1$

4 $\dfrac{x^2}{a^2} + \dfrac{y^2}{b^2} = 1$ **5** $\dfrac{x^2}{16} + y^2 = 1$ **6** $\dfrac{(x - 1)^2}{25} + \dfrac{(y - 3)^2}{9} = 1$

7 $x = 5 \cos t, y = 3 \sin t$ **8** $x = ct, y = c/t$ **9** $x = at^2, y = 2at$

10 $x = t^2, y = t^3$ **11** $(y - 1)^2 = (x + 2)^3$ **12** $y = \dfrac{x^2}{x + 2}$

*13 $x = a \cos^3 t, y = a \sin^3 t$

*14 A curve is given parametrically by

$$x = \sin t, \quad y = \cos^3 t, \quad -\pi < t \leqslant \pi$$

Show that
(a) $-1 \leqslant x \leqslant 1$ and $-1 \leqslant y \leqslant 1$

(b) $\dfrac{dy}{dx} = a \sin 2t$, where a is constant, and give the value of a

Find the value of $\dfrac{dy}{dx}$ when $x = 0$ and show that the curve has points of

inflexion where $t = -\dfrac{3\pi}{4}, -\dfrac{\pi}{4}, \dfrac{\pi}{4}$ and $\dfrac{3\pi}{4}$

Sketch the curve   (L)

Exercise 15.3

1 Obtain the equation of the normal to the curve $y^2 = x^3$ at the point (t^2, t^3). Show that the equation of the normal at the point where $t = \frac{1}{2}$ is $32x + 24y - 11 = 0$. Find the perpendicular distance from the point $(-1, 2)$ to this normal   (J)

2 A curve is given parametrically by the equations

$$x = 1 + t^2, \quad y = 2t - 1$$

Show that an equation of the tangent to the curve at the point with parameter t is

$$ty = x + t^2 - t - 1$$

Verify that the tangent at A(2, 1) passes through the point C(6, 5)

Show that the line $5y = x + 19$ passes through C and is also a tangent to the curve

Find also the coordinates of the point of contact of this line with the given curve   (L)

3 Prove that for all values of m, the line $y = mx - 2m^2$ is a tangent to the parabola $8y = x^2$

Find the value of m for which the line $y = mx - 2m^2$ is also a tangent to the parabola $y^2 = x$

The line PQ is a tangent to $8y = x^2$ at P and a tangent to $y^2 = x$ at Q. Find the coordinates of P and Q   (A)

4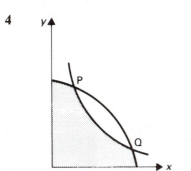

Part of the circle $x^2 + y^2 = 25$ and part of the hyperbola $xy = 12$ are shown in the diagram (not drawn to scale) in which P and Q are two of the points of intersection of the curves and O is the origin. Verify that P is the point (3, 4) and state (or obtain) the coordinates of Q. Without any calculating aids find
(a) the equation of PQ
(b) the tangent of the angle POQ
(c) the area of $\triangle$POQ   (J)

5 The vertex A of the triangle ABC is the point (7, 11) and equation of the side BC is $3x - 4y - 2 = 0$. The mid-point of BC has x-coordinate 2 and the area of the triangle ABC is 30 square units. Find the coordinates of B and C   (A)

● Exercise 15.4

For each of the following equations, state whether it represents a circle. If it does, find the centre and the radius of the circle:

1 $x^2 + y^2 = 9$ 　　　　　　　　　　**2** $4x^2 + 4y^2 = 1$ 　　**3** $4x^2 + y^2 = 9$

4 $(x - 1)^2 + (y + 2)^2 = 4$ 　　　　**5** $x^2 + y^2 - 2x - 4y = 0$

6 $x^2 - y^2 + 2x + 6y - 3 = 0$ 　　**7** $2x^2 + 2y^2 + 4x - y - 8 = 0$

8 $x^2 + y^2 + 6x + 4xy + 6 = 0$ 　　**9** $3x^2 + 3y^2 + x - 6 = 0$

10 $x(x + 3) + y(y - 4) = 0$

11 Find the equation of the diameter of the circle

$$x^2 + y^2 - 4x + 2y = 0$$

which passes through the origin

12 Find the equations of the tangent and the normal at the point $(1, -2)$ on the circle $x^2 + y^2 = 5$

13 Two circles C_1 and C_2 have equations

$$x^2 + y^2 + 4x - 3 = 0 \quad \text{and} \quad 2x^2 + 2y^2 + 4x - 6y + 3 = 0$$

respectively. Show that, for all values of λ, the equation

$$x^2 + y^2 + 4x - 3 + \lambda(2x^2 + 2y^2 + 4x - 6y + 3) = 0$$

respresents a circle passing through the points of intersection of C_1 and C_2

(a) For what particular value of λ does this circle pass through the origin?
(b) What happens when $\lambda = -\frac{1}{2}$?

14 Show that the circle given by

$$x^2 + y^2 - 2px - 2qy + p^2 = 0$$

touches the x-axis.

Two circles touch the x-axis and pass through the points $(3, 2)$ and $(2, 1)$. Find their equations

15 Sketch the circle whose equation is $(x - 3)^2 + (y - 4)^2 = 1$

Mark on your diagram the two tangents from the origin to this circle. Find the gradients of these tangents and the cosine of the acute angle between them.

16 $x^2 + y^2 + 2kx - 4y - 4k = 0$, $k \in \mathbb{R}$, $k > 0$ defines a set of circles. C_1 and C_2 belong to this set. C_1 has radius 5 and C_2 touches the x-axis. Find the value of k in each case
Show that C_1 and C_2 touch each other and find the equation of their common tangent

Exercise 15.5

1 Make rough sketches of the following curves:

(a) $\dfrac{x^2}{36} + \dfrac{y^2}{9} = 1$

(b) $\dfrac{x^2}{4} + y^2 = 1$

(c) $\dfrac{4x^2}{25} + \dfrac{4y^2}{9} = 1$

(d) $4x^2 + y^2 = 36$

(e) $\dfrac{(x-1)^2}{16} + \dfrac{(y-3)^2}{9} = 1$

(f) $y^2 = 12x$

(g) $y^2 + 8x = 0$

(h) $\dfrac{x^2}{9} - \dfrac{y^2}{25} = 1$

(i) $x = t^2, y = t^3, t \in \mathbb{R}$

(j) $x = t^2, y = t^3, t \geqslant 0$

(k) $x = 2\cos^3 \theta, y = 2\sin^3 \theta$

2 The points $P(8,3)$ and $Q(6,4)$ lie on an ellipse whose equation is $\dfrac{x^2}{a^2} + \dfrac{y^2}{b^2} = 1$. Find a and b.

The ellipse intersects the positive x-axis at A. Prove that AP is perpendicular to OQ
(J)

3 The tangent to the curve $4ay = x^2$ at the point P $(2at, at^2)$ meets the x-axis at the point Q. The point S is $(0, a)$.

(a) Prove that PQ is perpendicular to SQ
(b) Find a cartesian equation for the locus of the point, M, the mid-point of PS
(A)

4 A curve is given parametrically by the equations
$$x = a\cos^3 \theta, \quad y = a\sin^3 \theta, \quad 0 \leqslant \theta \leqslant 2\pi$$

Show that $\dfrac{dy}{dx} = -\tan \theta$ and give a sketch of the curve.

Find the equation of the tangent to the curve at the point T whose parameter is θ

The tangent meets the axes of x and y at the points P and Q respectively. Show that $PQ = a$. Find the Cartesian equation of the locus of the midpoint of PQ as T varies on the curve
(J)

5 The parametric equations of a curve are
$$x = 2t, \quad y = 2/t, \quad t \neq 0$$

The tangent to the curve at $P(2p, 2/p)$ meets the x-axis at A; the normal to the curve at P meets the y-axis at B. The point Q is the midpoint of AB. Find the parametric equations of the locus of Q as P moves on the given curve
(J)

Miscellaneous Exercise 15B

1 Find the distance from the origin to the centre of the circle

$$x^2 + y^2 + 6x + 8y - 24 = 0 \qquad\qquad \text{(L: 15.4)}$$

2 A curve is defined by the equations

$$x = at^2, \quad y = at^3$$

where a is a positive constant and t is a parameter. Sketch the curve and note on your sketch the set of values of t corresponding to each branch of the curve   (L: 15.5)

3 Find in any form an equation for the circle which passes through the points $O(0, 0)$, $P(6, 0)$ and $Q(0, 8)$. State the length of the radius and the coordinates of the centre of this circle   (L: 15.4)

4 Sketch the curve whose equation is $y = \dfrac{x - 2}{x + 2}$ and state the equations of its asymptotes. On the same diagram, sketch the curve whose equation is $x^2 + 4y^2 = 4$

Hence, or otherwise, find all real solutions of the equation

$$4(x - 2)^2 = (4 - x^2)(x + 2)^2 \qquad\qquad \text{(J: 12.2, 15.5)}$$

5 A curve is defined by the equations

$$x = t^2 - 1, \quad y = t^3 - t$$

where t is a parameter. Sketch the curve for all real values of t

Find the area of the region enclosed by the loop of the curve
    (L: 8.5, 11.3, 12.3)

6 The complex number $z = x + iy$ is such that $\dfrac{z + 3}{z - 4i} = \lambda i$, where λ is real. Show that $x(x + 3) + y(y - 4) = 0$.

If z is represented on an Argand diagram by the point P, deduce that, when λ varies, P moves on a circle. Find the radius of this circle and the complex number corresponding to its centre   (J: 14.2, 15.4)

7 The line $y = m(x - 1)$ $(m \neq 0)$ meets the parabola $y^2 = 4x$ at P and Q. Find, and simplify, a quadratic expression whose roots are the x coordinates of P and Q. Hence find, in terms of m, the coordinates of the midpoint M of PQ.

Given that m now varies, so that P and Q move on the parabola, find the Cartesian equation of the curve on which M moves.
    (J: 15.3, 10.11)

8 A point P moves so that the distance from the point $A(a, 3a)$ is twice the distance from the point $B(4a, -3a)$. Show that the locus of P is a circle of radius $2\sqrt{5a}$ and find the coordinates of its centre C.

The line through B at right angles to AB cuts the circle at D and E. Find the length of the chord DE   (A: 15.3, 15.4)

9 Sketch the curve given parametrically by

$$x = t^2 - 2, \quad y = 2t, \quad t \in \mathbb{R}$$

indicating on your sketch where (a) $t = 0$ (b) $t > 0,$ (c) $t < 0$

Calculate the area of the finite region enclosed by the curve and the y-axis   (L: 5.4, 15.5)

10 Find the radius and coordinates of the centre of the circle S given by the equation $x^2 + y^2 + 10x - 6y - 2 = 0$.

A chord AB of S is a tangent to the circle with the same centre but with radius equal to one half that of S. The tangents to S at A and B intersect at C. Prove that

(a) the $\triangle$ ABC is equilateral
(b) the point C is on the circle whose equation is

$$x^2 + y^2 + 10x - 6y - 110 = 0 \qquad\qquad \text{(J: 15.4)}$$

11 An ellipse is defined by the parametric equations

$$x = a \cos t, \quad y = b \sin t, \quad 0 \leqslant t < 2\pi$$

Show that the equation of the tangent to the curve at the point $P(a \cos p, \, b \sin p)$ is

$$\frac{x}{a} \cos p + \frac{y}{b} \sin p = 1$$

This tangent meets the curve defined by the parametric equations

$$x = 2a \cos \theta, \quad y = 2b \sin \theta, \quad 0 \leqslant \theta \leqslant 2\pi$$

at the points Q and R, which are given by $\theta = q$ and $\theta = r$, respectively. Show that p differs from each of q and r by $\pi/3$ (J: 15.3, 15.5)

12 Find the equation of the normal at the point P with parameter t on the curve with parametric equations

$$x = t^2, \quad y = 2t$$

Show that, if this normal meets the x-axis at G, and S is the point $(1, 0)$, then $SP = SG$.

Find also the equation of the tangent at P, and show that, if the tangent meets the y-axis at Z, then SZ is parallel to the normal at P   (L: 15.5)

13 Show that the curves whose equations are $y - 1 = x^3$ and $y + 3 = 3x^2$ intersect at a point on the x-axis and find the coordinates of this point. Show that the only other point at which the curves meet is $(2, 9)$ and that the curves have a common tangent here. Sketch the two curves on the same diagram. Show that the area of the finite region bounded by the curves is $27/4$ (J: 12.2, 5.4)

14 For the curve $y = (3 - 2x)(x^2 - 36)$, prove that $\dfrac{dy}{dx} > 0$ if and only if

$-3 < x < 4$

Sketch the curve $y = (3 - 2x)(x^2 - 36)$, clearly labelling with their coordinates

(a) the turning points
(b) the points where the curve intersects the coordinate axes

Hence show, in a separate sketch, the general shape of the curve

$$y^2 = (3 - 2x)(x^2 - 36) \qquad \text{(A: 12.3, 15.1)}$$

15 Two circles C_1 and C_2 have equations

$$x^2 + y^2 - 4x - 8y - 5 = 0 \quad \text{and} \quad x^2 + y^2 - 6x - 10y + 9 = 0$$

respectively. Find the x-coordinates of the points P and Q at which the line $y = 0$ cuts C_1 and show that this line touches C_2

Find the tangent of the acute angle made by the line $y = 0$ with the tangents to C_1 at P and Q. Show that, for all values of the constant λ, the circle C_3 whose equation is

$$\lambda(x^2 + y^2 - 4x - 8y - 5) + x^2 + y^2 - 6x - 10y + 9 = 0$$

passes through the points of intersection of C_1 and C_2. Find the two possible values of λ for which the line $y = 0$ is a tangent to C_3

(J: 15.4)

16 A curve is defined with parameter t by the equations

$$x = at^2, \quad y = 2at$$

The tangent and normal at the point P with parameter t_1 cut the x-axis at T and N respectively. Prove that $\text{PT/PN} = |t_1|$ (L: 15.3, 15.5)

17 Sketch the curve whose equation is $y = \dfrac{x+2}{x}$ and state the equations of its asymptotes.

By considering this sketch, or otherwise, show that the curve whose equation is $y = \ln\left(\dfrac{x+2}{x}\right)$ has no points whose x-coordinates lie in the interval $-2 \leqslant x \leqslant 0$. Sketch this curve on a separate diagram. Prove that the area bounded by the **second** curve, the x-axis and the lines $x = 1$ and $x = 2$ is $3 \ln 4/3$ (J: 5.4, 11.6, 12.2, 15.1)

18 The tangent to the parabola $y^2 = 4ax$ at the point $P(at^2, 2at)$ meets the y-axis at G. The normal at the point P meets the x-axis at the point H.

(a) Find the equation of the normal and show that the mid-point M of HG has coordinates $\left(a + \dfrac{at^2}{2}, \dfrac{at}{2}\right)$

(b) Find the cartesian equation of the locus of M as P moves on the parabola
     (A: 15.3, 15.5)

19 Sketch the curve whose equation is $y + 3 = \dfrac{6}{x-1}$

Find the coordinates of the points where the line $y + 3x = 9$ intersects the curve and show that the area of the region enclosed between the curve and the line is $\frac{3}{2}(3 - 4\ln 2)$

Determine the equations of the two tangents to the curve which are parallel to the line (J: 12.2, 15.3)

20 The points $A(-8, 9)$ and $C(1, 2)$ are opposite vertices of a parallelogram ABCD. The sides BC, CD of the parallelogram lie along the lines $x + 7y - 15 = 0$, $x - y + 1 = 0$, respectively. Calculate

(a) coordinates of D
(b) the tangent of the acute angle between the diagonals
(c) the length of the perpendicular from A to side CD
(d) the area of the parallelogram (J: 15.3)

*21 The tangents at the points $P(p^2, p^3)$ and $Q(q^2, q^3)$ to the curve $y^2 = x^3$ meets at the point R. Find the coordinates of R in terms of p and q. Given that the tangent to the curve at P is perpendicular to OQ, where O is the origin, prove that $pq = -\frac{2}{3}$. Find the Cartesian equation of the locus of R when P and Q move on the curve in such a way that the relation $pq = -\frac{2}{3}$ is always satisfied (J: 10.4, 15.5)

*22 The tangents to the parabola $y^2 = x$ at the variable points $P(p^2, p)$ and $Q(q^2, q)$ intersect at N. Given that M is the mid-point of PQ, prove that

$$MN = \tfrac{1}{2}(p - q)^2$$

Given that PN is always perpendicular to QN:

(a) show that the locus of N is a straight line, and write down its equation, and
(b) find the Cartesian equation of the locus of M (J: 15.5)

*23 A circle has centre (α, β). Show that if (α, β) lies on the line $2x = 3y + 7$ then the equation of the circle may be written in the form

$$x^2 + y^2 - (3\beta + 7)x - 2\beta y = k$$

If the circle also passes through the points $(1, 4)$ and $(2, -3)$, find its equation and show that the circle touches the y-axis

Calculate the length of a tangent from the point $(12, 6)$ to the circle (A: 15.4)

*24 A straight line parallel to the line $2x + y = 0$ intersects the x-axis at A and the y-axis at B. The perpendicular bisector of AB cuts the y-axis at C. Prove that the gradient of the line AC is $-\frac{3}{4}$

Find also the tangent of the acute angle between the line AC and the bisector of the angle AOB, where O is the origin (J: 15.3)

UNIT 16 NUMERICAL METHODS

16.1 Iterative Methods For Finding the Roots of Equations

Iterative Methods

An iterative method is a systematic method for producing successively better approximations to the requir-ed solution. It may be pictured as

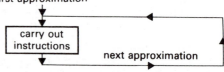

Each passage through the 'carryout instructions' box is called an **iteration**. We start with the first approximation and apply the first iteration to produce the second approximation. The second iteration produces the third approximation and so on. So, to apply an iterative method, we need to have:

(i)   a method of getting the first approximation
(ii) a set of instructions
(iii) an instruction that tells you when to stop.

(iii) is usually quite simple. It is often of the form 'perform three iterations' or 'find the fourth approximation' or 'stop when two successive approxima-tions agree to 3 decimal places'.

Location of Roots

1 Graphical Method

Example 1   Find the number and rough position of any roots of the equations (a) $\dfrac{x+1}{x} = 3$ (b) $\sin x = \dfrac{1}{2x}$, $-\pi \leqslant x \leqslant \pi$

> The roots of $\dfrac{x+1}{x} = 3$ are the x-coordinates of the points of intersection of $y = \dfrac{x+1}{x}$ and $y = 3$.

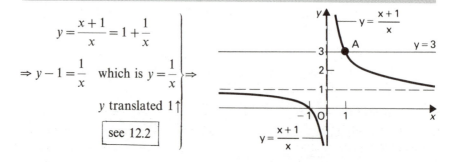

$$y = \frac{x+1}{x} = 1 + \frac{1}{x}$$

$$\Rightarrow y - 1 = \frac{1}{x} \quad \text{which is } y = \frac{1}{x}$$

y translated $1\uparrow$

see 12.2

$y = \dfrac{x+1}{x}$ and $y = 3$ intersect once, at A.

$\therefore$ the equation has one root and it lies in the interval $]0, 1[$

(b) The roots of $\sin x = \dfrac{1}{2x}$ are the x-coordinates of the points of intersection

of $y = \sin x$ and $y = \dfrac{1}{2x}$ (for $-\pi \leqslant x \leqslant \pi$)

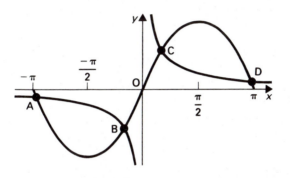

For $-\pi \leqslant x \leqslant \pi$, $y = \sin x$ and $y = \dfrac{1}{2x}$ intersect at A, B, C and D.

$\therefore$ the equation has four roots. One is a little less than π, one a little more than $-\pi$, one in the interval $] 0, \pi/2 [$ and one in the interval $] -\pi/2, 0[$.

2 Location of Roots by Change of Sign

If $f(a)$ is positive and $f(b)$ is negative and f is continuous between a and b, then $f(x) = 0$ for some value of x between a and b.

Example 2 Find the values of $f(x) = x^3 - 15$ for integral values of x from 1 to 4. Hence deduce the two consecutive integers between which the real root of $x^3 = 15$ lies.

The root of $x^3 = 15$ is the root of $f(x) = 0$.

x	1	2	3	4
$f(x)$	-14	-7	12	49

Change of sign between $x = 2$ and $x = 3$

Since $f(2)$ is negative and $f(3)$ is positive $f(x)$ must be zero for a value of x-between 2 and 3

Example 3 By considering the sign of $x^3 - x^2 - 5x + 3$ for integral values of x from -3 to 3, find three values of α such that the roots of $x^3 - x^2 - 5x + 3 = 0$ lie in the intervals $]\alpha, \alpha + 1[$ where $\alpha \in \mathbb{Z}$.

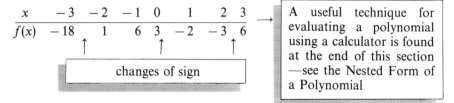

x	-3	-2	-1	0	1	2	3
$f(x)$	-18	1	6	3	-2	-3	6

changes of sign

A useful technique for evaluating a polynomial using a calculator is found at the end of this section —see the Nested Form of a Polynomial

Roots of $x^3 - x^2 - 5x + 3 = 0$ lie in $]-3, 2[$, $]0, 1[$, $]2, 3[$ that is, in $]\alpha, \alpha + 1[$ where $\alpha = -3, 0, 2$

Important Note Before considering the changes of sign, an equation must first be put into the form $f(x) = 0$.

Iterative Method No. 1: The Ten-Subdivision Method.

This is a systematic extension of the location of roots by change of sign.

Example 4 Use the ten-subdivision method to find a value for $\sqrt{2}$ correct to 2 decimal places.

We need to find the solution of $x = \sqrt{2}$ or $x^2 = 2$, or $x^2 - 2 = 0$. Let $f(x) \equiv x^2 - 2$.

Step 1:

x	0	1	2	3	4
$f(x)$	-2	-1	$+2$		

root

Notice that once you have located the root there is no need to find the rest of the values in the interval

Step 2: Divide [1, 2] into 10 parts

x	1	1.1	1.2	1.3	1.4	1.5	1.6	1.7	1.8	1.9	2
$f(x)$	-1	-0.8	-0.6	-0.3	-0.04	$+0.25$					

root

Step 3: Divide [1.4, 1.5] into 10 parts

x	1.40	1.41	1.42	1.43	1.44	1.45	1.46	1.47	1.48	1.49	1.50
$f(x)$	-0.4	-0.01	$+0.02$								

root

Step 4: Divide [1.41, 1.42] into 10 parts

x	1.410	1.411	1.412	1.413	1.414	1.415
$f(x)$	$-$	$-$	$-$	$-$	$-$	$+$

root

$\Rightarrow \sqrt{2}$ lies between 1.414 and 1.415
$\Rightarrow \sqrt{2} \approx 1.41$ to 2 decimal places

A Useful Technique: The Nested Form of a Polynomial

If you wish to evaluate the polynomial

$$2x^3 + 4x^2 + 3x + 1$$

for the value $x = -1.55$ you may find that your calculator does not handle powers of negative numbers. Even if it does, you will probably find that you need to use brackets, say

$$2(-1.55)^3 + 4(-1.55)^2 + 3(-1.55) + 1$$

which can be cumbersome.

To avoid such difficulties it is best to **nest** the expression:

$$2x^3 + 4x^2 + 3x + 1$$
$$\equiv [2x^2 + 4x + 3]x + 1$$
$$\equiv [(2x + 4)x + 3]x + 1$$

→ Note that the powers of x must be written in descending order first

This rearrangement avoids terms involving x^2 and x^3.

Example 6   Using the nested form of $2x^3 + 4x^2 + 3x + 1$, and a calculator, evaluate the polynomial for $x = 5$ and $x = -1.55$.

$$2x^3 + 4x^2 + 3x + 1 \equiv ((2x + 4)x + 3)x + 1$$

$x = 5$ A possible key sequence is

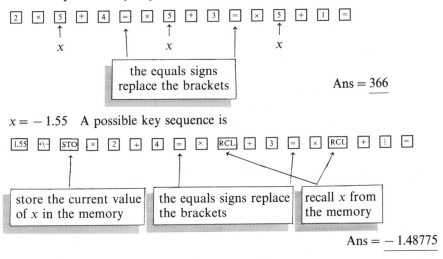

the equals signs replace the brackets

Ans $= 366$

$x = -1.55$ A possible key sequence is

store the current value of x in the memory

the equals signs replace the brackets

recall x from the memory

Ans $= -1.48775$

● Exercise 16.1   See page 348.

16.2 Further Iterative Methods

Iterative Method No. 2: Bisection

Bisection is a simpler, faster but reliable variation of the ten-subdivision method. Whereas the ten-subdivision method requires calculating up to nine values for each step, the bisection method only calculates one value. The mid point of each interval is calculated and, for each iteration, the half interval is used which contains the root.

Example 7   Use the bisection method to the find the value of the largest positive root of $x^3 - x^2 - 5x + 3 = 0$ which has a maximum error of less than 0.1.

Let $f(x) \equiv x^3 - x^2 - 5x + 3$ and let the approximation at each stage be the midpoint of the interval.

Step 1

x	-3	-2	-1	0	1	2	3
$f(x)$	$-18 \uparrow$	1	6	$3 \uparrow$	-2	$-3 \uparrow$	6

         root   root   largest positive root

$\Rightarrow$ 1st approximation $= x_1 = 2.5$ with maximum possible error 0.5

Step 2:

x	2	2.5	3
$f(x)$	-3	$-0.125 \uparrow$	6

             root

$\Rightarrow$ 2nd approximation $= x_2 = 2.75$ with max. poss. error 0.25

Step 3:

x	2.5	2.75	3
$f(x)$	$-0.125 \uparrow$	$+2.5$	6

          root

$\Rightarrow x_3 = 2.625$ with max. poss. error 0.125

Step 4:

x	2.5	2.625	2.75
$f(x)$	-0.125	$+1.07$	$+2.5$

$\Rightarrow x_4 = 2.5625$ with max. poss. error 0.0625 which is < 0.1

$\therefore$ largest positive root is ≈ 2.5625 with max. poss. error < 0.1

Note   For the ten-subdivision method and the bisection method the equation must be written in the form $f(x) = 0$.

Iterative Method No 3: Formula Iteration

In this method, the 'carry out instructions' box contains a formula connecting x_{n+1} and x_n.

$\xrightarrow{x_1} \boxed{x_{n+1} = f(x_n)} \xrightarrow{x_2} \boxed{x_{n+1} = f(x_n)} \xrightarrow{x_3} \boxed{x_{n+1} = f(x_n)} \xrightarrow{x_4}$

Example 8   Using $x_1 = 1$ and the iterative formula and $x_{n+1} = \frac{1}{2}(x_n + 2/x_n)$ calculate x_2, x_3 and x_4

$\xrightarrow{x_1 = 1} \boxed{x_2 = \frac{1}{2}\left(x_1 + \frac{2}{x_1}\right)} \xrightarrow{x_2 = 1.5} \boxed{x_3 = \frac{1}{2}\left(x_2 + \frac{2}{x_2}\right)} \xrightarrow{x_3 = 1.4167}$

$\boxed{x_4 = \frac{1}{2}\left(x_3 + \frac{2}{x_3}\right)} \xrightarrow{x_4 = 1.4142157}$

$\Rightarrow x_2 = 1.5, \; x_3 = 1.4167, \; x_4 = 1.4142157$

Example 9   To what equation does the iterative formula $x_{n+1} = \frac{1}{2}\left(x_n + \frac{2}{x_n}\right)$ give an approximate solution?

If x_n is the *exact* solution, for some n, then putting this in the iteration would give the same value for x_n.

$$\Rightarrow x_n = \frac{1}{2}\left(x_n + \frac{2}{x_n}\right)$$

$$\Rightarrow 2x = x + \frac{2}{x} \qquad \longleftarrow \boxed{\text{dropping the suffixes}}$$

$$\Rightarrow x = \frac{2}{x}$$

$$\Rightarrow x^2 = 2$$

Iterative Method No. 4: The Newton–Raphson Method

To find an approximate solution to an equation of the form

$$f(x) = 0$$

use the iterative formula

$$x_{n+1} = x_n - \frac{f(x_n)}{f'(x_n)}$$

Example 10   Show that the equation $x^3 + 2x^2 = 5x + 7$ has a root in the interval [2, 3]. Find the Newton–Raphson iteration formula for this equation and find the root correct to 4 decimal places.

Let
$$f(x) = x^3 + 2x^2 - 5x - 7 \qquad \boxed{\begin{array}{l}\text{The equation must be in the} \\ \text{form } f(x) = 0\end{array}}$$
$$f(2) = -1$$
$$f(3) = 23$$

Since f changes sign in the interval [2, 3], f has a root in that interval.

$$f'(x) = 3x^2 + 4x - 5$$

$\Rightarrow$ Newton–Raphson formula is

$$x_{n+1} = x_n - \frac{x_n^3 + 2x_n^2 - 5x_n - 7}{3x_n^2 + 4x_n - 5}$$

or

$$x_{n+1} = x_n - \frac{((x_n + 2)x_n - 5)x_n - 7}{(3x_n + 4)x_n - 5}$$

Let $\qquad x_1 = 2$ ←——————— | x_1 can be any value close to the root |

$$\Rightarrow x_2 = 2 - \frac{((2+2)2-5)2-7}{(3(2)+4)2-5} \simeq 2.06667$$

$$\Rightarrow x_3 = 2.06443$$

$$\Rightarrow x_4 = 2.06443$$

∴ to 4 d.p. the root is 2.0644

Note   Always work to as many decimal places as possible on your calculator. When possible, put the value you obtain for one iteration into the memory, to use in the next iteration. However you should only **record** values to one decimal place more than that required in the answer.

● **Exercise 16.2**   See page 349.

16.3 Permutations

A **permutation** is an ordered arrangement.

Permutations of Unlike Items

Example 11   In how many different ways can four books be arranged on a shelf?

No. of ways of choosing a book to go in the first position $= 4$
Then no. of ways of choosing a book to go in the second position $= 3$
and no. of ways of choosing a book to go in the third position $= 2$
and no. of ways of choosing a book to go in the last position $= 1$
$\Rightarrow$ No. of different ways of arranging the books $= 4 \times 3 \times 2 \times 1 = 24$

| Number of ways of arranging n items $= n!$ |

Example 12   In how many different ways can four books, chosen from 6, be arranged on a shelf?

No. of ways of choosing a book to go in the first position $= 6$
Then no. of ways of choosing a book to go in the second position $= 5$
and no. of ways of choosing a book to go in the third position $= 4$
and no. of ways of choosing a book to go in the last position $= 3$
$\Rightarrow$ No. of different ways of arranging 4 books chosen from 6

$$= 6 \times 5 \times 4 \times 3 = 360$$

Number of ways of arranging r items chosen from n
$$_nP_r = \underbrace{n(n-1)\cdots(n-r+1)}_{r \text{ terms}} = \frac{n!}{(n-r)!}$$

Permutations with Some Identical Items

Example 13   How many different arrangements are there of the letters in
the words (a)   PAPER   (b)   PUPPY   (c)   PEPPER

(a)   If we write PAPER as P_1AP_2ER then the number of different arrangements
is 5! But, without the suffixes, we will have duplications. For example
P_1AP_2ER and P_2AP_1ER will appear the same.

$$\Rightarrow \text{ number of different arrangements} = \frac{5!}{2} = \underline{60}$$

(b)   There are 5! different arrangements of $P_1UP_2P_3Y$ but

$$P_1UP_2P_3Y \qquad P_2UP_1P_3Y \qquad P_3UP_1P_2Y$$
$$P_1UP_3P_2Y \qquad P_2UP_3P_1Y \qquad P_3UP_2P_1Y$$

will appear the same without suffixes. The number of repeats is equal to the
number of arrangements of P_1, P_2, P_3—that is 3!

$$\Rightarrow \text{ number of different arrangements} = \frac{5!}{3!} = \underline{20}$$

(c)   Number of different arrangements of $P_1E_1P_2P_3E_2R$ is 6!

$$\Rightarrow \text{ number of different arrangements of PEPPER} = \frac{6!}{3!2!} = \underline{60}$$

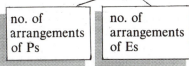

no. of arrangements of Ps	no. of arrangements of Es

> Number of arrangements of n objects of which p are alike of the first kind,
> q are alike of the second kind, r are alike of the third kind,...
>
> $$= \frac{n!}{p!q!r!\cdots}$$

Permutations when Repetition is Permitted

Example 14   There are four boxes of coins: one contains 2p coins, one
contains 5p coins, one contains 10p coins and one contains 20p coins. Find the
number of permutations of five coins chosen when repetition is permitted.

Since repetition is allowed, the no. of ways of choosing each coin is 4.
$\Rightarrow$ number of permutations is $4 \times 4 \times 4 \times 4 \times 4 = \underline{4^5}$

> Number of permutations of n unlike items taken r at a time, when
> repetition is allowed, is n^r

Closed (or Circular Permutations)

In these arrangements, there are no first or last places to consider. For example

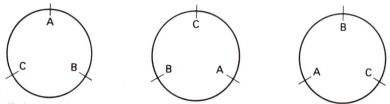

are all the same arrangement.

Example 15   In how many different ways can 4 children, chosen from 8, be seated round a circular table.

No. of ways of filling the first seat = 8
Then no. of ways of filling the second seat = 7
and no. of ways of filling the third seat = 6
and no. of ways of filling the fourth seat = 5

$\Rightarrow$ no. of ways $= 8 \times 7 \times 6 \times 5$

But, for any one arrangement in the four seats, the children can be moved clockwise four times and they will still be sitting in the same order. Hence the $8 \times 7 \times 6 \times 5$ ways is **four** times the ways of arranging them around a circular table.

$\therefore$ no. of different ways $= \dfrac{8 \times 7 \times 6 \times 5}{4} = \underline{420}$

Example 16   In how many different ways can 4 beads, chosen from 8, be threaded onto a circular wire.

The number of ways of arranging the 4 beads in a circle is

$\dfrac{8 \times 7 \times 6 \times 5}{4}$ $\leftarrow$ | as in example 15 |

but, a ring can be turned over.
For example.

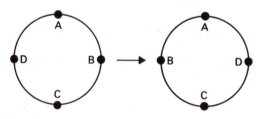

So the number of circular arrangements is **twice** the number of arrangements on a ring.

$\Rightarrow$ no. of arrangements of beads $= \dfrac{8 \times 7 \times 6 \times 5}{4 \times 2} = \underline{210}$

● **Exercise 16.3**   See page 350.

16.4 Combinations

A **combination** is a selection in which the order is irrelevant.

Example 17 In how many ways can 4 books be chosen from 4, if the order of choosing is not important?

Number of ways of choosing 4 books from $4 = 4! = 24$. But, if the order is unimportant, the number of ways $= \underline{1}$

Example 18 In how many ways can 4 books be chosen from 6, if the order of choosing is not important?

Number of ways of choosing 4 from $6 = 6 \times 5 \times 4 \times 3 = 360$. But, for each particular combination of 4 books there are $4! = 24$ arrangements.

$\Rightarrow$ Number of different sets (combinations) of books $= \dfrac{360}{24} = \underline{15}$

Number of combinations of r objects from n

$= \dfrac{\text{number of permutations of } r \text{ from } n}{\text{number of permutations of } r}$

$= \dfrac{n!}{(n-r)!\,r!} = {_n}C_r$

Example 19 How many different hands of 3 cards can be dealt from a pack of 52?

Number of hands $= \dfrac{52 \cdot 51 \; 50}{3!} = \underline{22100}$

Example 20 In how many ways can the England football team be selected from a squad of 14 players if the goal keeper has to be included.

This is equivalent to selecting 10 from 13.

No. of ways $= {_{13}}C_{10} = \dfrac{13 \cdot 12 \cdot 11}{3 \cdot 2 \cdot 1} = \underline{286}$

Division of a Group into Two Unequal Groups

Example 21 In how many ways can a group of 8 children be divided into a group of 6 and a group of 2.

If a group of 6 children is chosen this will automatically leave a group of 2.

$\therefore$ number of ways $= {_8}C_6 = \dfrac{8!}{6!\,2!} = \dfrac{8 \cdot 7}{1 \cdot 2} = \underline{28}$

Note You would have obtained the same number if you had chosen 2 from 8 $({_n}C_r = {_n}C_{n-r})$

Division of a Group into Two Equal Groups

Example 22   In how many ways can a group of 8 children be divided into two groups of 4?

Number of ways of choosing $4 = {}_8C_4 = \dfrac{8 \cdot 7 \cdot 6 \cdot 5}{4 \cdot 3 \cdot 2 \cdot 1} = 70$

But if the children are labelled ABCDEFGH, say, then the division into ABCD and EFGH is repeated later as EFGH and ABCD.
$\Rightarrow$ number of different combinations $= \frac{70}{2} = 35$

Independent Combinations

Example 23   In how many ways can you choose 8 apples and 3 oranges from 12 apples and 5 oranges.

The choice of apples and oranges are independent of each other
No. of combinations of apples $= {}_{12}C_8 = 495$
No. of combinations of oranges $= {}_5C_3 = 10$
$\Rightarrow$ No. of combinations of apples and oranges $= {}_{12}C_8 \times {}_5C_3 = 495 \times 10 = 4950$

- Exercise 16.4   See page 351.

16.5 Conditional Arrangements and Selections

Permutations with One Restriction

Example 24   How many permutations are there of the letters in the word TALLY where
(a)   the two Ls are next to each other
(b)   the two Ls are not next to each other.

(a)   T   A   (LL)   Y
      Consider the LL as one letter.
      No. of permutations with LL $= {}_4P_4 = 4! = 24$

(b)   No. of permutations where the Ls are separated
      $=$ no. of permutations $-$ no. of permutation with LL
      $= 5! - 4!$
      $= 120 - 24 = 96$

Example 25   How many numbers greater than 3000 can be made from the digits 2, 3, 4, 5 if each digit can be used only once?
Restriction: First digit must be 3, 4 or 5.

We start with the restricted item:
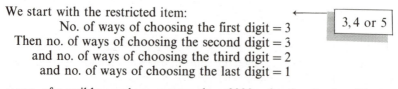
              No. of ways of choosing the first digit $= 3$
       Then no. of ways of choosing the second digit $= 3$
          and no. of ways of choosing the third digit $= 2$
           and no. of ways of choosing the last digit $= 1$

$\Rightarrow$ no. of possible numbers greater than $3000 = 3 \times 3 \times 2 \times 1 = 18$

Permutations with More than one Restriction

Example 26 How many **odd** numbers greater than 3000 can be made from 2, 3, 4, 5 if each digit can be used only once?
Restrictions: First digit must be 3, 4 or 5 and last digit must be 3 or 5.

> Since this is the most restrictive, this is the one we consider first

No. of ways of choosing the last digit $= 2$
Then no. of ways of choosing the first digit $= 2$
and no. of ways of choosing the second digit $= 2$
and no. of ways of choosing the third digit $= 1$

$\Rightarrow$ no. of possible odd numbers greater than $3000 = 2 \times 2 \times 2 \times 1 = 8$

Selection with Repetition

Example 27A A shop stocks 8 different packet soups. In how many different ways can you buy 4 packets of soup if (a) every packet you buy is different (b) at least two are the same.

(a) No. of selections $= {_8}C_4 = \dfrac{8 \cdot 7 \cdot 6 \cdot 5}{4 \cdot 3 \cdot 2 \cdot 1} = 70$

(b) These are 4 possibilities:

    (i) all 4 of the same variety
    (ii) 3 of the same variety and 1 other
    (iii) 2 of one variety and 2 of another
    (iv) 2 of one variety and 2 of other different varieties

    (i) No. of selections $= 8$
    (ii) There are 8 ways of choosing the three and 7 ways of choosing the one.
    (iii) There are 8 ways of choosing the first two and 7 ways of choosing the second two.
        Say the 8 kinds of soup are labelled A, B, C, D, E, F, G, H. Then, if you choose 2 packets of A followed by 2 packets of B, this is the same as choosing 2 packets of B followed by 2 packets of A.

$$\therefore \text{ no. of selections} = \frac{8 \times 7}{2} = 28$$

    (iv) There are 8 ways of choosing the two, 7 of choosing one and 6 of choosing one.
        But, choosing 2 of A, 1 of B and 1 of C is the same as choosing 2 of A, 1 of C and 1 of B.

$$\therefore \text{ no. of selections} = \frac{8 \times 7 \times 6}{2} = 168$$

$\therefore$ total no. of selections $= 8 + 56 + 28 + 168 = 260$

Example 28   How many different ways are there of selecting 4 letters from the letters in the word TATTERS?

No. of selections containing 3 Ts $= {}_4C_1 = 4$
No. of selections containing 2 Ts $= {}_4C_2 = 6$
No. of selections containing 1 T $= {}_4C_3 = 4$
No. of selections containing 0 Ts $= {}_4C_4 = 1$

Total no. of selections $= 4 + 6 + 4 + 1 = 15$

Permutations and Combinations

Example 29   How many different ways are there of arranging 4 letters from the letters in the word TATTERS?

No. of selections containing 3Ts $= {}_4C_1 = 4$

and each of these selections can be arranged $\dfrac{4!}{3!}$ ways

See example 13

$\therefore$ No. of arrangements $= 4 \times \dfrac{4!}{3!} = 16$

No. of selections containing 2Ts $= {}_4C_2 = 6$

and each of these selections can be arranged $\dfrac{4!}{2!}$ ways

$\therefore$ No. of arrangements $= 6 \times \dfrac{4!}{2!} = 72$

No. of selections containing, 1 T $= {}_4C_3 = 4$
and each of these selections can be arranged 4! ways
$\therefore$ No. of arrangements $= 4 \times 4! = 96$

No. of selections containing 0 Ts $= {}_4C_4 = 1$
and this can be arranged in 4! ways
$\therefore$ No. of arrangements $= 1 \times 4! = 24$

Total no. of arrangements $= 16 + 72 + 96 + 24 = 208$

• Exercise 16.5   See page 352.

16.6 Approximate Methods for Evaluating Definite Integrals

Several methods have been devised for the approximate evaluation of the definite integral $\int_a^b f(x)\,dx$ when the indefinite integral $\int f(x)\,dx$ is difficult or impossible to find.

The first two methods we consider here use the relationship

$$\int_a^b f(x)\,dx = \text{area under the curve } y = f(x) \text{ and between the ordinates } x = a \text{ and } x = b,$$

provided f is continuous and $f(x) \geqslant 0$ for all $x \in [a, b]$

1 The Trapezium Rule

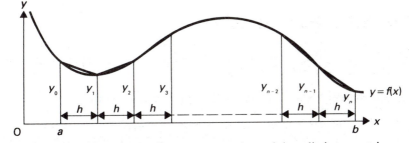

The area under the curve between $x = a$ and $x = b$ is split into n strips, each of width h, by taking $(n-1)$ equally spaced ordinates between $x = a$ and $x = b$. Each strip is approximately a trapezium of width h where $h = (b - a)/n$.

Let $y_0 = f(a)$, $y_1 = f(a + h)$, $y_2 = f(a + 2h), \ldots, y_n = f(b)$

$$\text{Area under curve} = \int_a^b f(x)\,dx \approx \frac{h}{2}[y_0 + y_1] + \frac{h}{2}[y_1 + y_2] + \cdots + \frac{h}{2}[y_{n-1} + y_n]$$

$$\Rightarrow \boxed{\int_a^b f(x)\,dx = \frac{h}{2}[y_0 + 2y_1 + 2y_2 + 2y_3 + \cdots + 2y_{n-1} + y_n] \\ \text{where } h = (b - a)/n \text{ and } n \text{ is the number of intervals}}$$

2 Simpson's Rule

In this method, the area under the curve is divided into $2n$ strips of equal width h.

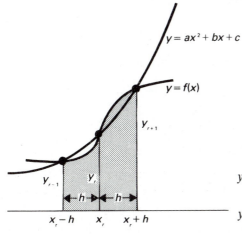

Consider a pair of strips bounded by the ordinates y_{r-1}, y_r, y_{r+1}. A parabola $y = ax^2 + bx + c$ can be found passing through the three points $(x_r - h, y_{r-1}), (x_r, y_r), (x_r + h, y_{r+1})$. Simpson's Rule uses the area under the parabola as an approximation for the area under the curve $y = f(x)$.

Since the three points lie on the parabola $y = ax^2 + bx + c$ then,

$$\left. \begin{aligned} y_{r-1} &= a(x_r - h)^2 + b(x_r - h) + c \\ y_r &= ax_r^2 + bx_r + c \\ y_{r+1} &= a(x_r + h)^2 + b(x_r + h) + c \end{aligned} \right\} \quad (1)$$

Shaded area is

$$\int_{x_r-h}^{x_r+h} (ax^2 + bx + c)\,dx = \frac{a}{3}[(x_r+h)^3 - (x_r-h)^3] + \frac{b}{2}[(x_r+h)^2 - (x_r-h)^2]$$
$$+ c[(x_r+h) - (x_r-h)]$$

Simplifying this expression and using the equation (1) gives the shaded area as

$$\int_{x_r-h}^{x_r+h} (ax^2 + bx + c)\,dx = \frac{h}{3}[y_{r-1} + 4y_r + y_{r+1}]$$

Extending this argument to cover all $2n$ strips

$$\Rightarrow \qquad \int_a^b f(x)\,dx \approx \frac{h}{3}[y_0 + 4y_1 + 2y_2 + 4y_3 + 2y_4 + \cdots + 4y_{2n-1} + y_{2n}]$$

where $h = (a-b)/2n$ and $2n$ is the number of intervals

Note   that Simpson's Rule can only be applied to an *even* number of strips, i.e. an *odd* number of ordinates.

3   Using a Series Expansion

If $f(x) = a_0 + a_1 x + a_2 x^2 + a_3 x^3 + \cdots$

See 9.5

then $\int_a^b f(x)\,dx = \int_a^b (a_0 + a_1 x + a_2 x^2 + \cdots)\,dx$

Note   This equation only holds if the expansion on the R.H.S. converges, i.e. for the values of x for which the particular expansion is valid—see 9.5

Example 30   Tabulate, to 3 decimal places, the values of $f(x) = \dfrac{1}{1+\sqrt{x}}$ for $x = 1$, $x = 25$ and three equally spaced values of x in between.

(a)   Use these values to estimate the value of $\displaystyle\int_1^{25} \frac{1}{1+\sqrt{x}}\,dx$

   to 3 decimal places using (i)   the trapezium rule (ii)   Simpson's rule

(b)   Use the substitution $u = \sqrt{x}$ to find the exact value of this integral.
(c)   Evaluate your answer to (b) to 3 decimal places.

First we need to divide the interval $[1, 25]$ into four equal parts that is

$$\overset{\longleftrightarrow h}{\underset{1 \quad 7 \quad 13 \quad 19 \quad 25}{\mid \quad \mid \quad \mid \quad \mid \quad \mid}}.$$

Note that $h = (25 - 1)/4 = 6$

x	1	7	13	19	25
$f(x)$	0.5	0.274	0.217	0.187	0.167

(a) (i) $\displaystyle\int_1^{25} \frac{1}{1+\sqrt{x}} dx \approx \frac{h}{2}[y_0 + 2y_1 + 2y_2 + 2y_3 + y_4]$

$$= 3[0.5 + 2(0.274 + 0.217 + 0.187) + 0.167]$$

$$\Rightarrow \int_1^{25} \frac{1}{1+\sqrt{x}} dx \approx 6.069$$

(ii) $\displaystyle\int_1^{25} \frac{1}{1+\sqrt{x}} dx \approx \frac{h}{3}[y_0 + 4y_1 + 2y_2 + 4y_3 + y_4]$

> We say that the factors of $y_0, y_1 y_2, y_3$ and y_4 are $1, 4, 2, 4, 1$ respectively, and we use them in the following table.

r	x	y_r	factor f	$y_r \times f$
0	1	0.5	1	0.5
1	7	0.274	4	1.096
2	13	0.217	2	0.434
3	19	0.187	4	0.748
4	25	0.167	1	0.167

$$\Sigma(y_r \times f) = 2.945$$

$$\int_1^{25} \frac{1}{1+\sqrt{x}} dx \approx \tfrac{6}{3} \times \Sigma(y_r \times f) = 5.89$$

$$\Rightarrow \int_1^{25} \frac{1}{1+\sqrt{x}} dx \approx 5.89$$

(b) Let $u = \sqrt{x} \Rightarrow \qquad u^2 = x$

$$\Rightarrow \qquad 2u\,du = dx$$

and when $x = 1, u = 1$
when $x = 25, u = 5$

$$\therefore \int_1^{25} \frac{1}{1+\sqrt{x}} dx = \int_1^5 \frac{2u}{1+u} du \rightarrow \boxed{\text{Integrand is top heavy}}$$

$$= 2\int_1^5 \left(1 - \frac{1}{1+u}\right) du$$

$$= 2[u - \ln(1+u)]_1^5 = \underline{8 - 2\ln 3}$$

(c) $8 - 2\ln 3 \approx \underline{5.80}$ to 3 d.p.

Example 31   Using the first three terms of the series expansion for xe^x, estimate the value of $\displaystyle\int_0^{0.02} xe^x\,dx$ to 3 significant figures.

$$e^x = 1 + x + \frac{x^2}{2!} + \cdots \longleftarrow \boxed{\text{Series is valid for all } x}$$

$$\Rightarrow xe^x \approx x + x^2 + \frac{x^3}{2}$$

Hence
$$\int_0^{0.02} xe^x\,dx \approx \int_0^{0.02} \left(x + x^2 + \frac{x^3}{2} \right) dx$$

$$\approx \left[\frac{x^2}{2} + \frac{x^3}{3} + \frac{x^4}{8} \right]_0^{0.02}$$

$$\Rightarrow \int_0^{0.02} xe^x\,dx \approx 0.000203 \text{ (to 3.s.f.)}$$

- Exercise 16.6   See page 353.

16.7 + A Level Questions

- Miscellaneous Exercise 16B   See page 354.

Unit 16 EXERCISES

Exercise 16.1

1   Write the following polynomials in nested form:
  (a) $2x^2 - 3x + 4$ (b) $3x^3 - 2x^2 + x - 3$
  (c) $x^4 - x^2 + 2$ (d) $4x^5 - 3x^2 + x$
  (e) $x^3 - 5x^2 + 3x$

2   Find the value of each of the following, giving your answers to six significant figures:
  (a) $2x^2 - 3x + 4$ when $x = 3.56$
  (b) $4x^5 - 3x^2 + x$ when $x = 3.56$
  (c) $x^3 - 5x^2 + 3x$ when $x = -0.105$

3   Draw sketches to find the number and the rough position of any roots of the equations:

  (a) $\ln x = \dfrac{1}{x}$ (b) $e^x = \tan x, \quad -\pi/2 \leqslant x \leqslant \pi/2$

  (c) $\cos x = 2^x, \quad -\pi \leqslant x \leqslant \pi$ (d) $3 - x = \dfrac{-3}{x}$

4   Given that the equation $x^4 - 2x - 1 = 0$ has two real roots, find the two pairs of consecutive integers between which these roots lie, and use the ten subdivision method to find an approximation for the positive root which is correct to two decimal places

5   Use the ten-subdivision method to find an approximate value, to three decimal places, for the solution of $e^x = \tan x$ [Remember to put your calculator into radian mode]

6   Use the ten sub-division method to find a value for $\sqrt{5}$ correct to 2 decimal places.

*7   How many positive roots are there of the equation $2^{-x} = \sec x$? How many negative roots are there?

*8   Use the graph of $f(\theta) = \sin \theta$ to find a rough approximation for the root of $\theta = \sin \theta + \pi/3$

*9   By sketching a pair of graphs find rough approximate values for the roots of $(3 - 2x)\cos x = 2$, where $-\pi \leqslant x \leqslant 2\pi$

Exercise 16.2

1   Given that $x^3 - 2x^2 + 3x - 4 = 0$ has a root in the interval $1 < x < 2$, use the bisection method to find the value of this root with an error of less than 0.1

2   Use the bisection method to find an approximation to the largest positive root of $x^3 - 4x^2 + x + 1 = 0$ with an error of less than 0.05

3   (a)   Using $x_1 = 1$ and the iterative formula

$$x_{n+1} = \frac{x_n^2 + 1}{2x_n - 1}$$

find x_2, x_3, x_4, x_5, x_6, giving your answers to five decimal places. Also give, to five decimal places, the value to which this series is converging

(b)   Use $x_1 = -1$ and the same iterative formula. To what value does this process give an approximation to five decimal places?

(c)   The answers to (a) and (b) are approximate values of the roots of an equation. What is the equation?

4   Two iterative rearrangements of the equation $x^2 - x - 1 = 0$ are

(a) $x_{n+1} = x_n^2 - 1$ (b) $x_{n+1} = \dfrac{x_n + 1}{x_n}$

One of these converges very slowly to a root of the equation. Use $x_1 = 1.5$ and investigate the values produced by both iterations. Decide which one converges slowly and state what happens to the other one

5   Verify that $2x^3 - 3x^2 + 2x + 1 = 0$ has a solution in the interval $[-1, 0]$. Find the Newton–Raphson iteration formula for this equation. Taking $x_1 = -0.5$ find the solution correct to three decimal places

6 Verify that $3 \sin x + 2\pi = 3x$ has a solution in the interval $[2,3]$. Find the Newton–Raphson formula for this equation and find the solution correct to four decimal places [Make sure your calculator is in radian mode!]

7 By considering the signs of $xe^x - 3$ for positive integral values of x find two integers between which this root of $e^x = 3/x$ lies. Using the smaller of these two integers as x_1, apply the Newton–Raphson process to obtain the root correct to 3.d.p.

8 Show graphically, or otherwise, that the equation $\ln x = 4 - x$ has only one real root and prove that this root lies between 2.9 and 3

By taking 2.9 as a first approximation to this root and applying the Newton–Raphson process once to $\ln x - 4 + x = 0$ or otherwise, find a second approximation giving your answer to 3 significant figures   (L)

Exercise 16.3

Find the number of ways of arranging:

1 7 items on a shopping list

2 3 different coins in a row

3 5 different coloured electric light bulbs in 3 sockets

4 6 houseplants, chosen from 8, in a line on the window sill

5 6 houseplants, chosen from 8, in a circle

6 5 people round a circular table

7 3 beads threaded on a circular ring

8 3 beads, chosen from 5, threaded on a circular ring

9 Evaluate (a) $_{15}P_2$ (b) $_7P_3$ (c) $_7P_4$ (d) $_3P_3$

10 Write down expressions in terms of n for
   (a) $_nP_2$ (b) $_{n-1}P_3$ (c) $_nP_{n-1}$

11 How many three letter words can be formed using
   (a) 3 letters without repetition
   (b) 3 letters with repetition
   (c) 8 letters without repetition
   (d) 8 letters with repetition?

Find the number of ways of arranging:

12 The letters x, y, z, z

13 3 red balls, 1 white ball, 1 green ball in a row

14 The letters of the word FRENZY

15 The letters of the word FRENETIC

16 The letters of the word AARDVARK

17 3 green counters, 4 white counters, 1 red counter in arrow

18 3 red counters, 4 green counters, 1 white counter in a circle

19 3 red counters, 1 black counter in a row

20 3 red counters, 1 black counter in a circle

*21 In how many ways can 10 girls be arranged in a line if the shortest girl must be at the right hand end?

*22 How many different 6 digit numbers can be made using $2, 3, 3, 4, 4, 4$? How many of these are even?

Exercise 16.4

1 Evaluate: (a) $_7C_3$ (b) $_5C_1$ (c) $_4C_4$

2 Verify that $_7C_3 = {}_7C_4$

3 In how many ways can:
   (a) 3 photographs be chosen from 8 proofs
   (b) 7 cards be dealt from 10
   (c) a team of 11 cricketers be selected from 12?

4 How many different combinations of 3 letters can be chosen from the letters U, V, W, X, Y, Z?

5 In how many ways can U, V, W, X, Y, Z can be divided into two groups of 4 letters and 2 letters?

6 In how many ways can U, V, W, X, Y, Z be divided into two groups, each containing 3 letters?

7 In how many ways can U, V, W, X, Y, Z be divided into 3 groups, containing 3, 2, 1 members.

8 A tutor group of 24 children have been invited to send a team of 4 children to take part in a T.V. quiz show.
   (a) In how many ways can the team be chosen?
   (b) The tutor group decide that Mary, who has just come out of hospital, should be in the team. In how many ways can the team now be chosen?

9 In how many ways can 3 white and 1 black keys be simultaneously pressed on a keyboard with 50 white and 35 black keys?

10   In how many ways can a tutor group committee of 4 boys and 4 girls be formed from 12 boys and 11 girls?

11   In how many ways can a group of letters containing 1 capital letter, 3 consonants and 2 vowels be chosen from 3 capital letters, 6 consonants and 4 vowels?

12   Tent A holds 2 people. Tent B holds 3 people. Tent C holds 4 people. In how many ways can 9 people be assigned to tents A, B or C?

13   A test consists of 30 questions and each question has only two possible answers, just one of which is correct. Calculate the number of different ways in which it is possible to answer

(a)   exactly 5 questions correctly
(b)   exactly 25 questions correctly   (L)

14   There are 3 post-boxes marked P, Q and R. Find the number of ways in which 8 different letters can be posted so that 4 of them are in box P, 2 of them are in box Q, and 2 of them are in box R.   (L)

Exercise 16.5

1   How many four digit numbers can be made from the digits 2, 3, 4, 5 if:
(a)   repetition is allowed
(b)   repetition is not allowed
(c)   repetition is not allowed and the number is even
(d)   repetition is not allowed and the number must be even and less than 5000?

2   In how many arrangements of the letters of the word ASSEMBLY do the two Ss always come together?

3   How many different combinations of 3 letters taken from P, P, Q, Q, R, R, S are there?

4   (a)   How many different ways are there of selecting 3 letters from the letters of the word CALCULUS?
(b)   How many different ways are there of arranging 3 letters from the letters of the word CALCULUS?

5   How many different permutations are there of the letters in the word PARALLEL in which the two As are *not* next to each other?

6   A student must answer exactly 7 out of 10 questions in an examination. Given that she must answer *at least* 3 of the first 5 questions, determine the number of ways in which she may select the 7 questions.   (L)

7   Find how many four digit numbers can be formed from the six digits 2, 3, 5, 7, 8 and 9, without repeating any digit

Also find how many of these numbers

(a)   are less than 7000   (b)   are odd   (L)

8 A small chess club consists of 3 married couples, 5 unmarried men and 2 unmarried women. Calculate the number of ways in which a team of four may be chosen in each of the following cases:

(a) When a team is to contain exactly one married couple
(b) When a team is to contain at least one man and at least one women

[In each case, teams with the same members but chosen in a different order are to be regarded as identical.]

*9 Calculate the greatest possible number of points of intersection when 3 circles and 4 ellipses are drawn on a sufficiently large piece of paper

Exercise 16.6

Use the trapezium rule to find approximate values of the following definite integrals, giving your answers to 3 significant figures:

1 $\displaystyle\int_0^{1.2} \sqrt{(1+x)}\,dx;$ 6 strips

2 $\displaystyle\int_1^{2.2} xe^x\,dx$; 7 ordinates

3 $\displaystyle\int_1^2 \log_{10} x\,dx$; 10 strips

4 $\displaystyle\int_{\pi/6}^{\pi/2} \frac{\sin x}{x}\,dx$; 4 strips

5 $\displaystyle\int_0^1 10^x\,dx$; 6 ordinates

Use Simpson's rule to find approximate values of the following definite integrals giving your answers to 3 significant figures:

6 $\displaystyle\int_0^1 (1+xe^{2x})\,dx;$ 3 ordinates

7 $\displaystyle\int_0^2 \sqrt{(1+x^2)}\,dx;$ 4 strips

8 $\displaystyle\int_0^\pi \sqrt{\sin x}\,dx$; 5 ordinates (J)

9 $\displaystyle\int_0^\pi x\sin dx$; 7 ordinates

10 $\displaystyle\int_0^{10} \frac{1}{1+x^3}\,dx$; 11 ordinates

Using the first three terms of the appropriate series expansions, estimate the values of the following definite integrals, giving your answers to 3 significant figures:

11 $\displaystyle\int_0^{0.5} \ln(1-x)\,dx$ **12** $\displaystyle\int_0^1 e^{x^2}\,dx$

13 $\displaystyle\int_0^{0.2} \frac{1}{\sqrt{(1+x^2)}}\,dx$ **14** $\displaystyle\int_0^{\pi/2} \frac{\sin x}{x}\,dx$ (using $\pi \approx 3.142$)

15 The area enclosed by the curve $y = e^{-x^2/2}$, the x-axis and the ordinates $x = 3$ and $x = -3$ is rotated about the x-axis through 2π radians. Use Simpson's method with seven ordinates to find an estimate of the volume of the solid of revolution so formed in terms of e and π

16 Sketch the curve $y = \ln x$

The region between the curve, the x-axis and the line $x = 7$ is rotated about the x-axis through 2π radians. Showing all your working in the form of a table, use Simpson's method with six intervals to estimate the volume of revolution so formed, to two decimal places

Miscellaneous Exercise 16B

[In this exercise, all permutation and combination questions will be cross-references to 16.5, when they are in fact covered by 16.3, 16.4 or 16.5. It is up to the student to decide which techniques are applicable]

1 Find the number of different permutations of the 10 letters of the word *STATISTICS* (L: 16.5)

2 Some values of a function f are given in the table below. Use the trapezium rule to estimate $\displaystyle\int_0^{0.5} f(x)\,dx$ to 2 decimal places

x	0	0.1	0.2	0.3	0.4	0.5
$f(x)$	0.10	0.23	0.45	0.52	0.44	0.21

(L: 16.6)

3 Show that the equation $\sin x - \ln x = 0$ has a root lying between $x = 2$ and $x = 3$

Given that this root lies between $a/10$ and $(a + 1)/10$, where a is an integer, find the value of a.

Estimate the value of the root to 3 significant figures (L: 16.1)

4 A committee of 4 people is to be selected from a group consisting of 8 men and 4 women. Determine the number of ways in which the committee may be formed it it is to contain

(a) exactly one woman
(b) at least one woman

(L: 16.5)

5 Tabulate, to three places of decimals, the values of $(1 + x^4)^{1/4}$, for $x = 0$, 0.2, 0.4, 0.6, 0.8

Using Simpson's rule with five ordinates estimate to 3 significant figures, the value of

$$\int_0^{0.8} (1 + x^4)^{1/4}\, dx$$

By expanding $(1 + x^4)^{1/4}$ to powers of x as far as, and including, the term in x^8, obtain to 3 significant figures, a second estimate for the value of this integral (L: 16.6)

6 Show that the equation $f(x) = 0$, where $f(x) \equiv x^3 + x^2 - 2x - 1$, has a root in each of the intervals $x < -1$, $-1 < x < 0$, $x > 1$

Use the Newton–Raphson procedure, with an initial value 1, to find two further approximations to the positive root of $f(x) = 0$, giving your final answer to two decimal places (L: 16.1, 16.2)

7 Write down the expansion of $(1 + x^2)^{1/2}$, in ascending powers of x up to the term in x^4. Hence, by integrating the series term by term, find an approximate value of $\int_0^{1/2} \sqrt{(1 + x^2)}\, dx$, giving your answer to 3 decimal places.

Show that Simpson's rule with two equal intervals leads to the same approximation for $\int_0^{1/2} \sqrt{(1 + x^2)}\, dx$ (A: 16.6)

8 Four visitors Dan, May, Nan and Tom arrive in a town which has five hotels. In how many ways can they disperse themselves among the five hotels

(a) if four hotels are used to accommodate them
(b) if three hotels are used to accommodate them in such a way that May and Nan stay at the same hotel? (L: 16.5)

9 Show that the equation $e^x \cos 2x - 1 = 0$ has a root between 0.4 and 0.45

Taking 0.45 as a first approximation to this root, apply the Newton–Raphson process once to obtain a second approximation, giving your answer to 3 significant figures (L: 16.1, 16.2)

10 From eight persons, including Mr and Mrs Smith, a committee of four persons is to be chosen. Mr Smith will not join the committee without his wife, but his wife will join the committee without her husband. In how many ways can the committee be formed? (L: 16.5)

11 By considering the roots of the equation $f'(x) = 0$ or otherwise. Prove that the equation $f(x) = 0$ where $f(x) \equiv x^3 + 2x + 4$, has only one real root. Show that this root lies in the interval $-2 < x < -1$

Use the iterative procedure

$$x_{n+1} = -\tfrac{1}{6}(x_n^3 - 4x_n + 4), \quad x_1 = -1$$

to find two further approximations to the root of the equation giving your final answer to 2 decimal places (L: 16.1, 16.2)

12 Find the number of different permutations of the 8 letters of the word *SYLLABUS*.

Find the number of different selections of 5 letters which can be made from the letters of the word *SYLLABUS*   (L: 16.5)

13 Tabulate, to 3 decimal places, the function $f(x)$ where

$$f(x) \equiv \sqrt{(1 + x^2)}$$

for values of x from 0 to 0·8 at intervals of 0·1. Use these values to estimate I, to 3 decimal places, where

$$I = \int_0^{0.8} f(x)\,dx$$

by Simpson's rule.

Obtain a second estimate of the value of I, also to 3 decimal places, by using the first 3 terms in the binomial expansion of $\sqrt{(1 + x^2)}$
                                                                        (L: 16.6)

14 Whole numbers are formed from digits selected from 1, 2, 3, 4, 5, 6 without repetition. Calculate how many of these numbers are less than 2350
                                                                        (J: 16.5)

15 Given that $x > 0$ and that $y = \dfrac{\ln x}{x}$, find $\dfrac{dy}{dx}$

State the set of values of x for which $\dfrac{dy}{dx} > 0$ and the set of values of x for which $\dfrac{dy}{dx} < 0$. Hence show that y has a maximum value of $1/e$

Sketch the graph of $y = \dfrac{\ln x}{x}$ for $1 \leqslant x \leqslant 5$

Find the area of the finite region R bounded by the curve $y = \dfrac{\ln x}{x}$, the x-axis and the line $x = 5$

The region R is rotated completely about the x-axis to form a solid of revolution S. Use Simpson's rule, taking ordinates at $x = 1, 2, 3, 4$ and 5 to estimate, to 2 significant figures, the volume of S.   (A: 16.6)

16 The registration number of a car consists of 3 letters of the alphabet followed by an integer between 1 and 999 inclusive followed by a letter of the alphabet, e.g. *ABC 123 D*, *XYZ 78 A*, *PQR 5 S*. Given that *all* 26 letters of the alphabet may be used and that any letter or digit may be repeated, find the total number of different registration numbers which could be formed.

Find also in how many of these registration numbers the letters and digits are all different [Answers may be left in factor form]
                                                                        (L: 16.5)

17 Given that $(1 + x)y = \ln x$, show that, when y is stationary

$$\ln x = (1 + x)/x$$

Show, graphically, or otherwise, that this latter equation has only one real root and prove that this root lies between 3.5 and 3.8

By taking 3.5 as a first approximation to this root and applying the Newton–Raphson process once to the equation $\ln x - (1 + x)/x = 0$, find a second approximation to this root, giving your answers to 3 significant figures. Hence find an approximation to the corresponding stationary value of y (L: 16.1, 16.2)

*18 Find how many distinct numbers greater than 5000 and divisible by 3 can be formed from the digits 3, 4, 5, 6 and 0, each digit being used at most once in any number (J: 16.5)

*19 Given that $f(x) \equiv (1 - x^2/9)^{1/2}$, tabulate values of $f(x)$ to 3 decimal places for values of x between 0 and 3 at intervals of 0.5

With these seven ordinates, use Simpson's rule to calculate an approximate value for I, where

$$I = \int_0^3 \left(1 - \frac{x^2}{9}\right)^{1/2} dx$$

Hence estimate, to 1 decimal place the area of the region enclosed by the ellipse $\dfrac{x^2}{9} + \dfrac{y^2}{4} = 1$

Use the substitution $x = 3 \sin \theta$ to evaluation I in terms of π and hence estimate π to 1 decimal place (L: 11.1, 11.3, 15.5, 16.6)

*20 A pack of 20 cards consists of 5 red, 5 blue, 5 green and 5 yellow cards, and the cards in each set of a particular colour are numbered 1, 2, 3, 4 and 5, respectively. In a certain game a hand consists of 5 cards. Find numerical expression for

(a) the number of different hands which contain cards of at least two different colours
(b) the number of different hands which contain three cards marked with one number and two cards marked with a second number (J: 16.5)

*21 Find an approximate value for $\displaystyle\int_0^{\pi/3} \sqrt{(\sec x)} \, dx$, by using Simpson's Rule with five equally spaced ordinates.

Write down the least value of $\sec x$ for $0 \leqslant x < \dfrac{\pi}{2}$

Hence show that $\displaystyle\int_0^{\pi} \sqrt{(\sec x)} \, dx > \alpha$ for $0 < \alpha < \dfrac{\pi}{2}$ (A: 16.6)

ANSWERS*

Unit 1

Exercise 1.1

1 $x = 1, y = 2, z = 1$ 2 $x = 4, y = 1, z = 1$ 3 $x = 2, y = -1, z = 3$

4 $x = -2, y = -1, z = 1$ 5 $u = 3, v = 1, w = -1$ 6 $x = -2, y = 4, z = 3$

7 $\left.\begin{array}{l} x = 3 \\ y = 9 \end{array}\right\}$ and $\left.\begin{array}{l} x = -2 \\ y = 4 \end{array}\right\}$ 8 $\left.\begin{array}{l} x = 1 \\ y = 1 \end{array}\right\}$ and $\left.\begin{array}{l} x = -3 \\ y = 9 \end{array}\right\}$ 9 $\left.\begin{array}{l} x = 4 \\ y = 2 \end{array}\right\}$ and $\left.\begin{array}{l} x = -4 \\ y = -2 \end{array}\right\}$

10 $\left.\begin{array}{l} x = 1 \\ y = 4 \end{array}\right\}$ and $\left.\begin{array}{l} x = -2 \\ y = -2 \end{array}\right\}$ 11 $\left.\begin{array}{l} x = 1 \\ y = 2 \end{array}\right\}$ and $\left.\begin{array}{l} x = 4 \\ y = -4 \end{array}\right\}$ 12 $\left.\begin{array}{l} x = 0 \\ y = -\frac{1}{2} \end{array}\right\}$ and $\left.\begin{array}{l} x = -1 \\ y = -1 \end{array}\right\}$

13 (a) $\left.\begin{array}{l} x = 3 \\ y = 0 \end{array}\right\}$ and $\left.\begin{array}{l} x = 0 \\ y = 3 \end{array}\right\}$ (b) $x = 1\frac{1}{2}, y = 1\frac{1}{2}$ (c) $\left.\begin{array}{l} x = 3 \\ y = 0 \end{array}\right\}$ and $\left.\begin{array}{l} x = 1 \\ y = 0 \end{array}\right\}$

14 $a = -2, b = 10$ 15 $p = r = 0, q = -1$ 16 $m = -1, n = -2, p = -3$

17 $\left.\begin{array}{l} x = -1 \\ y = 3 \end{array}\right\}$ and $\left.\begin{array}{l} x = 2 \\ y = -3 \end{array}\right\}$ 18 $\left.\begin{array}{l} x = 5 \\ y = -5 \end{array}\right\}$ and $\left.\begin{array}{l} x = 0 \\ y = -10 \end{array}\right\}$ 19 $\left.\begin{array}{l} x = 1 \\ y = \frac{1}{2} \end{array}\right\}$ and $\left.\begin{array}{l} x = \frac{1}{5} \\ y = -\frac{7}{10} \end{array}\right\}$

20 $\left.\begin{array}{l} x = -2 \\ y = 1 \end{array}\right\}$ and $\left.\begin{array}{l} x = 3 \\ y = -1 \end{array}\right\}$ 21 $x = 60, y = 40, z = 20$

Exercise 1.2

1 $36°, 18°, 135°, 20°, 22.5°, 67.5°, 54°, 120°, 270°, 86°$

2 $\pi^c/10, 2\pi^c/5, 9\pi^c/10, 4\pi^c/3, 2\pi^c, 17\pi^c/36, 5\pi^c/18, \pi^c/12, \pi^c/9, 4\pi^c$

3 $0.5, 0.924, 0.901, 1, 0.5, 0.766, 0, -0.309, -0.5, 1$

4 (a) $6\pi/5\,\text{cm}$ (b) $2\pi\,\text{cm}$ (c) $2\,\text{cm}$ (d) $7.5\,\text{cm}$ (e) $16\pi\,\text{cm}$

5 (a) $18\pi/5\,\text{cm}^2$ (b) $4\pi\,\text{cm}^2$ (c) $10\,\text{cm}^2$ (d) $56.25\,\text{cm}^2$ (e) $160\,\pi\,\text{cm}^2$

6 1.25^c

7 $5.2\,\text{cm}$

8 (a) $10.27\,\text{cm}^2$ (b) $22.11\,\text{cm}^2$

9 $20\,\text{cm}^2$

10 $5.80\,\text{cm}^2$ and $195.26\,\text{cm}^2$

11 2.86^c

12 $35\pi/4\,\text{cm}^2$

13 $5\pi\,\text{cm}$

14 $2.715\,\text{m}^3$

Exercise 1.3

1 (a) $2\sqrt{2}$ (b) $2\sqrt{3}$ (c) $2\sqrt{5}$ (d) $5\sqrt{2}$ (e) $4\sqrt{5}$ (f) $3\sqrt{2}$

   (g) $6\sqrt{2}$ (h) $5\sqrt{5}$ (i) $10\sqrt{2}$ (j) $15\sqrt{2}$ (k) $20\sqrt{5}$ (l) $5\sqrt{3}$

2 (a) $7\sqrt{3}$ (b) not possible   (c) $4\sqrt{11}$ (d) $3\sqrt{2} - 2$ (e) $3\sqrt{2} - 4$

   (f) $5 - \sqrt{5}$

3 (d) $4\sqrt{3}$ (b) $15\sqrt{6}$ (c) $6\sqrt{2}$ (d) 2 (e) $\frac{14}{3}$ (f) $\sqrt{5}$

4 (a) $9\sqrt{3}$ (b) 0 (c) 3 (d) $9 - \sqrt{3}$ (e) -5 (f) $6\sqrt{2}$

5 (a) 1 (b) $5 - 3\sqrt{3}$ (c) $8 + 3\sqrt{5}$ (d) $6 + 3\sqrt{3} - 2\sqrt{2} - \sqrt{6}$

   (e) $x^2 - 2$ (f) $y + 4\sqrt{3y} + 9$

6 (a) $3 - 2\sqrt{2}$ (b) $3 + 2\sqrt{2}$ (c) $x - 1$ (d) $4x + 1 - 4\sqrt{x}$

   (e) $x - y$ (f) $a + b - 2\sqrt{ab}$

7 (a) $\sqrt{3}/3$　(b) $\sqrt{2}/4$　(c) $\sqrt{2}/4$　(d) $3\sqrt{2}/2$　(e) $3\sqrt{3}$

8 (a) $\sqrt{2}-1$　(b) $\dfrac{\sqrt{3}+1}{2}$　(c) $2-\sqrt{3}$　(d) $5-2\sqrt{5}$

(e) $\dfrac{4\sqrt{3}+6}{3}$　(f) $6+3\sqrt{3}$　(g) $\dfrac{2\sqrt{3}-\sqrt{2}}{10}$　(h) $-\left(\dfrac{2\sqrt{7}+7}{3}\right)$

9 (a) $7+4\sqrt{3}$　(b) $\dfrac{5-3\sqrt{2}}{7}$　(c) $2-\sqrt{3}$　(d) $\sqrt{5}-2$

10 (a) $2\sqrt{2}$　(b) $\dfrac{4}{4-x}$　(c) $\dfrac{x-\sqrt{x}}{x-1}$　(d) $\sqrt{x}-\sqrt{y}$　(e) $\dfrac{x+2\sqrt{x}}{x}$

(f) $x-\dfrac{1}{x}$　(g) $\dfrac{3}{1-x}$

Exercise 1.4

1 $5,2,p^3+2,\cos x+2$　**2** $12,6,2,\dfrac{12}{y}$

3 $4,2,0,-2,10,4-2x,4-4x,4-6y$
4 (a) $0,0,2\sqrt{6}$　(b) many–one
5 (a) $6,6,-2$　(b) $\{-2,1,6\}$　(c) $\{-3,-2\}$　(d) one–one
(e) many–one.

6 (a) $x\to x+3$　(b) $x\to 2x$　(c) $x\to\dfrac{x}{100}$

7 (a) $\{2,3,4\}$　(b) $x\to x-1$　(c) domain $\{2,3,4\}$, range $\{1,2,3\}$

8 (a) $x\to\dfrac{x+3}{5}$　(b) $x\to\dfrac{x}{5}+3$　(c) $x\to\dfrac{1}{2}\left(\dfrac{x}{3}-1\right)$　(d) $x\to\frac{1}{2}(3x-1)$

(e) $x\to 3-x$　(f) $x\to\dfrac{6}{x}$

9 (a) $5,9,1\frac{1}{2}$　(b) 4
10 (a) Both one–one　(b) $0,1,0.5$　(c) $\sin(x^2),\sin(x-1),\sin(x+3),\sin(x+2)$
(d) $3,y+2,p^2+2,\cos x+2,\sin x+2$

Exercise 1.5

1 $60°$　**2** $14.5°$ and $165.5°$　**3** $\pi/4$ and $5\pi/4$　**4** $210°$ and $330°$
5 $40°,320°$
6 (a) $\pm50.5°$　(b) $\pm129.5°$　(c) $60°,-120°$　(d) $-60°,120°$
(e) $-30°,-150°$
7 (a) $45°,225°$　(b) $135°,315°$　(c) $197.5°,342.5°$　(d) $107.5°,252.5°$
(e) $0°,180°,360°$
8 $19.5°,160.5°,199.5°,340.5°$
9 $\pi/8,5\pi/8$
10 $10°,50°,130°,170°$
11 $\pm120°,$
14 $33.7°,213.7°$
15 $116.6°,-63.4°$
16 (a) $100.7,169.3,280.7,349.3$　(b) $31.6,121.6,211.6,301.6$　(c) $221°$
(d) $324°$　(e) $20°,100°,260°,340°$　(f) $0°$

Exercise 1.6

1

Angle	$\sin x$	$\cos x$	$\tan x$	$\text{cosec}\, x$	$\sec x$	$\cot x$
0			0		1	∞
$\pi/6$			$1/\sqrt{3}$	2	$2/\sqrt{3}$	$\sqrt{3}$
$\pi/4$	$1/\sqrt{2}$			$\sqrt{2}$	$\sqrt{2}$	1
$\pi/3$	$\sqrt{3}/2$	$\frac{1}{2}$	$\sqrt{3}$	$2/\sqrt{3}$	2	$1/\sqrt{3}$
$\pi/2$		0		1	∞	0
$2\pi/3$	$\sqrt{3}/2$	$-\frac{1}{2}$	$-\sqrt{3}$	$2/\sqrt{3}$	-2	$-1/\sqrt{3}$
$3\pi/4$	$1/\sqrt{2}$	$-1/\sqrt{2}$	-1	$\sqrt{2}$	$-\sqrt{2}$	-1
$5\pi/6$	$\frac{1}{2}$	$-\sqrt{3}/2$	$-1/\sqrt{3}$	2	$-2/\sqrt{3}$	$-\sqrt{3}$
π			0	∞	-1	∞

2 (a) $1 - 2\sin^2\theta$ (b) $2 - 2\sin^2\theta - 3\sin\theta$
3 (a) $\sec^2\theta$ (b) $\cot\theta$ (c) $\tan^4\theta$ (d) 1 (e) $\sec^2\theta$ (f) $\tan\theta$
4 $\pm 1/\sqrt{2}$
5 $14.5°, 90°, 165.5°$
6 $54.7°, 125.3°$
7 $0°, 120°$
8 $45°, 63.4°$
9 $60°, 180°$
10 $60°, 180°$
11 $90°$

Exercise 1.7
3 (a) $x = 0$ (b) $(0, -1)$ (c) -1 (d) $[-1, 3]$
4 (a) $x = 2$ (b) $(2, 9)$ (c) 9 (d) Domain $[-1, \infty[$ [Range] $-\infty, 9]$
5 (a) $(0, 0)$ (b) $(0, -1)$ (c) $(1, 0)$ (d) $(-3, 0)$ (e) $(1, 0)$ (f) $(0, 3)$
6 (a) $1, -2$ (c) $]-2, 1[$ (d) $x = -\frac{1}{2}$
7 (a) $(1\frac{1}{2}, 1\frac{1}{2})$ (b) $(1, 0), (-2, 3)$ (c) $(1, 3)$
8 (a) $]0, 2[$ (b) $[-2, 2]$ (c) $[1, 5]$ (d) $]-\infty, 1[\cup]3, \infty[$
   (e) $]-1, \frac{1}{3}[$ (f) $]-\infty, -3] \cup [-\frac{1}{2}, \infty[$
9 (a) $(\pi, 0), (2\pi, 0)$ (b) $(3\pi/2, 0)$

Exercise 1.8

1	2.41 and -0.41	12	1	23 2 imaginary roots
2	-5.46 and 1.46	13	2 and $-\frac{1}{2}$	24 2 distinct real roots
3	4.303 and 0.697	14	-1 and $-\frac{2}{3}$	25 Line does not intersect
4	0.653 and -7.653	15	-4.697 and -0.321	the curve
5	-0.158 and 3.158	16	3.097 and -0.431	27 $k = 12$
6	-3 and $-\frac{1}{2}$	17	-1 and 0.4	28 $p = \pm 10$
7	1.309 and 0.191	18	3.213 and 0.187	29 $k = \pm 40$
8	-1 and $-\frac{1}{3}$	19	2 distinct real roots	
9	$a \pm \sqrt{a^2 - b}$	20	2 imaginary roots	
10	4.791 and 0.209	21	2 distinct real roots	
11	1.162 and -5.162	22	2 equal real roots	

Exercise 1.9
1 (a) $(a - b)(a^2 + ab + b^2)$ (b) $(x - 3)(x^2 + 3x + 9)$
  (c) $(p + 2)(p^2 - 2p + 4)$ (d) $(1 + y)(1 - y + y^2)$

2 (a) $A = 2, B = -4$ (b) $A = 2, B = 1$
3 $A = 5, B = 1$
4 $a = 2, b = 8$
5 $a = 2, b = -3; (x^2 - x + 2)(x - 1)(x + 3)$
6 $4(x - 1)(x - 2) - (x - 3)(x - 1) + 2(x - 2)(x - 3)$
7 $a = -2, b = 1$
8 (a) $A = -1, B = 2$ (b) $A = B = 2$ (c) $A = B = 1$
9 (a) $p = \frac{3}{2}, q = -\frac{1}{2}$ (b) $p = 3, q = 1$ (c) $p = -1, q = 1$
10 $l = 2, m = -1, p = 6$

11 $1 - \dfrac{\frac{1}{3}}{x + 1} + \dfrac{\frac{1}{3}x - \frac{2}{3}}{x^2 - x + 1}$ **12** $A = 1, B = -1, C = -2$

13 $\dfrac{1}{x + 2} - \dfrac{4}{(x + 2)^2} + \dfrac{4}{(x + 2)^3}$

Miscellaneous Exercise 1A

1 $x = 2, y = -1, z = \frac{1}{2}$ **2** $\left. \begin{matrix} x = 5 \\ y = 0 \end{matrix} \right\}$ and $\left. \begin{matrix} x = -3 \\ y = 4 \end{matrix} \right\}$
3 $90°, 360°, 30°, 60°, 18°, 45°, 135°$
4 $\pi^c/10, 2\pi^c/5, \pi^c, 3\pi^c/2, 5\pi/6.$
5 20.61 cm
6 50 cm^2
7 $4 : 3$

8 (a) $\sqrt{15} - 3$ (b) $3\sqrt{3}$ (c) 1 (d) $10\sqrt{2} - 9$ (e) $\dfrac{\sqrt{10}}{2}$

   (f) $\dfrac{5 + 2\sqrt{3}}{13}$

9 (a) $45°$ (b) $121°$ (c) no solution in the range   (d) $28.8°, 151.2°$
   (e) $67.4°, 112.6°$ (f) $25°$
11 $35.8°, 125.8°, 215.8°, 305.8°$
12 (a) $2\cos^2 \theta - 1$ (b) $3 - 7\cos^2 \theta$
13 TTFTTF
14 (a) $30°, 150°, 270°.$ (b) $0°, 360°, 48.2°, 311.8°$
15 $15°, 75°, 135°, 21.1°, 81.1°, 141.1°$
16 (a) $(5 - r)(25 + 5r + r^2)$ (b) $(3a + 4b)(9a^2 - 12ab + 16b^2)$
17 (a) 3.41 and 0.59 (b) 2.37 and 0.63
18 (a) -0.21 and -4.79 (b) 2.69 and -0.19

19 (a) 5 (b) $2x^2 - 5$ (c) $x \to \dfrac{x + 5}{2}$ (d) 3

20 (b) $]-\infty, -3[\cup]2, \infty[$ (c) $(-0.5, -6.25)$
21 (a) $A = \frac{3}{5}, B = -\frac{3}{5}, C = -\frac{6}{5}$ (b) $A = \frac{3}{4}, B = \frac{3}{4}, C = -\frac{3}{2}$
   (c) $A = 2, B = 0$
   $C = -2, D = 3$

Unit 2

Exercise 2.1

1 (a) $\frac{5}{2}$ (b) 4 (c) 0 (d) $\frac{4}{3}$ (e) ∞ (f) -1
2 (a) $(2, 4\frac{1}{2})$ (b) $(2\frac{1}{2}, -3)$ (c) $(0, 5)$ (d) $(-2\frac{1}{2}, 0)$ (e) $(-1, 2)$
   (f) $(\frac{9}{2}, \frac{9}{2})$

3 (a) $\sqrt{13}, \sqrt{17}, \sqrt{2}$ not rt $\angle$ d.

(b) $\sqrt{168}, \sqrt{136}, \sqrt{68}$ rt $\angle$ d.

(c) $\sqrt{13}, \sqrt{37}, \sqrt{20}$ not rt $\angle$ d.

4   (a) $\frac{5}{2}$ (b) $\frac{1}{3}$ (c) $\frac{3}{2}$ (d) $\frac{7}{4}$ (e) a/b

5   (a) $/\!/$el   (b) $\perp^r$ (c) $/\!/$el   (d) $/\!/$el   (e) neither   (f) $\perp^r$

6   13

8 $2, -\frac{1}{2}, \frac{2}{11}$, rt $\angle$ at B

10   (a) 1   (b) -1 (c) ∞

11 $\alpha = \pm\sqrt{\dfrac{5}{2}}$

12 $6b = 4a + 13$

14   1

15 $(-3\frac{1}{2}, -\frac{1}{2}), 17.5$

16   AB $= 5$ units, $m_{AB} = -\frac{4}{3}$

17   B is $(-3, -1)$

Exercise 2.2

1   (a) $y = 2x + 1$ (b) $3x + y = 0$ (c) $4y + 13 = x$ (d) $y + px + p = p^2$

2   (a) $y = 3x + 1$ (b) $y + 2x + 1 = 0$ (c) $5y + x = 17$ (d) $y + x = p$
   (e) $y + x = p + q$

3 $y = 3x - 5$

4 $2y = x - 3$

5 $3y = 5x - 6$

6 $y + 2 = 4x$

7 $y = x$

8 $y + x = 4$

9 $2y + 3x = 4; (\frac{22}{13}, -\frac{7}{13})$

10 $y = 3x + 1, 5y + x = 13, (\frac{1}{2}, 2\frac{1}{2})$

11   (a) $2, 1$ (b) $y = x$ (c) $y = 2x - 1$ (d) $(1, 1)$

12   (a) $y = 2x + 18$ (b) $2y + x + 4 = 0$ (c) $(-8, 2)$

13 $3y + 4x = 16$ $(16, -16)$

Miscellaneous Exercise 2A

1   (a) $\dfrac{4}{\sqrt{13}}$ (b) $\dfrac{18}{\sqrt{29}}$ (c) $\dfrac{4}{\sqrt{29}}$ (d) 1

2   6

3 $\dfrac{9}{\sqrt{29}}$ **4** (a) $(1, 4)$ (b) $(1, 1)$

5 $2y + x + 3 = 0$

6   (a) Yes   (b) No   (c) Yes.

7   (a) $/\!/$ el   (b) neither   (c) $\perp^r$

8   14.5

9 $(3\frac{1}{2}, 3\frac{1}{2})$

11 $q^2 + 9 = 6p$

12 $y = x - 4$

13 $3\sqrt{2}$

14 $y + 2x = 10$;
   $y = x - 5$

15 $(3, -1)$

16 $4y = 3x + 5$
   $13y = 15x + 4$
   $5y = 9x - 6$
   $(\frac{7}{3}, 3)$

Unit 3

Exercise 3.1

1 $\dfrac{QR}{PR} = 6.3$ 6.03   6.003   6.0003   6.00003

2 $\dfrac{dy}{dx} = 4x$ **3** $\dfrac{dy}{dx} = 2x$ **4** $\dfrac{dy}{dx} = 4$ **5** $\dfrac{dy}{dx} = 3x^2$

6 $\dfrac{dy}{dx} = -\dfrac{1}{x^2}$ **7** $\dfrac{dy}{dx} = 6 - 2x$ **8** $\dfrac{dy}{dx} = 4x^3$ **9** $\dfrac{ds}{dt} = -1 + 6t$

10 $\dfrac{dv}{dr} = 6r + 2$

Exercise 3.2

1 (a) $5x^4$ (b) $11x^{10}$ (c) $-3x^{-4}$ (d) $\frac{1}{3}x^{-2/3}$ (e) $\dfrac{-4}{x^5}$

2 (a) 2 (b) $\dfrac{-3}{x^7}$ (c) $\frac{5}{2}x^{3/2}$ (d) 0 (e) -3

3 (a) $12x^3$ (b) $\frac{5}{2}x^4$ (c) $-\dfrac{3}{x^2}$ (d) $\frac{21}{2}x^{1/2}$ (e) $\frac{1}{4}x^{-3/4}$

   (f) $-\frac{1}{2}x^{-3/2}$

4 (a) $2 - 3x^2$ (b) $3 - 10x$ (c) $-\dfrac{1}{x^2}$ (d) $18x - 6$ (e) $12x^3 + \dfrac{2}{x^3}$

   (f) $2x - \dfrac{2}{x^3}$ (g) $2x - \dfrac{4}{x^2}$ (h) $-\dfrac{1}{x^2} + \dfrac{14}{x^3} - \dfrac{3}{x^4}$

   (i) $\frac{3}{2}x^{1/2} + \frac{1}{2}x^{-1/2}$ (j) $\frac{1}{2}x^{-1/2} + \frac{1}{2}x^{-3/2}$ (k) $2x + 2$ (l) $8x$

5 (a) $3 - 10t$ (b) $6t^2 - 2t$ (c) $2 + \dfrac{1}{t^2}$

6 5 **13** $(2, 6)$ **20** $(4, -23\frac{2}{3}), (-3, 26\frac{1}{2})$.
7 48 **14** $(1, 2), (-1, 8)$
8 $1/2\sqrt{3}$ **15** $(3, 5)$
9 $-\frac{1}{16}$ **16** $(\frac{1}{3}, \frac{4}{27}), (-1, 0)$
10 -5 **17** $(1, 1)$
11 4 **18** $(2, 3\frac{1}{2})(-2, 2\frac{1}{2})$
12 $-\frac{3}{2}$ **19** $0, 6$

Exercise 3.3
1 (a) $y = 12x - 12$ (b) $y = 5x + 1$ (c) $y + x + 2 = 0$ (d) $y = 4x - 4$
   (e) $y = 20x - 35$
2 (a) $12y + x = 146$ (b) $5y + x = 5$ (c) $y = x$ (d) $4y + x = 1$
   (e) $20y + x = 503$
3 $y = 12x - 1$
4 $y = 2x + 3, \quad (-3, -3)$
5 $(-9, -5)$ **6** $(0, 7)$ **7** $3y = x - 1; 3y + x = 4, \quad (\frac{5}{2}, \frac{1}{2})$
8 $(2, 4) \quad y = 4x - 4$
9 $2y = x - 6$
10 $y = -\frac{25}{3}$
11 $-\frac{1}{27}$ or $\frac{55}{27}$
12 $3\sqrt{10}, \sqrt{10}$.

Exercise 3.4

1 (a) 0,   (b) 3
   (c) ± 2
2 (a) -25 (b) $-2, 2$
   (c) $-\frac{7}{9}, \frac{25}{9}$
3 Max at $(\frac{3}{2}, \frac{9}{4})$
4 Min at $(2, -1)$
5 Min at $(3, 0)$
6 Min at $(-\frac{1}{4}, -\frac{9}{8})$
7 Min at $(3, -18)$
   Max at $(-3, 18)$
8 Min at $(2, 12)$
9 12
10 1

Exercise 3.5

1 P of I at $(0, 0)$
2 Min at $(0, 0)$
3 P of I at $(0, 5)$
4 Min at $(0, -2)$
5 P of I at $(0, 1)$
6 $\frac{9}{8}$ 7 0
8 (a) P of I at $(\frac{1}{2}, \frac{1}{2})$—not st. pt.
   (b) P of I at $(1, 15)$ is a st. pt.
        P of I at $(-1, -1)$—not st. pt.
9 Min at $(2, -14)$, Max at $(-2, 18)$
10 P of I at $(-1, 2)$
11 Max at $(1, 1)$, Min at $(2, 0)$
12 P of I at $(2, -3)$
13 P of I at $(3, 1)$
14 Min at $(1, 3)$, Max at $(-1, 1)$

Exercise 3.6

1 $v = 2\frac{1}{4}$ m/s; $a = -3$ m/s^2; $t = 1$ s
2 $-10 + 50t; \frac{1}{5}s, a = 0$
3 0.2 m/s, -9.6 m/s
4 52 rad/s.
5 $t = 1$, acc$^n = -5$
6 112.5 m/s^2

7 Min value $= 0$;   Max value $= \dfrac{4000\pi}{27}$

8 Max value $= 3456$;   Min value $= 0$

9 $A = 200x - \dfrac{x^2}{2}$;   Max area $= 20\,000$ m^2

10 $x = \sqrt[3]{400/\pi}$
11 $2048 \, \pi/81$
12 4.5 cm, 9 cm, 6 cm

13 $r = \dfrac{30}{4 + \pi}$

14 2^c

Miscellaneous Exercise 3A

1 (a) $\dfrac{dy}{dx} = 2$ (b) $\dfrac{dx}{dt} = -1$ (c) $-\dfrac{dV}{dr} = 4$ or $\dfrac{dV}{dr} = -4$

2 (a) Let $y = (1 - 3x)^2 = 1 - 6x + 9x^2$
    $y + \delta y = 1 - 6(x + \delta x) + 9(x + \delta x)^2$
    $\delta y = -6\delta x + 18x\delta x + 9(\delta x)^2$

    $\dfrac{\delta y}{\delta x} = -6 + 18x + 9\delta x$

    $\therefore \dfrac{dy}{dx} = \lim_{\delta x \to 0} (-6 + 18x + 9\delta x) = -6 + 18x$

(b)   Let $y = \dfrac{1}{1-x}$

$$y + \delta y = \dfrac{1}{1-x-\delta x}$$

$$\delta y = \dfrac{1}{1-x-\delta x} - \dfrac{1}{1-x}$$

$$= \dfrac{\delta x}{(1-x-\delta x)(1-x)}$$

$$\dfrac{\delta y}{\delta x} = \dfrac{1}{(1-x-\delta x)(1-x)}$$

$$\therefore \dfrac{dy}{dx} = \dfrac{1}{(1-x)^2}$$

3 (a) $20p^3 - 8p - \dfrac{4}{p^2}$ (b) $5p^{3/2} - \frac{1}{2}p^{-1/2}$ (c) $4p^3 - 4p^{-5}$

   (d) $\frac{3}{4}(p^{-2} - p^{-4})$

4 $4, -3, 12$

5 $\frac{21}{8}$

6 Tgt: $y = (3-2p)x + p^2$ Normal: $y(2p-3) = x - 10p + 9p^2 - 2p^3$

7 0

8 When $x = 1$, grads are 2 and -2; when $x = -1$, grads are -2 and 2

9 Max at $(1, 13)$; Min at $(-2, -14)$

10 $(-1, \frac{71}{4})$

11 P is $\left(\dfrac{2p^3}{3p^2 - 2}, 0\right)$, Q is $(0, -2p^3)$, PQ $= \dfrac{2p^3}{3p^2 - 2}[9p^4 - 12p^2 + 5]^{1/2}$

12 $76°, 72.5°$

13 $a = \frac{3}{2}, \quad b = -1, \quad c = \frac{7}{6}$

Unit 4

Exercise 4.1

1 (a) $\sin 104°$ or $\sin 76°$ (b) $\sin 4°$ (c) $\cos 52°$ (d) $\cos 63°$
   (e) $\sin 3x$ (f) $\cos \pi/3$ (g) $\cos 2x$

2 (a) $-\sin A$ (b) $-\cos A$ (c) $\cos A$ (d) $-\sin A$

4 (a) $\cos(x-y)$ (b) $\sin(3\theta - \varphi)$ (c) $\sin(\varphi + 20°)$
   (d) $\sin(A - 2B)$ (e) $\cos 5\theta$ (f) $-\cos(C + D)$
   (g) $\cos 3\theta$ (h) $\sin 2\theta$ (i) $\sqrt{3}/2$ or $\sin 60°$
   (j) $\tan 90°$ or ∞

5 (a) $2\sin\theta\cos\phi$ (b) $2\cos C \cos D$ (c) $-2\sin\alpha\sin\beta$

6 (a) $\frac{3}{5}$ (b) $\frac{4}{3}$ (c) $\frac{5}{13}$ (d) $\frac{12}{13}$ (e) $\frac{63}{65}$ (f) $\frac{33}{65}$
   (g) $\frac{16}{65}$ (h) $\frac{56}{65}$ (i) $\frac{63}{16}$ (j) $\frac{33}{56}$

7 $\frac{56}{65}, -\frac{33}{65}, \frac{36}{325}, -\frac{323}{325}, -\frac{116}{845}, -\frac{837}{845}$

8 (a) $\frac{117}{125}$ (b) $\frac{4}{5}$ (c) $-\frac{7}{25}$ (d) $\frac{117}{44}$ (e) $-\frac{24}{7}, -\frac{44}{117}$

Exercise 4.2

1 (a) $18.4°$ (b) $8.1°$ (c) $8.1°$ (d) $8.1°$ (e) $81.9°$

2 (a) $52.5°, 232.5°$ (b) $7.4°, 187.4°$ (c) $37.9°, 217.9°$

Exercise 4.3

1 (a) $\sin 28°$ (b) $\tan 70°$ (c) $\cos 52°$ (d) $\cos \pi/3$
   (e) $\cos 8\theta$ (f) $\tan 6\theta$ (g) $\sin 4\theta$ (h) $\frac{1}{2}\sin 2\theta.$

2 (a) $2\sin\theta\cos\theta$ (b) $\dfrac{2\tan 4\theta}{1-\tan^3 4\theta}$ (c) $2\cos^2 2\theta - 1$

   (d) $\sin\alpha$ (e) $\frac{1}{2}\sin\varphi$ (f) $\tan 45°$ (g) $\cos\pi/6$
   (h) $-\cos 2\phi$ (i) $1+\sin 2A$ (j) $\frac{1}{2}\sin 3\theta$

3 (a) $\frac{24}{25}, -\frac{7}{25}$ (b) $\frac{336}{625}, \frac{527}{625}$ (c) $\frac{120}{169}, -\frac{119}{169}$

4 $\frac{8}{15}$

5 $\frac{3}{4}, \frac{24}{7}$

6 (a) $\frac{1}{2}(1+\cos 2A)$ (b) $\frac{1}{2}(1-\cos 2A)$ (c) $2\sin^2 A$

   (d) $2\cos^2 A$ (e) $2\sin^2\dfrac{\beta}{2}$ (f) $2\cos^2\dfrac{\beta}{2}$

7 $\pm 1/\sqrt{5}$

8 $\pm\frac{1}{2}$

9 $\sqrt{2}-1$

10 $\frac{12}{13}, \frac{119}{169}, \pm 5/\sqrt{26}$

Exercise 4.4

1 $0°, 180°, 30°, 150°$ 2 $30°, 90°, 150°$ 3 $0°, 180°, 82.8°$ 4 $0°, 120°$
6 (a) $x(1-y^2)=2y$ (b) $x=2y^2-1$
   (c) $xy^2=y^2-2$ (d) $y=1-2x^2$
7 $\cos 4A = 8\cos^4 A - 8\cos^2 A + 1$ 8 $30°, 150°, 270°$
9 $90°, 270°, 210°, 330°$ 10 $60°, 300°$
11 $35.3°, 144.7°, 215.3°, 324.7°$ 12 $90°, 270°, 45°, 225°$

Exercise 4.5

1 $\hat{A}=29.9°$ 2 $\hat{Z}=50.6°$ 3 $\hat{Q}=43.2$ 4 $c=8.11\,\text{cm}$
5 $1.79\,\text{cm}$ 6 $p=56.41\,\text{cm}$ 7 $a=7.86\,\text{cm}$ 8 $x=5.94\,\text{cm}$
9 $e=26.17\,\text{cm}$ 10 $\hat{Y}=70.2$ 11 $108.2°$

12    $\hat{B}=152.9°, \quad \hat{C}=7.1°, \quad c=1.08\,\text{cm}$ ⎫
     or $\hat{B}=27.1°, \quad \hat{C}=132.9°, \quad c=6.43\,\text{cm}$ ⎭

13 $4.26\,\text{cm}$ or $2.63\,\text{cm}$
14 $\hat{C}=59°, \quad a=42.17\,\text{cm}, \quad c=36.95\,\text{cm}$
15 $a=9.71\,\text{cm}, \quad \hat{B}=61.9°, \quad \hat{C}=39.9°$
16 $\hat{Z}=31.7°, \quad x=4.01\,\text{cm}, \quad z=5.63\,\text{cm}$
17 $\hat{P}=18.8°, \quad \hat{Q}=150.6°, \quad \hat{R}=10.6°$
18 $\hat{A}=59.7°, \quad \hat{C}=11.1°, \quad c=33.23\,\text{m}$
21 $6\,\text{km}$
22 $6.43\,\text{cm}$ and $9.42\,\text{cm}$

Exercise 4.6

1 (a) $55\,\text{cm}^2$ (b) $1074.3\,\text{cm}^2$ (c) $3\sqrt{3}$
2 $5.08\,\text{cm}$
3 $a=5.10\,\text{cm}, \quad b=2.40\,\text{cm}, \quad \text{Area}=4.32\,\text{cm}^2, \quad h=3.6\,\text{cm}$
4 $137.6\,\text{cm}^2$ 5 $1.22\,\text{cm}$ 6 $70\,\text{m}$
7 $YZ=140\,\text{m}, \quad \text{ht}=28.9\,\text{m}$
8 $28.76\,\text{km}$
9 $82.8°, 27.9°, 69.3°$
10 (a) $58.6°$ and $121.4°$ (b) $14.45\,\text{cm}$ (c) $73.1°$ and $106.9°$
11 $71.74\,\text{cm}^2$
12 (a) $10.95\,\text{cm}$ (b) $5.62\,\text{cm}$
13 $27.45\,\text{cm}$

Exercise 4.7

1 (a) $12\sqrt{3}$ cm (b) $35.3°$ (c) $54.7°$
2 (a) 4.04 m (b) $38.9°$ (c) 11.2 m (d) $19.9°$
3 $15°$
4 (a) 11.3 cm 12.4 cm (b) $23.8°$ (c) $32°$
5 40 m
6 (a) 10 cm (b) 7.81 cm (c) 9.43 cm (d) $70.2°$
7 (a) $h\tan 65°$ (b) $h\tan 55°$ (c) 23.3 m
8 (a) $2\sqrt{7}$ (b) $72°$
9 40.4 m

Miscellaneous Exercise 4A

1 (a) $0°, 67.4°$ (b) $130.4°, -153°$ (c) $36.8°, 126.8°$
2 $82.9°$
3 $-\frac{12}{5}, \frac{12}{13}, -\frac{5}{13}$
4 $\frac{21}{29}, \frac{840}{841}, \frac{41}{841}, \pm 2/\sqrt{29}, \pm 5/\sqrt{29}$
5 $7, 7\sqrt{2}/10, \sqrt{2}/10$
6 $\frac{1}{2}[2-\sqrt{3}]^{1/2}, \frac{1}{2}[2+\sqrt{3}]^{1/2}$
7 $2\sin 2A \cdot \cos 2A$

Miscellaneous Exercise 4B

1 $120°, 300°$ 2 (a) $30°, 90°, 150°$ (b) $120°$ 3 $33.4°$
4 $\tan\phi = \dfrac{1-\tan\theta}{1+\tan\theta}$ 7 $10\sqrt{3}, 25\sqrt{3}, 142°$
8 $\tan(45+\theta) = \dfrac{1+\tan\theta}{1-\tan\theta}$ (a) $2+\sqrt{3}$ (b) $\dfrac{2}{\cos 2\theta}$
9 $AC = 2$, $BC = 2\sqrt{3}$, $AB = 2\sqrt{7}$; $\sqrt{\frac{3}{7}}$
10 $\tan\frac{1}{2}\theta = 1/k$ 11 (a) $204a^2$ (b) $24a$ (c) $35.8°$ (d) $36.9°$
12 $k \leqslant x \leqslant 2k; 2\sqrt{3}k$
13 $AP^2 = a^2 + c^2 - ac[\cos A\hat{B}C - \sqrt{3}\sin A\hat{B}C]$; $\lambda = \frac{1}{2}$, $\mu = 2\sqrt{3}$
14 D is $(7, 7)$; centre $(4, 3)$, radius 5; $A\hat{B}C = 71.57°$; 31.3 sq. units

Unit 5

Exercise 5.1

1 (a) $\cos^2 x - 1$ (b) $\cos(x^2 - 1)$ (c) $\cos(\cos x)$
2 (a) $2x^2$ (b) $4x^2$ (c) $x^2 + 1$ (d) $2(x+1)^2$
  (e) x^4 (f) $4x^2 + 1$ (g) $4x + 2$
3 $y = u^2$, $u = x - 3$ 4 $y = \sin u$, $u = x^2$ 5 $y = u^2$, $u = \sin x$
6 $y = \cos u$, $u = x - 1$ 7 $y = \cos u, u = \sin v, v = x^2$
8 $y = e^u$, $u = x + 3$ 9 $y = \log u$, $u = v^2$, $v = x - 3$
10 $y = \log u$, $u = \cos v$, $v = \sin w$, $w = x + 1$
11 $y = u^2$, $u = 2x + 1$
12 (a) $y = u^3$ (b) $y = \sqrt{u}$ (c) $y = \dfrac{1}{u}$ (d) $y = \frac{1}{2}u$ (e) $y = \frac{1}{4}u^2$
13 (a) $6(2x+1)^2$ (b) $\dfrac{1}{\sqrt{2x+1}}$ (c) $\dfrac{-2}{(2x+1)^2}$ (d) 1 (e) $(2x+1)$
14 (a) $220(4x+3)^{10}$ (b) $-72x(1-4x^2)^2$ (c) $64x(2x^2-1)^3$

15 (a) $3(x+5)^2$ (b) $10(2x-1)^4$ (c) $-18(1-3x)^5$

(d) $24x(3x^2-6)^3$ (e) $\dfrac{3}{2(1+3x)^{1/2}}$ (f) $\dfrac{5x}{(1+5x^2)^{1/2}}$

(g) $\frac{1}{3}(3-2x)(3x-x^2)^{-2/3}$ (h) $\dfrac{10x^4}{3}(x^5+1)^{-1/3}$

16 $4t-1, 3, \dfrac{4t-1}{3}$

17 (a) -1 (b) -2 (c) $\frac{1}{2}$

Exercise 5.2

1 $5(x-1)^4$ **2** $8(2x-1)^3$ **3** $-15(2-3x)^4$

4 $2(3x^2+x)(6x+1)$ **5** $6x(x^2+1)^2$ **6** $\dfrac{-1}{(x+1)^2}$

7 $\dfrac{-2}{(2x+1)^2}$ **8** $\dfrac{-2(2x-1)}{(x^2-x)^3}$ **9** $\dfrac{-24x}{(x^2-1)^4}$

10 $\dfrac{1}{\sqrt{(2x+1)}}$ **11** $\dfrac{2x-3}{3(x^2-3x)^{2/3}}$ **12** $\dfrac{-1}{2(x-1)^{3/2}}$

13 $\dfrac{2x}{3(4-x^2)^{4/3}}$ **14** $\dfrac{3(1+\sqrt{x})^2}{2\sqrt{x}}$ **15** $\dfrac{-1}{4\sqrt{x}(1+2\sqrt{x})^{5/4}}$

16 $\dfrac{-15}{2\sqrt{x}(3+5\sqrt{x})^4}$ **17** $\dfrac{4(2x-3)^3}{\sqrt{\{(2x-3)^4-1\}}}$

18 $4\{(x^2+1)^5-2x\}\{5(x^2+1)^4x-1\}$

19 $\dfrac{-4x}{[4+\sqrt{(4x^2+1)}]^2(4x^2+1)^{+1/2}}$ **20** $\dfrac{-6\{(x+1)^2-4(3x-1)\}}{\{(x+1)^3-2(3x-1)^2\}^3}$

21 $-12, y+12x+11=0$ **22** $2y+\sqrt{5}x=4\sqrt{5}$

23 $1, -2$ **24** $1-\dfrac{4}{(t+1)^2}, 3$

Exercise 5.3

1 (a) $y=x^5+c$ (b) $y=\dfrac{x^5}{5}+c$ (c) $y=\dfrac{3x^5}{5}+c$

2 $y=\dfrac{5x^4}{4}-\dfrac{1}{4}$

3 $s=3t-\dfrac{t^3}{3}+5$

4 $\dfrac{x^6}{6}+c, \dfrac{x^{12}}{12}+c, \dfrac{x^{-3}}{-3}+c, \dfrac{-1}{2x^2}+c, \dfrac{2x^{5/2}}{5}+c, -2x^{-1/2}+c, \dfrac{5x^{6/5}}{6}+c,$

$\dfrac{-2}{x}+c, \dfrac{-1}{2x^6}+c, \dfrac{-1}{15x^3}+c, \dfrac{-5}{3x^3}+c, \frac{2}{3}x^{3/2}+c, 2x^{1/2}+c, x^{1/2}+c$

5 $x^4-\dfrac{x^3}{3}+c$

6 $3x + \dfrac{1}{2x^2} + c$

7 $x^3 + \dfrac{1}{x} + \dfrac{3x^2}{2} + c$

8 $\dfrac{5x^4}{4} - \dfrac{x^{-4}}{4} + 5x + c$

9 $-\dfrac{1}{x^3} + \dfrac{1}{x} - \dfrac{x^2}{2} + c$

10 $\dfrac{2x^{3/2}}{3} - \dfrac{2x^{5/2}}{5} + c$

11 $\dfrac{4x^{5/2}}{5} + \dfrac{2x^{-3/2}}{3} + c$

12 $-2x^{-2} - \dfrac{x^{-3}}{3} + 2x + c$

13 $-\tfrac{2}{3}x^3 - x + \dfrac{3x^2}{2} + c$

14 $\dfrac{x^2}{2} + \dfrac{1}{x} + c$

15 $\dfrac{x^3}{3} + 2x - \dfrac{1}{x} + c$

16 $3x + c$

17 $\dfrac{3x^5}{5} + c$

18 $\dfrac{81x^5}{5} + c$

19 $\tfrac{3}{5}(x+1)^5 + c$

20 $\tfrac{1}{15}(3x+1)^5 + c$

21 $\dfrac{x^3}{3} + c$

22 $\dfrac{(x+1)^3}{3} + c$

23 $\dfrac{(x+3)^3}{3} + c$

24 $\dfrac{(2x+1)^3}{6} + c$

25 $\dfrac{(5x+4)^3}{15}$

26 $\dfrac{(2+3x)^5}{15} + c$

27 $\dfrac{(3-x)^4}{-4} + c$

28 $-\tfrac{8}{3}(2-x)^{3/2} + c$

29 $\tfrac{10}{3}(3+x)^{3/2} + c$

30 $\tfrac{3}{8}(1+2x)^{4/3} + c$

31 $\dfrac{-1}{2(2x+3)} + c$

32 $-\tfrac{2}{3}(1-3x)^{1/2} + c$

33 $\dfrac{(qx+r)^{p+1}}{(p+1)q} + c$

34 $y = \dfrac{x^3}{3} - \dfrac{x^2}{2} + x - 50$

35 $\tfrac{8}{3}x^{3/2} + \tfrac{1}{6}(4x+1)^{3/2} + \tfrac{2}{5}(1-2x)^5 + c$

36 $-\tfrac{2}{3}(2-x)^{3/2} - 2(2-x)^{1/2} - \dfrac{1}{2-x} + c$

Exercise 5.4

1 $\tfrac{1}{4}$

2 42

3 $17\tfrac{1}{3}$

4 $\tfrac{1}{2}$

5 $\tfrac{2}{3}$

6 6

7 8

8 $6\tfrac{2}{3}$

9 $-\tfrac{1}{4}$

10 $6\tfrac{2}{3}$

11 (a) 18 (b) $\dfrac{7\sqrt{2}}{3}$

12 (a) 80 (b) $\tfrac{4}{3}$ (c) $2\tfrac{2}{3}$ (d) $\tfrac{8}{3}(3\sqrt{3}-1)$

13 (a) $(2, 0)$ (b) $\tfrac{4}{3}$

14 $4\frac{1}{2}$ **16** $4\frac{2}{3}$ **18** $1\frac{3}{4}$ **20** $\left(3\frac{1}{12}\right)$
15 $\frac{1}{3} < x < 2, \frac{125}{54}$ **17** $27\frac{3}{4}$ **19** $\frac{1}{24}$ **21** $13\frac{1}{6}, -\frac{25}{6}$

Exercise 5.5

1 24π **3** $8\pi/3$ **5** $\dfrac{128\pi}{3}$ **7** $4\pi/3$ **9** $(3,0), 27\pi/4$ **11** $\dfrac{192\pi}{5}$

2 10π **4** $4\frac{2}{3}\pi$ **6** $68\pi/3$ **8** $16\pi/9$ **10** $34\frac{2}{15}\pi$ **12** $\frac{1}{3}\pi r^2 h$

Miscellaneous Exercise 5A

1 $\frac{4}{15}$ **2** $\frac{4}{3}$ **3** $\frac{4}{3}$ **4** $\dfrac{16\sqrt{2}}{15}$

5 (a) $28(4x+9)^6$ (b) $9x^2(x^3+1)^2$

  (c) $\dfrac{-4x}{3}(3-2x^2)^{-2/3}$ (d) $\dfrac{-1}{\sqrt{x}(1+\sqrt{x})^2}$

  (e) $\dfrac{-3}{2x^2(1+3x^{-1})^{1/2}}$ (f) $\dfrac{16}{(1-2x)^3}$

6 (a) $y = u^2$,   (b) $y = \dfrac{3}{u}$ (c) $y = \frac{1}{3}u$ (d) $y = u+7$

7 (a) $x^3 + x^2 - x + c$ (b) $\frac{2}{3}x\sqrt{x} + 8\sqrt{x} + c$ (c) $-\dfrac{1}{3x} + \dfrac{x^2}{2} + x^3 + c$

8 (a) $\dfrac{3x^2}{2} + 4x - \dfrac{1}{x} + c$ (b) $\frac{14}{3}$ (c) $4\frac{2}{3}$ (d) -2

  (e) $\dfrac{a}{3}(b^3 - a^3) + \dfrac{b}{2}(b^2 - a^2)$

9 $\frac{16}{3}$

10 $\dfrac{c^2}{2}; 2b^{1/2}c^{3/2}(\sqrt{2}-1); 1:16(\sqrt{2}-1)^2$

11 $k = c/2$
12 $1\frac{13}{24}$
13 $32\frac{3}{4}$
14 $\frac{37}{96}$
15 10 units^2

Miscellaneous Exercise 5B

1 $\frac{2}{3}$ **2** $y = 6x + 15; \dfrac{408\pi}{5}$ **3** $\frac{a}{3}$ **4** $h = 4,\ r = 2$

6 Max at $x = 2$; min at $x = \frac{2}{3}$; $2\frac{5}{6}$ **7** $1:2$
8 $2ab$ **9** $3\sqrt{3}$

Unit 6

Exercise 6.1
1 (a)(b)(d)(f) are A.P.s.
2 (a) $31, 5n - 4$ (b) $0, \frac{7}{2} - \frac{1}{2}n$ (c) $-9, 5 - 2n$ (d) $p + 12q, p + (n-1)2q$
3 (a) 27 (b) 26 (c) 36
4 (a) 750 (b) 450 (c) -18 (d) 203
5 3375
6 1482
7 41, 675
8 6, 4 **14** 2
9 3, 27 **15** $-6, 640$
10 14 **16** $-2, 0, 2$
11 116 **17** $12, 2n + 4$
12 57 **18** $167, \frac{14}{83}, 11$
13 $7, 4, -3$ **19** $15, 23, 31, 39, 47$

Exercise 6.2
1 (a)(b)(e) are G.P.s.
2 (a) $3, 2 \cdot 3^9$ (or 39366), $2 \cdot 3^{n-1}$ (c) $x, 3x^{10}, 3x^n$

(b) $\dfrac{2}{3}, \dfrac{2^9}{3^6}, \dfrac{2^{n-1}}{3^{n-4}}$ (d) $\dfrac{1}{3}, \dfrac{1}{2 \cdot 3^9}, \dfrac{1}{2 \cdot 3^{n-1}}$

3 (a) ± 10 (b) $\pm 9, \pm 1$ (c) $\pm 6, \pm 1.5, 0.75$ (d) $12, -24$
4 (a) 7 (b) 8 (c) 9

5 (a) 189 (b) 342 (c) 3906 (d) 9.75 (e) 11.1111 (f) $\dfrac{x^{17} - x}{x^2 - 1}$

6 $a = \frac{1}{2}$, $r = 2$ and $a = \frac{1}{2}$, $r = -2$
7 2,279 **8** $\frac{375}{16}$ **9** $\frac{16}{27}, \frac{-27}{8}$

10 (a) $\frac{3}{4}$ (b) $\frac{16}{3}$ (c) $\frac{10}{9}$ (d) $\frac{1}{3}$ (e) $\dfrac{1}{1 - x}$ (f) $\sec^2 \theta$

11 (a) $|x| < 1$ (b) $x^2 > 1$ (c) $|x| < \frac{1}{3}$
12 $\frac{2}{3}$

13 (a) $\dfrac{x}{1 - x}$ (b) $\dfrac{1}{1 + y}$

14 $|z| < 3, \dfrac{3z}{3 - z}$

15 5120 **16** 2 or $\frac{1}{3}$

Exercise 6.3
1 $1 + 2 + 3 + 4$ **2** $3 + 12 + 27$ **3** $1 \cdot 0 + 2 \cdot 1 + 3 \cdot 2 + 4 \cdot 3 + 5 \cdot 4$
4 $3^3 + 4^3 + \cdots + 9^3$ **5** $\frac{1}{2} + \frac{1}{3} + \frac{1}{4} + \cdots + \frac{1}{8}$ **6** $\frac{1}{11 \cdot 13} + \frac{1}{12 \cdot 14} + \frac{1}{13 \cdot 15} + \frac{1}{14 \cdot 16}$
7 $1^4 + 2^4 + 3^4 + \cdots + n^4$
8 $4 \cdot 7 + 5 \cdot 8 + 6 \cdot 9 + \cdots + (n+1)(n+4)$ **9** $x + x^2 + \cdots + x^n$

10 $\frac{1}{80}$ **12** $3n; n^2 + n + 2$ **14** $\displaystyle\sum_{1}^{30} \frac{1}{r}$ **16** $\displaystyle\sum_{1}^{14} (2r - 1)^3$

11 1.3 **13** $\displaystyle\sum_{1}^{20} 2r$ **15** or $\begin{array}{c} \displaystyle\sum_{0}^{14} (3r + 1) \\ \displaystyle\sum_{1}^{15} (3r - 2) \end{array}$

17 $\displaystyle\sum_{3}^{21} r(r+1)$ or $\displaystyle\sum_{1}^{19}(r+2)(r+3)$ **22** $\displaystyle\sum_{1}^{\infty}(-1)^r\frac{1}{(2r-1)}$ **27** 54

18 $\displaystyle\sum_{1}^{\infty}\frac{1}{r^2}$ **23** $\frac{1}{20}$ **28** 368

19 $\displaystyle\sum_{4}^{99}\frac{1}{r(r+1)}$ or $\displaystyle\sum_{1}^{96}\frac{1}{(r+3)(r+4)}$ **24** n^2-2n **29** $1,3,5,7,9$

20 $\displaystyle\sum_{1}^{2n}\frac{r+1}{(r+2)(r+3)}$ **25** $1\frac{1}{12}$ **30** $\dfrac{3n}{2}(n-1)$

21 $\displaystyle\sum_{1}^{11}(-1)^{r+1}2r$ or $\displaystyle\sum_{1}^{11}(1-1)^{r-1}2r$ **26** 14   **31** 31.02

32 9.83

33 3

34 (a) 81   (b) 7   **35** (a) -1 (b) 12   **36** (a) $\frac{1}{10}$ (b) 7

37 (a) $(\frac{1}{4})^{n-1}$ (b) an infinite no.   **38** $-1, \dfrac{3}{x}, -\dfrac{9}{x^2}, \dfrac{27}{x^3}$

39 $\dfrac{2x}{1-x}$ **40** $\displaystyle\sum_{1}^{\infty}(-1)^r rx^r$

Exercise 6.5

1 4900	**10** $\frac{1}{3}n(2n+1)(4n+1)$	**18** $(3r-2)^2; \dfrac{n}{2}(6n^2-3n-1)$
2 405	**12** $n(n+2)$	
3 715	**13** $\frac{1}{3}n(n+1)(n+2)$	**19** $27r^3; \frac{27}{4}n^2(n+1)^2$
4 14175	**14** $\frac{1}{6}n[2n^2-9n+13]$	
5 3630	**15** $4r; 2n(n+1)$	**20** $16r^2-1; \dfrac{n}{3}(16n^2+24n+5)$
6 42075	**16** $(2r-1)^2; \frac{1}{3}n(4n^2-1)$	
7 7500	**17** $(r+1)(r+2); \frac{1}{3}n(n^2+6n+11)$	**21** $3,7,11; 4r-1$
8 (a) 650		
(b) 221 00		
(c) 845 000		

Exercise 6.6

1 9	**11** $\frac{729}{1000}$	**21** $\frac{9}{4}$	**31** x
2 4	**12** $\frac{4}{9}$	**22** 6	**32** y
3 3	**13** -125	**23** 3	**33** p^5
4 $\frac{1}{3}$	**14** $\frac{9}{4}$	**24** 1	**34** q
5 $\frac{5}{9}$	**15** 0.5	**25** $\frac{3}{4}$	**35** $\dfrac{x}{x-1}$
6 5	**16** $\frac{10000}{81}$	**26** 10	**36** $\dfrac{x+6}{(x+3)^{5/2}}$
7 $\frac{1}{4}$	**17** $\frac{5}{7}$	**27** 1	**37** $\dfrac{x^2-x-1}{x-1}$
8 1728	**18** $\frac{8}{5}$	**28** 8	**38** 2
9 $\frac{1}{243}$	**19** $\frac{2}{5}$	**29** $\frac{1}{8}$	**39** $\dfrac{\frac{1}{2}y+1}{(y+1)^{4/3}}$
10 8	**20** 1	**30** $\frac{5}{3}$	**40** $\dfrac{2p^2-3q^2}{(p+q)(p-q)^{1/2}}$

Exercise 6.7

1
```
                    1
                 1     1
              1     2     1
           1     3     3     1
        1     4     6     4     1
     1     5     10    10    5     1
  1     6     15    20    15    6     1
1     7     21    35    35    21    7     1
```

2 $1 + 7x + 21x^2 + 35x^3 + 35x^4 + 21x^5 + 7x^6 + x^7$
3 $1 + 8x + 24x^2 + 32x^3 + 16x^4$
4 $1 - 9x + 27x^2 - 27x^3$

5 $1 - \dfrac{3x}{2} + \dfrac{3x^2}{4} - \dfrac{x^3}{8}$

6 $16 + 32x + 24x^2 + 8x^3 + x^4$
7 $32 + 240x + 720x^2 + 1080x^3 + 810x^4 + 243x^5$
8 $16x^4 - 160x^3 + 600x^2 - 1000x + 625$

9 $t^4 + 4t^3 + 6 + \dfrac{4}{t^2} + \dfrac{1}{t^4}$

10 $a^6 - 18a^5b + 135a^4b^2 - 540a^3b^3 + 1215a^2b^4 - 1458ab^5 + 729b^6$
11 $x^{14} - 7x^{12}y^3 + 21x^{10}y^6 - 35x^8y^9 + 35x^6y^{12} - 21x^4y^{15} + 7x^2y^{18} - y^{21}$

12 $x^6 - 3x^2 + \dfrac{3}{x^2} - \dfrac{1}{x^6}$

13 $20x^3 + \dfrac{160}{x} + \dfrac{64}{x^5}$

14 $10 + 6\sqrt{3}$ 16 14 18 $46\sqrt{3} - 1$
15 $124 - 32\sqrt{15}$ 17 $a = 362, \quad b = 209$

Exercise 6.8
1 (a) 30   (b) 60   (c) 132   (d) 66   (e) 126   (f) 15
2 (a) 6   (b) 35   (c) 1716   (d) 1   (e) $-\frac{5}{128}$ (f) $\frac{21}{32}$
3 $1 + 4x + 6x^2 + 4x^3 + x^4$
4 $1 - 5y + 10y^2 - 10y^3 + 5y^4 - y^5$
5 $1 + 12x + 60x^2 + 160x^3 + 240x^4 + 192x^5 + 64x^6$

6 $1 - x + \dfrac{x^2}{3} - \dfrac{x^3}{27}$

7 $1 + \frac{1}{2}x - \dfrac{x^2}{8} + \dfrac{x^3}{16} + \cdots$

8 $1 - 2x + 3x^2 - 4x^3 + \cdots$

9 $1 - \frac{1}{2}x + \dfrac{3x^2}{8} - \dfrac{5x^3}{16} + \cdots$

10 $1 + \dfrac{3x}{2} + \dfrac{15x^2}{8} + \dfrac{35x^3}{16} + \cdots$

11 $1 - 16x + 160x^2 - 1280x^3 + \cdots; |x| < \frac{1}{4}$

12 $1 - x - x^2 - \dfrac{5x^3}{3} + \cdots; |x| < \frac{1}{3}$

13 $1 + \dfrac{2x}{9} - \dfrac{x^2}{81} + \dfrac{4x^3}{2187} + \cdots; |x| < 3$

14 $1 - 2x + 4x^2 - 8x^3 + \cdots; |x| < \frac{1}{2}$
15 729 729

16 $1 - x + x^2 - x^3 + x^4 + \cdots$

17 $1 + x + x^2 + x^3 + x^4 + \cdots$ **19** $1 - x^2 + x^4 - x^6 + x^8 + \cdots$
18 $1 + 2x + 3x^2 + 4x^3 + 5x^4 + \cdots$ **20** $(1 - 2x)^{-1} = 1 + 2x + 4x^2 + 8x^3 + 16x^4 + \cdots$

Exercise 6.9

1 $32 + 80x + 80x^2 + 40x^3 + \cdots$; all x

2 $\sqrt{2}\left[1 - \dfrac{x}{16} - \dfrac{3x^2}{512} - \dfrac{7x^3}{8192} + \cdots\right.$; $|x| < 4$

3 $\dfrac{1}{\sqrt{3}}\left[1 - \dfrac{x}{3} + \dfrac{x^2}{6} - \dfrac{5x^3}{54} + \cdots\right.$; $|x| < \frac{3}{2}$

4 $1 + 6x + 27x^2 + 108x^3 + \cdots$; $|x| < \frac{1}{3}$.

5 495	**12** 0.2495	**20**	$2 - 3x + 4x^2 - 5x^3 + \cdots$
6 10 206	**14** $\frac{7}{16}$	**21**	$160x^3$
7 $\frac{30618}{125}$	**15** $\frac{40}{3}$	**22**	$-61236x^5$
8 60	**16** $\frac{4}{243}$	**23**	-20
9 1.0121	**17** $\frac{1}{16}p^{-5/2}$	**24**	5670
10 1.0149	**18** $5 + 4x + 14x^2 + 22x^3 + \cdots$	**26**	$A = 220, \quad B = 284$
11 0.9900	**19** $1 + 2x + 2x^2 + 2x^3 + \cdots$		

Miscellaneous Exercise 6A

1 (a) $-9; 6\sqrt{2}$ (b) $\frac{5}{2}x, 2x$
2 4, 16
3 (a) $\frac{3}{2}$ (b) $\frac{2}{3}$ (c) $\frac{4}{3}$ (d) 128

4 (a) $\dfrac{2p + q}{(p + q)^2 p^{1/2}}$ (b) $\dfrac{5q^{4/3}}{p^{11/2}}$

5 47   **9** 25, 67   **13** 10
6 -1.328 **10** $a = 2, \quad d = 3$ **14** 1287 286
                         11 $108°$ **15** (a) 4 m
7 $\dfrac{3n}{5} - \dfrac{8}{5}; 575$ **12** 7 terms   (b) 4.1 cm
                                                                    (c) $23\frac{1}{3}$ m
8 (a) 400   (b) 7327.5

Miscellaneous Exercise 6B

1 15 350   **13** $k = 2, \quad p = \frac{1}{2}; -\frac{5}{8}; \quad |x| < \frac{1}{2}$
2 (a) -116 (b) 72   **14** $50(r + 1), 257 500$
3 $\frac{2}{27}$ **15** $\frac{1}{3}$
4 63   **16** $\frac{1}{2}, \frac{9}{16}, -\frac{27}{64}$
5 (a) $1 + \frac{1}{3}px - \frac{1}{9}p^2x^2$ **17** (a) $q = -\frac{1}{8}$ (b) $-\frac{1}{3}$ (c) $5\frac{1}{2}$
     (b) $1 + qx - q^2x^2$
6 2660   **18** $\dfrac{55}{1728}$
7 $a = 1.5, \quad b = 6.25$
8 $a = 15, \quad n = 17$ **19** $\dfrac{3^{5/4}a}{2^{n-1}}$
9 $a + 3b$
10 $\frac{55}{53}$ **20** $-\dfrac{1}{2}; \dfrac{2a}{3}$
11 $-2\frac{1}{2}, \frac{1}{2}; 205$
12 (a) $1 + \frac{1}{2}x - \frac{1}{8}x^2 + \frac{1}{16}x^3 + \cdots$
      (b) $1 - \frac{1}{4}x + \frac{1}{16}x^2 - \frac{1}{64}x^3 + \cdots$

22 $1 + \frac{1}{2}y - \frac{1}{8}y^2; a_0 = 1, \quad a_1 = \frac{1}{2}, \quad a_2 = -\frac{1}{8}; 0.0103$

23 $\dfrac{3l}{5}, l, \dfrac{7l}{5}$

25 $a = 1, \quad b = -1, \quad c = \frac{3}{2}, \quad d = -\frac{5}{2}; 2.9704$
26 $\frac{1}{2}[1 + (1 - 4a)^{1/2}]$ and $\frac{1}{2}[1 - (1 - 4a)^{1/2}]; 0.969$ and 0.031

Unit 7

Exercise 7.1

1 $2\sin 2A \cdot \cos A$
2 $2\cos 3A \cos 2A$
3 $-2\sin 2A \cdot \sin A$
4 $2\cos 4A \cos 3A$

5 $2\cos\dfrac{5A}{2}\cdot\sin\dfrac{A}{2}$

6 $2\sin 3A \cdot \sin A$
7 $2\cos(A+B)\cdot\cos(A-B)$

8 $2\cos\left(\dfrac{5A+3B}{2}\right)\cdot\sin\left(\dfrac{5A-3B}{2}\right)$

9 $2\sin 45° \cdot \cos 15°$
10 $-2\sin 60° \cdot \sin 5°$
11 $\cos 5\theta + \cos \theta$
12 $\sin 5\theta + \sin 3\theta$
13 $\cos 6\theta - \cos 2\theta$
14 $\cos 2\theta - \cos 4\theta$
15 $\frac{1}{2}(\cos 6\theta + \cos 4\theta)$

16 $\frac{1}{2}(\sin 90° - \sin 30°)$

17 (a) $\sqrt{\dfrac{3}{2}}$ (b) $-\dfrac{1}{\sqrt{2}}$

18 $45°, 90°, 135°$
19 $0°, 30°, 90°, 150°, 180°$
20 $0°, 45°, 90°, 135°,$
    $180°, 60°, 120°$
21 $30°, 90°, 150°$
22 $0°, 30°, 60°, 90°,$
    $120°, 150°, 180°, 20°,$
    $100°, 140°$

23 $\tan\dfrac{3x}{2}$

24 $-\cot\dfrac{3x}{2}$

25 $-\cot\theta$

26 $\cot 5\theta$

27 $\dfrac{\sin\theta}{\cos 2\theta}$

28 $-\tan\left(\dfrac{x-y}{2}\right)$

33 $22\frac{1}{2}°, 90°, 112\frac{1}{2}°$
34 $0°, 30°, 60°, 90°,$
    $120°, 150°, 180°,$
    $40°, 80°, 160°$
35 $30°, 150°, 45°,$
    $135°$
36 $30°, 90°, 150$
37 $10°, 130°$

Exercise 7.2

1 $\sqrt{2}\cos(x-45°)$
2 $\sqrt{6}\cos(x+35.3°)$
3 $\sqrt{5}\sin(x-26.6°)$
4 Max 5, Min -5
13 $25\cos(\theta - 73.7°)$

5 $30°, 390°$
6 $-53.1°, 126.9°$
7 $119.5°, 346.7°$
8 $19.4°, 270°$

11 $\sqrt{5}\sin(2x$
    $+26.6°); 0°, 63.4°, 180°$
12 $79.3°, 115.8°$

   (a) Max 22, $\theta = 73.7°$;   Min -28, $\theta = 253.7°$
   (b) Max 625, $\theta = 73.7, 253.7°$;   Min 0, $\theta = 163.7°, 343.7°$
   (c) Max ∞, $\theta = 163.7°, 343.7°$;   Min $\frac{1}{25}$, $\theta = 73.7, 253.7°$

Exercise 7.3

1 $\theta = n\pi + (-1)^n\pi/4$
2 $\theta = 2n\pi \pm \pi/3$

3 $\theta = n\pi + \pi/4$

4 $\theta = n\pi + (-1)^{n+1}\pi/3$

5 $\theta = 2n\pi \pm \dfrac{3\pi}{4}$

6 $\theta = n\pi + \pi/6$
7 $\theta = n\pi + (-1)^n\pi/6$
8 $\theta = n\pi - \pi/3$
9 $\theta = 2n\pi \pm \pi/4$
10 $\theta = n\pi + (-1)^n 0.25$

11 $\theta = 2n\pi \pm 0.32$
12 $\theta = n\pi + 1$

13 $\theta = \dfrac{n\pi}{2} + (-1)^n\dfrac{\pi}{12}$

14 $\theta = \frac{2}{3}n\pi \pm \dfrac{\pi}{18}$

15 $\theta = \dfrac{n\pi}{4} - \dfrac{\pi}{16}$

16 $\theta = 3n\pi + (-1)^n\pi/2$
17 $\theta = \frac{2}{3}n\pi$
18 $\theta = (2n-1)\pi/2$
    or $(2n+1)\pi/2$

19 $\theta = (2n-1)\pi/4$
    or $(2n+1)\pi/4$
20 $\theta = n\pi \pm \pi/6$
21 $\theta = n\pi \pm \pi/3$
22 $\theta = n\pi + (-1)^n\pi/6$
    and $\theta = n\pi + (-1)^4\pi/2$

23 $\theta = \dfrac{n\pi}{3}$ and
    $\theta = \dfrac{n\pi}{3} + (-1)^n\pi/6$

24 $\theta = n\pi + \pi/4$

25 $\theta = \dfrac{n\pi}{2} + (-1)^n\pi/6;$ **26** $\theta = n360° \pm 90°;$ **27** $\theta = \dfrac{n\pi}{2} + \dfrac{\pi}{8}$ and

$\theta = \pi/6, \pi/3$ $\theta = \pm 90°, \pm 270°$ $\theta = \dfrac{n\pi}{2} + 0.554$

Exercise 7.4

1 $\theta = \dfrac{n\pi}{2 - (-1)^n}$

2 $\theta = \dfrac{n\pi}{2}$ [This includes all values of $\theta = n\pi$]

3 $\theta = \dfrac{n\pi}{2}$

4 $\theta = \dfrac{n\pi}{5 - (-1)^n 3}$

5 $\theta = \dfrac{\pi}{8} - \dfrac{n\pi}{2}$

6 $\theta = \dfrac{n\pi}{3} + \dfrac{\pi}{6}$

7 Either $\theta = \dfrac{n\pi + (-1)^n\pi/2}{3 + (-1)^n}$ or

$\theta = \dfrac{n\pi}{2} + \dfrac{\pi}{8}$ and $\theta = \dfrac{\pi}{4} - n\pi$

8 $\theta = (5n + (-1)^n)\pi/15;$
$\theta = \pi/15, 4\pi/15, 11\pi/15, 14\pi/15$

9 $\theta = (6n \pm 1)\dfrac{2\pi}{9};$ $\theta = \dfrac{2\pi}{9}$

10 $\theta = \dfrac{n\pi}{5}$ and $\dfrac{n\pi}{2};$
$\theta = 0, \dfrac{\pi}{5}, \dfrac{2\pi}{5}, \dfrac{3\pi}{5}, \dfrac{4\pi}{5}, \pi, \dfrac{\pi}{2}$

11 $\theta = (2n + 1)\dfrac{\pi}{8};$ $\theta = \pi/8, 3\pi/8, 5\pi/8, 7\pi/8$

12 $\theta = n90° + (-1)^n 30°;$
$\theta = 30°, 60°, 210°, 240$

13 $\theta = n90° - 30°;$ $\theta = 60°, 150°, 240°, 330°$

14 $\theta = 360°n - 30°$ and $\theta = 120°n - 10°;$
$\theta = 110°, 230°, 330°, 350°$

15 $\theta = \dfrac{7\pi}{36}, \dfrac{11\pi}{24}, \dfrac{19\pi}{36}, \dfrac{23\pi}{24}, \dfrac{31\pi}{36}$

16 $\theta = \pi/10, \pi/2, 9\pi/10$

17 $70°, 250°$

18 $\theta = 15°, 135°, 255°, 315°$

Miscellaneous Exercise 7A

1 (a) $149.5°, 340.5°$
  (c) $120°, 240°, 70.5°, 289.5°$
  (e) $26.6°, 206.6°, 116.6°, 296.6°$
  (b) $120°, 240°, 41.4°, 318.6°$
  (d) $30°, 150°, 19.5°, 160.5°$
  (f) $135°, 315°, 18.4°, 198.4°$

2 (a) $120°, 240°$
  (c) $90°, 30°, 150°, 270°$
  (e) $0°, 180°, 360°, 60°, 120°, 240°, 300°$
  (b) $41.8°, 138.2°, 270°$
  (d) $120°, 240°, 0°, 360°$
  (f) $210°, 330°, 48.6°, 131.4°$

3 $\pm 90°, 19.5°, 160.5°$

4 $45°, -135°, 63.4°, -116.6°$

5 $\pm 120°, 0°, \pm 180°, \pm 90°$

6 $15°, 75°, -105°, -165°, \pm 30°, \pm 90°$

7 $110.2°, 26.2°$

8 $180°, -46.4°$

9 $-18.4°, 161.6°, 63.4°, -116.6°$

10 $\pm 70.5°, \pm 101.5°$

11 $0°, \pm 180°, \pm 30°, \pm 150°$

12 $36.9°, 90°, -90°, -143.1°$

13 $40.8°, 139.2°, -8.8°, -171.2°$

14 $0°, 54°, 126°, 162°, -18°, -90°$

15 $45°, 114.3°, -24.3°, -135°$

16 $\dfrac{n\pi}{2 - (-1)^n}$

17 Either $\dfrac{n\pi + (-1)^n\pi/2}{2 + (-1)^n}$ or

$\left[\dfrac{2}{3}n\pi + \dfrac{\pi}{6}\quad \text{and}\quad \dfrac{\pi}{2} - 2n\pi\right]$

18 $(2n + 1)\dfrac{\pi}{8}$

19 $2n\pi \pm \dfrac{\pi}{2}$ and $\dfrac{2}{5}n\pi \pm \dfrac{\pi}{10}$

20 Either $[2n\pi + \pi/2$ and $2n\pi - \pi/6]$
  or $n\pi + (-1)^n\pi/6 - \pi/3$

21 $2n\pi \pm 2.2097^c - 0.46^c$

Miscellaneous Exercise 7B

1 (a) $1 - \theta^2/8$ (b) $1 - 9\theta^2/2$ (c) θ (d) $\dfrac{1}{\sqrt{2}}\left(\theta + \dfrac{\theta^2}{2} - 1\right)$

2 $\sqrt{2}\sin(x - \pi/4); 2n\pi + \pi/2$ and $(2n + 1)\pi$
3 $\pi/8, \pi/2, 5\pi/8, 3\pi/2, 9\pi/8, 13\pi/8$
4 $0, \pi, 7\pi/6, 11\pi/6, 2\pi$

5 $\dfrac{2t}{1 + t^2}, \dfrac{1 - t^2}{1 + t^2}; 72°$ and $220°$

6 $2\sin(\theta - 30°); 70°, 190°, 310°, 330°$
7 $0, \pi/4$
8 $40°, 80°, 90°, 160°$
9 $3.2, 18.4°; 108.4°$ or $288.4°$

10 (a) $\dfrac{n\pi}{3} - (-1)^n\dfrac{\pi}{18}$ (b) $360n + 36.9$ or $360n - 270$

11 $36.9°$; Max when $\theta = -36.9°$, Min when $\theta = 143.1°$, zero when $\theta = 0°$ and $-73.8°$

12 (a) $\dfrac{n\pi}{6}$ (b) $\frac{2}{3}n\pi \pm \dfrac{\pi}{6}$ and $\dfrac{n\pi}{2} + (-1)^n\dfrac{\pi}{12}$

13 $f(\theta) = \cos(\theta + \pi/6) + \cos\pi/6$;   Max is $1 + \sqrt{3}/2$ for $\theta = -\pi/6$
    Min is $-1 + \sqrt{3}/2$ for $\theta = 5\pi/6$; $a = \sqrt{3}$, $b = -\frac{1}{2}$, $c = -\sqrt{3/4}$
14 (a) $\pi/3, \pi, 5\pi/3$ (b) $0, \pi/6, 5\pi/6, \pi, 7\pi/6, 11\pi/6, 2\pi$ (c) $5\pi/6, 3\pi/2$
15 $18, 0; 15°, 75°$ 16 0.8416

18 $\dfrac{2t}{1 - t^2}; \dfrac{1 + t^2}{1 - t^2}; 18°$

19 $2\tan 2\theta$
20 $20\sin(\theta + 53.1°)$; Max perimeter is 20, when $\theta = 36.9°$
21 $2n\pi \pm \pi/3$ and $n\pi + \pi/4$
22 (a) $7 - 4\sqrt{3}$ (b) $90°, 270°$
23 $90°n + (-1)^{n+1}19°; 161°$
24 $-\sqrt{2} \leqslant a \leqslant \sqrt{2}; 45°, 225°$
25 (c) $p = 2$, $q = \frac{3}{4}$ (d) $r = 1$, $s = \frac{3}{2}$

26 $\sqrt{2}\cos(\theta + \pi/4)$; (a) $\dfrac{7\pi}{8}, \dfrac{15\pi}{8}$ (b) (i) $\text{Max}\left(\dfrac{3\pi}{4}, 1 + \sqrt{2}\right), \text{Min}\left(\dfrac{7\pi}{4}, 1 - \sqrt{2}\right)$

    (ii) $\text{Max}\left(\dfrac{\pi}{4}, \infty\right)$ and $\left(\dfrac{5\pi}{4}, \infty\right)$, $\text{Min}\left(\dfrac{3\pi}{4}, \dfrac{-1}{\sqrt{2}}\right)$

Unit 8

Exercise 8.1
1 $2\cos 2x$
2 (a) $3\cos\theta$ (b) $-4\sin\theta$ (c) $\frac{1}{5}\sec^2\theta$ (d) $2\cos\theta + \sin\theta$

   (e) $3\cos\theta - \sec^2\theta$ (f) $2\sin\theta$ (g) $\dfrac{\sec^2\theta - 3\sin\theta}{2}$

3 (a) $-2\cos x + c$ (b) $\dfrac{\sin x}{3} + c$ (c) $3\tan x + c$

   (d) $-3\cos x - \sin x + c$ (e) $-\cos x + 2\tan x + c$

4 (a) 1   (b) 1   (c) 1   (d) −7   (e) −1
5 (a) −5 cos t   (b) −4 cos t + 5 sin t   (c) 3 sin t

6 (a) $1 - \dfrac{5\sqrt{3}}{2}$ (b) $\dfrac{35 + \sqrt{3}}{2}$

7 (a) 0   (b) 1   (c) 4

8 (a) $\dfrac{\sqrt{3} - 1}{2}$ (b) 3

9   $\dfrac{9\pi}{\sqrt{2}}$ 10 $y = \sqrt{2}[x - \pi/4 + 2]$, $\sqrt{2}y + x = 4 + \pi/4$

11 (a) 1   (b) −2   (c) $-\frac{4}{3}$

12 (a) $\frac{1}{2}\sin x$ (b) $-\frac{1}{2}\sin x$ (c) $x - \sin x + c$ (d) $\frac{1}{2}(x + \sin x) + c$

Exercise 8.2

1 (a) $6 \cos 6x$ (b) $-\frac{1}{2}\sin\dfrac{x}{2}$ (c) $3 \sec^2 3x$ (d) $-0.2 \sin(0.2x)$

   (e) $\frac{1}{5}\cos\dfrac{x}{5}$ (f) $20 \sec^2 10x$ (g) $-15 \sin 5x$ (h) $\cos 2x$

   (i) $2 \sec^2\dfrac{x}{2}$ (j) $\frac{1}{9}\cos\dfrac{x}{3}$

2 (a) $-\frac{1}{6}\cos 6x + c$ (b) $2 \sin x/2 + c$ (c) $\frac{1}{3}\tan 3x$

   (d) $-\frac{2}{3}\cos\dfrac{3x}{2} + c$ (e) $4 \tan\dfrac{x}{2} + c$ (f) $\sin 4x + c$

   (g) $-\cos\dfrac{x}{5} + c$ (h) $-\frac{1}{9}\cos 3x + c$ (i) $2 \tan 4x + c$

   (j) $\sin(0.3x) + c$

3 (a) $\cos(x + 1)$ (b) $-2 \sin(2x - 1)$ (c) $6 \cos(3x - 1)$

   (d) $-\sin(3x + 2)$ (e) $-4 \cos(1 - 2x)$ (f) $-2 \sin\left(\dfrac{x}{2} + 1\right)$

   (g) $-18 \sec^2(1 - 6x)$ (h) $\pi \cos(\pi x - \pi/4)$ (i) $-\dfrac{\pi}{2} \cdot \sin\left(\dfrac{\pi x}{2} + \dfrac{\pi}{3}\right)$

   (j) $-\pi^2 \sec^2(2 - \pi x)$

4 (a) $-\frac{1}{3}\cos(3t + 2) + c$ (b) $\frac{1}{2}\sin(2t - 1) + c$ (c) $2 \tan\left(\dfrac{t}{2} + 3\right) + c$

   (d) $2 \cos(4 - t) + c$ (e) $[\sin(\pi t + \pi/4)/\pi] + c$
5 (a) $2 \sin x \cdot \cos x$ (b) $-3 \cos^2 x \cdot \sin x$ (c) $5 \tan^4 x \cdot \sec^2 x$
   (d) $12 \sin^3 x \cdot \cos x$ (e) $-2 \cos^3 x \cdot \sin x$ (f) $4 \sin 2x \cdot \cos 2x$

   (g) $-12 \cos^2 4x \cdot \sin 4x$ (h) $6 \tan^3 x \cdot \sec^2 3x$ (i) $3 \sin^3\dfrac{x}{4} \cdot \cos\dfrac{x}{4}$

   (j) $\dfrac{\cos x}{2\sqrt{\sin x}}$ (k) $\dfrac{-\sin x}{2\sqrt{\cos x}}$ (l) $\dfrac{\sec^2 x}{2\sqrt{\tan x}}$

   (m) $\dfrac{-2 \cos x}{\sin^3 x}$ (n) $\dfrac{6 \sin 2x}{\cos^4 2x}$ (o) $\tan^{1/2} 4x \cdot \sec^2 4x$

6 (a) $\frac{1}{2}$ (b) $\frac{2}{3}$ (c) $\pi/2$ (d) $-\frac{1}{3}$ (e) π (f) -3

7 $4, y - 2\sqrt{3} = 4x - \dfrac{2\pi}{3}$ **10** 1 **14** (a) $\dfrac{4\sqrt{2}}{\pi}$ (b) π

                                    11 $\pi/2$

8 $-9/\sqrt{2}$ **12** $\sqrt{2}y = x + 1 - \pi/8$ **15** (a) $\pi/2$ (c) $\pi/2$

9 1 **13** $0, \pi, \pi/3$

Exercise 8.3

1 $\cos x - x \sin x$ **2** $2x \sin x + x^2 \cos x$ **3** $\tan x + x \sec^2 x$ **4** $\dfrac{2}{(1 + x)^2}$

5 $\dfrac{6x - 3x^2 - 1}{(1 - x)^2}$ **6** $\cos^2 x - \sin^2 x$ **7** $\dfrac{x \sec^2 x - \tan x}{x^2}$

8 $\dfrac{\cos x + x \sin x}{\cos^2 x}$ **9** $\dfrac{-x \sin x - 3 \cos x}{2x^4}$ **10** $\dfrac{4}{(1 - 2x)^3}$

11 $x \sin x + 1 - \cos x$ **12** $\sin 2x + 2x \cos 2x$ **13** $2x \cos x - x^2 \sin x - 1$

14 $3 - 2x^3 \sec^2 2x - 3x^2 \tan 2x$ **15** $(x + 3)(2 \cos x + \sin x) + 2 \sin x - \cos x$

17 (a) $-\operatorname{cosec}^2 \theta$ (b) $-\operatorname{cosec} \theta \cdot \cot \theta$

18 (a) $0, \frac{2}{3}$ (b) 2

19 (a) $-\dfrac{4}{(2 + t)^3}$ (b) $6 \cos t - 3t \sin t$ (c) $\dfrac{2 \sin t}{t^2} + \dfrac{2 \cos t}{t^3} - \dfrac{\cos t}{t}$

20 $\sqrt{2}\left(1 - \dfrac{3\pi}{4}\right)$

22 $4 \cos 2x \cos 5x - 10 \sin 5x \sin 2x$

23 $\cos^2 x - 2x \cos x \cdot \sin x$

24 $\cos^3 x - 3(2 + x) \cos^2 x \cdot \sin x$

21 $\dfrac{4x \cos 2x - 2 \sin 2x}{x^2}$ **25** $(x - 2)/[2(x - 1)^{3/2}]$

27 $\text{Max}\,(2, -1); \text{Min}\,(-2, -\frac{1}{9})$

Exercise 8.4

1 $\dfrac{x}{3y}$ **5** $\dfrac{1}{y^2}$ **8** $-\dfrac{y^2}{x^2}$

2 $\dfrac{x^2}{y^2}$ **6** $-\dfrac{y}{x}$ **9** $\dfrac{3 - 3x^2 y^2}{2x^3 y}$

3 $\dfrac{-2 - x}{y}$ **7** $\dfrac{y - 4x}{5 + 2y - x}$ **10** $-\sqrt{\dfrac{y}{x}}$

4 $\dfrac{4}{y}$

11 (a) $\dfrac{r \tan \theta}{2}$ (b) $r\left[\dfrac{\cos \theta + 2 \sin 2\theta - 2 \cos 2\theta}{\sin 2\theta + \cos 2\theta - 1 - \sin \theta}\right]$

12 (a) 1 (b) $-\frac{1}{4}$

14 -114

15 $\infty, 0; \frac{1}{2}; 90°, 36.9°$

16 $(1, 1)$ and $(-1, -1); \tan^{-1} 3$ or $71.6°$

17 $5y = 2\sqrt{2}x - 10; \ 2\sqrt{2}y + \sqrt{25}x = 29\sqrt{2}$

18 $\text{Max} - 3, \text{Min} \frac{3}{2}$

19 $\theta = n180° + 45°$

Exercise 8.5

1 $(0, 3), (\sqrt{2}, 3/\sqrt{2}), (2, 0), (1 - 3\sqrt{3}/2), (-\sqrt{3}, \tfrac{3}{2})$

2 $12\sqrt{5}$ or 26.83

3 -3

4 $\left(\dfrac{1}{4}, \dfrac{3\sqrt{3}}{2}\right), \left(\dfrac{1}{\sqrt{2}}, \sqrt{2}\right), \left(\dfrac{3\sqrt{3}}{4}, \dfrac{1}{2}\right), (-2, 0)$

5 $\tfrac{6}{25}$

6 $(1, 2)$

8 $y^2 = 12x$

9 $xy = 9$

10 $\dfrac{x^2}{4} + \dfrac{y^2}{9} = 1$

11 $\dfrac{x^2}{25} + \dfrac{(y - 1)^2}{9} = 1$

12 $(2 - x)^3 = y - 3$

13 $3y + 2x = 13$

14 $\left(\dfrac{x - 3}{2}\right)^2 + (1 - y)^2 = 1$

15 $y + x = 2$

16 $x^2 = y^2 + 8$

17 $y = 2x^2 - 1$

18 $x^{2/3} + y^{2/3} = 1$

19 $y + tx = 4t + 2t^3$

20 $t^2 y + x = 2ct$

21 $y + tx = 2at + at^3; 4a^2(1 + t^2)^2$

Exercise 8.6

1 (a) $\pi/6$ (b) $-\pi/4$ (c) $\pi/3$ (d) $\pi/3$ (e) $-\pi/4$ (f) $4\pi/6$
    (g) $-\pi/6$ (h) $\pi/3$ (i) $\pi/3$

4 $\dfrac{2}{\sqrt{1 - 4x^2}}$

5 $\dfrac{3}{9 + x^2}$

6 $\dfrac{-4}{\sqrt{1 - 16x^2}}$

7 $\dfrac{1}{2\sqrt{x - x^2}}$

8 $\dfrac{-2x}{\sqrt{1 - x^4}}$

9 $\dfrac{-1}{x^2 + 1}$

10 $\dfrac{2}{4 + x^2}$

11 $\dfrac{1}{\sqrt{2x - x^2}}$

12 $\dfrac{1}{(x - 1)\sqrt{x^2 - 2x}}$

13 $\dfrac{-1}{x\sqrt{x^2 - 1}}$

14 $\dfrac{1}{1 + x^2}$

15 $\dfrac{-2}{(2 + x)\sqrt{2x}}$

16 $\dfrac{1}{x\sqrt{4x^2 - 1}}$

17 $\dfrac{-\sin x}{1 + \cos^2 x}$

18 -1

19 $\dfrac{2\cos x}{\sqrt{1 - 4\sin^2 x}}$

21 -2

22 $-2(1 - x^2)^{1/2}$

Exercise 8.7

1 $\sin^{-1}\dfrac{x}{2} + c$

2 $\sin^{-1}\dfrac{x}{\sqrt{3}} + c$

3 $\tfrac{1}{2}\tan^{-1}\dfrac{x}{2} + c$

4 $\pi/4$

5 $\tfrac{1}{4}\sin^{-1} 4x + c$

6 $\tfrac{1}{5}\sin^{-1} 5x + c$

7 $\dfrac{1}{\sqrt{2}}\sin^{-1}\sqrt{2}x + c$

8 $\pi/12$

9 $\tfrac{1}{4}\tan^{-1}\dfrac{x}{A} + c$

10 $\tfrac{1}{12}\tan^{-1}\dfrac{3x}{4} + c$

11 $\tfrac{1}{15}\tan^{-1}\dfrac{5x}{3} + c$

12 $\tfrac{1}{2}\tan^{-1} 2x + c$

13 $\dfrac{1}{\sqrt{3}}\tan^{-1}\sqrt{3}x + c$

14 $\tfrac{1}{3}\sin^{-1}\dfrac{3x}{4} + c$

15 $\dfrac{1}{\sqrt{3}}\sin^{-1}\dfrac{\sqrt{3}x}{\sqrt{2}} + c$

16 $\dfrac{1}{4} + \dfrac{\pi}{8}$

17 $\dfrac{\pi}{2}$

18 $\sqrt{3} - 1 - \dfrac{\pi}{12}$

19 $\dfrac{3\pi}{8} - \dfrac{1}{4}$

20 $-\pi/2$

21 $- \operatorname{cosec} x(\operatorname{cosec} x + \cot x)$ **27** $\pi/4$

22 $x \sec x \cdot \tan x + \sec x$ **28** $\pi/3$

23 $\sec^7 x + \operatorname{cosec} x \cdot \cot x$

24 $-2 \cot x \cdot \operatorname{cosec}^2 x$

25 $3 \sec^3 x \tan x$ **29** $\dfrac{\pi}{6\sqrt{15}}$

26 $\dfrac{-2 \operatorname{cosec}^3 x}{(1 - \cot x)^2}$

Miscellaneous Exercise 8A

1 Let $y = \cos x/2$

$$y + \delta y = \cos\left(\frac{x + \delta x}{2}\right)$$

$$\delta y = \cos\left(\frac{x + \delta x}{2}\right) - \cos\frac{x}{2}$$

$$\frac{\delta y}{\delta x} = \frac{-2\sin\left(\frac{x}{2} + \frac{\delta x}{4}\right)\sin\frac{\delta x}{4}}{\delta x} = -\tfrac{1}{2}\sin\left(\frac{x}{2} + \frac{\delta x}{4}\right)\left[\frac{\sin\left(\frac{\delta x}{4}\right)}{\frac{\delta x}{4}}\right]$$

As $\delta x \to 0$, $\dfrac{\delta y}{\delta x} \to \dfrac{dy}{dx}$ and $\left[\dfrac{\sin\left(\frac{\delta x}{4}\right)}{\delta x/4}\right] \to 1$

$$\therefore \frac{dy}{dx} = \lim_{\delta x \to 0} \frac{\delta y}{\delta x} = -\tfrac{1}{2}\sin\frac{x}{2}$$

2 (a) $6\cos x(1 + 3\sin x)$ (b) $4\sin 2x \cos 2x - 9\cos^2 3x \cdot \sin 3x$

    (c) $\tfrac{3}{2}\sqrt{\sin x \cdot \cos x}$ (d) $-4\tan 2x \sec^2 2x$

4 (a) 4 (b) 2 (c) $\dfrac{3\pi}{8} - \sqrt{2} + \tfrac{7}{4}$

5 Max value $= \dfrac{\pi}{2} - 1$;   Min value $= \dfrac{3\pi}{2} + 1$

6 $2\cos 2t, -2\sin t; \dfrac{dy}{dx} = \dfrac{-\sin t}{\cos 2t}; \quad t = n\pi$

7 $y = x + \pi - 4$

8 $8, 16, \pi/4$

9 4

10 2

11 $-\tfrac{1}{3}$

12 (a) $x(\cos^2 x - \sin^2 x) + \sin x \cdot \cos x$ (b) $\dfrac{x\cos x}{x - 1} - \dfrac{\sin x}{(x - 1)^2}$

13 $-2t/(1 - t^2)$

14 $\pi\left[\dfrac{3\pi}{4} - 2\right]$

15 Max value $= 1/4$;   Min value $= 1$

Miscellaneous Exercise 8B

1 $\dfrac{x^2 - y}{x - y^2}$ **3** $11y + 7x = 18$ **4** (a) $-\frac{1}{4}$ (b) 4

5 Max is 2 at $x = \pi/6$, Min is -2 at $x = 2\pi/3$; $\dfrac{5\pi}{12}$ and $\dfrac{11\pi}{12}$; $\dfrac{\pi}{2}[\pi + \sqrt{3}]$

7 $\dfrac{6x}{(2x^2 + 1)^2}$; $x > 0$; $y = 0$, $y = -1$

8 $a = 3$, $b = 5$, $c = 2$
9 $\sec\theta(\tan\theta + \sec\theta)$, $-\csc\theta(\cot\theta + \csc\theta)$

10 (a) $\dfrac{-n^2\sin n\theta\cos\theta + \sin\theta \cdot n\cos n\theta}{\cos^2\theta}$

11 $(\frac{1}{7}, \frac{15}{7})$

12 $\dfrac{1 - 2\sin^2\theta - \sin\theta}{\cos\theta}$; $\left(\dfrac{1}{2}, \dfrac{3\sqrt{3}}{4}\right)$, $\left(\dfrac{1}{2}, \dfrac{-3\sqrt{3}}{4}\right)$, $(1, 0)$; $\dfrac{13\pi}{10}$

14 $\cos x$, $-\sin x$
16 $(\pi/2, 0)$, $(\pi, 0)$

17 R is $(2ap^2, 3ap)$; $4y + 6px = 12ap + 15ap^3$; $\dfrac{3ap^3}{\sqrt{16 + 36p^2}}$

Exercise 9.1
1 (a) $\log_3 9 = 2$ (b) $\log_{10} 10\,000 = 4$ (c) $\log_4 8 = \frac{3}{2}$
   (d) $\log_8 512 = 3$ (e) $\log_{144} 1728 = \frac{3}{2}$ (f) $\log_8 \frac{1}{2} = -\frac{1}{3}$
   (g) $\log_6 1 = 0$ (h) $\log_5 \frac{1}{25} = -2$ (i) $\log_{27} 9 = \frac{2}{3}$
   (j) $\log_q p = 3$ (k) $\log_x z = y$ (l) $\log_m 3 = n$
2 (a) $125 = 5^3$ (b) $100 = 10^2$ (c) $81 = 3^4$
   (d) $3° = 1$ (e) $8 = (1/2)^{-3}$ (f) $125^{1/3} = 5$
   (g) $p° = 1$ (h) $q^3 = p$ (i) $5^y = x$
   (j) $a^c = 3$ (k) $p^r = q$ (l) $r^q = p$
3 (a) $\log p + \log r$ (b) $\log p - \log q$ (c) $\log p + \log q + \log r$
   (d) $\log p + \log q - \log r$ (e) $2\log q$ (f) $2\log p - \log q$
   (g) $3\log q + 2\log p$ (h) $-\log p$ (i) $1/2\log p$
   (j) $\frac{1}{2}(\log q - \log r)$ (k) $\log r + \frac{1}{2}\log p$ (l) $4\log p - \log r - \frac{1}{2}\log q$
4 (a) $\log 15$ (b) $\log 4$ (c) $\log 6$ (d) $\log 125$ (e) $\log 50$
   (f) $\log 4$ (g) $\log 3$ (h) 0   (i) $\log 3/8$ (j) 0
5 (a) 3   (b) 4   (c) 2   (d) $\frac{1}{2}$ (e) 0   (f) $\frac{2}{3}$
   (g) -1 (h) 3   (i) -2 (j) 2   (k) $\frac{1}{2}$ (l) $-\frac{1}{3}$
6 (a) $1 + \log_{10} a$ (b) $2 - 2\log_{10} b$ (c) $\frac{1}{2}\log_{10} b - 3$
7 (a) $\log_{10} 2$ (b) $\log_{10} 400$ (c) $\log_{10} 8$ (d) $\log_4 36$
   (e) $\log_6 4$ (f) $\log_4 8$
8 (a) 2   (b) 3   (c) $\log(x - 1)$ (d) 0
9 3   (b) q (c) 1   (d) 5   (e) 9

Exercise 9.2
1 (a) 1.861   (b) 1.771   (c) 3.907   (d) 1.464   (e) 2.792
   (f) 1.131
2 (a) 1.594   (b) 1.465   (c) 0.256   (d) -1.113
3 (a) $-1, 0$ (b) ± 1 (c) $0, 1$ (d) 3
4 4 and 8
5 3

6 $x = 3,$ $y = 9$
7 $x = 0,$ $y = 0$ and $x = \frac{1}{3},$ $y = 1$
8 $x = 3$ and 6
9 1

10 3
11 $x = 1,$ $y = 0$ and $x = y = 2$
12 -1

Exercise 9.3

3 (a) $\dfrac{1}{x}$

(b) $\dfrac{1}{x + 5}$

(c) $\dfrac{3}{3x - 1}$

(d) $\dfrac{-1}{2 - x}$ or $\dfrac{1}{x - 2}$

(e) $\dfrac{2x - 1}{x^2 - x + 1}$

(f) $\dfrac{-3x^2}{1 - x^3}$ or $\dfrac{3x^2}{x^3 - 1}$

(g) $-\tan x$

(h) $2 \cot 2x$

(i) $\dfrac{\sec^2 x}{\tan x}$

4 (a) $3e^x$ (b) $-e^{-x}$ (c) $-5e^{-5x}$ (d) $1/2e^{1/2x}$

(e) $3x^2 e^{x^3}$ (f) $\sec^2 x e^{\tan x}$ (g) $e^x e^{e^x}$

(h) $e^x - \dfrac{1}{e^x}$ (e) $2e^{2x} - 2e^{-2x}$

5 (a) $\dfrac{1}{2(2 + x)}$ (b) $\dfrac{4}{4 - x^2}$ (c) $\dfrac{12}{3x - 1}$

(d) $2 \cot x$ (e) $\dfrac{-1}{x}$ (f) $\dfrac{-5}{2(2 + x)(1 + 3x)}$

(g) $\dfrac{2x + 3}{x(x + 1)}$ (h) $\dfrac{x^2}{x^3 - 1}$ (i) $2 \tan x$

6 $x(2 \ln x + 1)$
8 2

10 Stat value $= \dfrac{1}{e}$; Max.

12 (a) $2 + 2\ln(2x - 1)$

(b) $\dfrac{e^x}{1 + e^x}$

(c) $e^x \left[\dfrac{1}{x} + \ln 3x \right]$

(d) $-3e^{-3x}[\cos 3x + \sin 3x]$

(e) $2 \sec x$

(f) $\dfrac{-(1 + x)}{(1 + x)(1 + x^2)}$

(g) $\dfrac{1}{\sqrt{x(1 - x)}}$

(h) $e^{x \sin x}(x \cos x + \sin x)$

Exercise 9.4
1 (a) $4 \ln|x| + c$ (b) $\frac{1}{4} \ln|x| + c$ (c) $x - \ln|x| + c$
(d) $\ln|x + 2| + c$ (e) $\frac{1}{2} \ln|2x - 3| + c$ (f) $\frac{1}{3} \ln|3x + 1| + c$
(g) $-\frac{1}{2} \ln|1 - 2x| + c$ (h) $\frac{2}{3} \ln|x - 4| + c$ (i) $\frac{3}{5} \ln|5x - 2| + c$
(j) $-\ln|3 - x| + c$ (j) $-\ln|5 - 2x| + c$ (k) $-\frac{4}{3} \ln|1 - 3x| + c$

2 (a) $\frac{1}{3} e^{3x} + c$ (b) $-\frac{1}{3} e^{-3x} + c$ (c) $-\dfrac{1}{e^x} + c$ (d) $3e^x + c$

(e) $2e^{2x} + c$ (f) $-\frac{3}{4} e^{-4x} + c$ (g) $e^{x^3} + c$ (h) $e^{\sin x} + c$

3 (a) $x - 2 \ln|x| - \dfrac{1}{x} + c$ (b) $\dfrac{x^3}{3} + 2x - \dfrac{1}{x} + c$ (c) $\dfrac{e^{2x}}{2} + 2x - \dfrac{e^{-2x}}{2} + c$

4 (a) $\ln|x^2 - 1| + c$ (b) $\ln|x^3 - x^2 + 1| + c$ (c) $\frac{1}{2}\ln|x^2 + 2x + 3| + c$
(d) $\frac{1}{4}\ln|x^4 - 3| + c$ (e) $-\ln|1 + \cos x| + c$ (f) $\frac{1}{6}\ln|x^6 + 2| + c$
(g) $-\frac{1}{4}\ln|\cos 4x| + c$ (h) $\frac{1}{3}\ln|\sin 3x|c$ (i) $\ln|1 + \tan x| + c$

5 (a) $1 - \dfrac{1}{e^2}$ (b) $\dfrac{1}{2}\left(e - \dfrac{1}{e}\right)$ (c) $\dfrac{1}{2e^{1/2}}$ (d) $\frac{1}{3}\ln 4$ (e) $\frac{1}{2}\ln\frac{5}{4}$

(f) $\frac{1}{2}\ln 2$
6 $2\ln\frac{3}{4}$
7 $3\ln 3$
8 $\ln|\tan x| + c$
9 $\ln|\sec x + \tan x| + c$

10 Max is $\left(\dfrac{1}{\sqrt{2}}, \dfrac{1}{\sqrt{2e}}\right)$; Min is $\left(-\dfrac{1}{\sqrt{2}}, -\dfrac{1}{\sqrt{2e}}\right)$

Exercise 9.5

1 $1 + 5x + \frac{25}{2}x^2 + \frac{125}{6}x^3 + \cdots; \dfrac{5^r x^r}{5!}$; valid all x

2 $-6\left[x + \dfrac{x^3}{3} + \cdots\right]; 3[-1 + (-1)^r]\dfrac{x^r}{r}$; valid $|x| < 1$

3 $\ln 3 - \dfrac{x}{3} - \dfrac{x^2}{16} - \dfrac{x^3}{81} + \cdots; \dfrac{-x^r}{r3^r}$; valid for $-3 \leqslant x < 3$

4 $x - 2x^3 + \dfrac{2x^5}{3}$

5 $1 - \dfrac{x^2}{2} + \dfrac{x^3}{3}$

6 $x + \dfrac{x^3}{3!} + \dfrac{x^5}{5!}$

7 $1 + x - x^2$

8 (a) $1 + x + \dfrac{3x^2}{2} + \frac{5}{2}x^3 + \cdots$ (b) $1 + x + \frac{3}{2}x^2 + \frac{7}{6}x^3$; $k = \frac{4}{3}$

9 (a) $px - \frac{1}{6}p^3 x^3 + \frac{1}{120}p^5 x^5 + \cdots$ (b) $x - qx^2 + \frac{1}{2}q^2 x^3 + \cdots$; $p = -2$,
$q = b$, $r = \frac{104}{3}$
10 $a = \frac{2}{3}$, $b = \frac{2}{5}$; 2.00 671

Exercise 9.6 Note: Results from graphs are only approximate.
1 (a) $w = \frac{5}{2}R + 50$ (b) $l = 0.02w + 9.4$
2 $R \approx 2.3V + 96$ **3** $\theta \approx 8.2 - 0.4T$
4 $p = 2$, $q = 0$ are in error; $a \approx 0.5$, $b \approx -2.5$
5 $f \approx 10$
6 $\alpha \approx 2$, $\beta \approx -5$
7 $a = 2.0$ or 1.9, $b = 0.50$ or 0.51; $y = 5.0(\pm 0.1)$
8 $a \approx 12.6$, $b \approx 1.12$

9 $\dfrac{y}{x^2} = z = ax + b$; $a \approx 4$, $b \approx 2$

Miscellaneous Exercise 9A

	Y	X	m	C
1	$\dfrac{1}{y}$	x^2	a	b
2	$\dfrac{y^3}{x}$	x	a	ab
3	$\dfrac{1}{y}$	$\dfrac{1}{x}$	$-\dfrac{a}{b}$	$\dfrac{2}{b}$
4	$\log y$	x	$\log b$	$-\log a$
5	$\log s$	t	$-\log b$	$\log a$
6	$y^2 - x$	y	p	$-q$
7	x	$\dfrac{e^y}{x}$	a	b
8	$x^2 - y$	x	$a + b$	$-ab$

9 (a) $\log_5 125 = 3$ (b) $\log_8 32 = \frac{5}{3}$
10 (a) $16 = 2^4$ (b) $9° = 1$ 11 $\log x + 3 \log y - \frac{1}{2} \log z$
12 (a) $\log 3$ (b) 3   (c) x 13 (a) 6   (b) $\frac{1}{2}$ (c) 2
14 (a) 1.189   (b) 2.33   15 0   16 $\frac{1}{5}$ and 25

17 (a) $\dfrac{2x}{x^2 + 3}$ (b) $\cot x$ (c) $5e^{5x}$ (d) $-\frac{1}{2}e^{-1/2x}$ (e) xe^{x^2}
    (f) $\sec x$

18 (a) $\dfrac{6}{2x + 1}$ (b) $\dfrac{-1}{2(x - 1)}$

19 (a) $2 \ln x + c$
    (b) $-2e^{-x} + c$

    (c) $3 \ln x + \dfrac{2}{x} + c$

    (d) $\ln(x + 1)$
    (e) $\frac{3}{2}e^{2x} + c$
    (f) $\frac{1}{2} \ln(3x^2 + 2x - 1) + c$
    (g) $e^{x^3} + c$
20 (a) $2e^2 - 2$ (b) $\frac{2}{3} \ln 28$ (c) $-\ln 2$ or $\ln \frac{1}{2}$

Miscellaneous Exercise 9B
1 (a) $e^{-x}(\cos x - \sin x)$ (b) $2 \tan x$

2 (a) $-\dfrac{(x + 1)}{x^2}e^{-x}$ (b) $\cos x/(1 + \sin x)$

3 (a) $x^2(1 + 3 \ln x)$ (b) $-(1 + \cos x + x \sin x)/x^2$

4 (a) $6 \sin 3x \cdot \cos 3x$ (b) $\dfrac{\sec x \tan x}{1 + \sec x}$

5 $x = 3$, $y = 9$ and $x = 9$, $y = 3$
6 -0.39
7 8.7   12 $\frac{1}{2}[\ln(1 + x) - 2x - \ln(1 - x)]; 0$
10 $\frac{1}{2}n(n + 1)$ 13 $a = \ln 2$, $b = 4$

11 $k = 8$ 14 $\dfrac{3\pi}{4}, \dfrac{7\pi}{4}, \dfrac{11\pi}{4}, \dfrac{15\pi}{4}$

15 $|r| < 1$; $x < 0$; $\dfrac{1}{1 + e^x}$

16 $x = 27$, $y = 27$ and $x = 9$, $y = 81$

17 (a) $x = 3^{+1.5}$ (b) $y = 21$

18 (a) $3x + 2y = 4$ (b) $y \ln 6 = \ln x + \ln 5$

19 (a) $\dfrac{1}{p} + \dfrac{1}{q}$ (b) $\dfrac{q}{p}$

20 $1 - \dfrac{x^2}{6} + \dfrac{x^4}{120} + \cdots$; $-y - \dfrac{y^2}{2} - \dfrac{y^3}{3} - \cdots$; $A = -\frac{1}{6}$, $B = -\frac{1}{180}$

21 $\frac{1}{2}[\log_3 x + \log_3 y]$; $x = \frac{1}{3}$, $y = 3^6$ and $x = 3^6$, $y = \frac{1}{3}$

23 (a)   125 or $\frac{1}{125}$ (b) $\pm \dfrac{1}{\sqrt{2}}$.

Unit 10

Exercise 10.1

1 (a) $2x^3 + 5x^2 + 2x + 15$ (b) $6x^4 - 7x^3 + 12x^2 - 9x + 2$
  (c) $3x^4 + 14x^3 - 10x^2 + 7x - 2$ (d) $5x^5 - 7x^3 + 4x^2 + 2x - 4$
  (e) $9x^5 - 4x^3 + 18x^2 + 3x - 6$ (f) $x^6 - 1$

2 (a) $-16, -5$ (b) $4, -2$ (c) $-5, -3$ (d) $15, 5$

3 (a) $\dfrac{x - 13}{(x^2 - 1)(2x - 5)}$ (b) $\dfrac{x + 3}{x^2 - 3x + 9}$ (c) $x^2 + x + 1$ (d) $x^2 + 4$

4 (a)   20   (b)   103   (c)   12   (d) -17

5 (a) $3x + 1, 6$ (b) $2x^2 + x + 4, 15$ (c) $x^2 + 3x + 9, 36$
  (d) $x^4 - 2x^2 + 3x + 1, 2$ (e) $x^2 - x, 4x - 4$ (f) $2x^2 - x + 5, 2 - x$
  (g) $3x - 5a, 2a^2$ (h) $a^2 - 2ab - 2b^2, 0$

6 (a) $\frac{19}{16}$ (b) $-\frac{7}{27}$ (c) $5 + 2a^2 - a^3$ (d) $b + ac + c^2$

Exercise 10.2

1 (a)   Yes   (b)   No   (c)   No   (d)   Yes   (e)   Yes   (f)   No

2 (a) $(x - 1)^2(x + 3)$ (b) $(x - 1)(x - 2)(x^2 + 4)$
  (c) $(x - 3)(x + 3)(x^2 + x + 1)$ (d) $(x + 1)(x - 2)(2x - 1)$
  (e) $(x - 1)(3x + 1)(x^2 + 2)$ (f) $(x - y)(x^2 + xy + y^2)$
  (g) $(2x - 3)(4x^2 + 6x + 9)$

3 $a = 2$ **4** $x + 4$ **5** $t = -1$ **6** 3

7 $a = \pm 3$; $3(x - 3)(x + 1)^2$ and $-3(x + 3)(x - 1)^2$

Exercise 10.3

1 1   **2** $3\frac{1}{4}$ **3** 3   **4** $p = 1$, $q = -3$

5 $x + 2$ **6** $2, 5, 1$

7 (a)   no repeated factors   (b) $x - 2$ (c) $x - 2$, $x + 2$

8 $10x + 7$

9 $-48, -45, 80$ **10** $2x + 1$ **11** $m = 1$, $n = 3$

Exercise 10.4

1 (a) $5, 3$ (b) $-\frac{2}{3}, -\frac{5}{3}$ (c) $3, -1$
  (d) $-1, -10$ (e) $-3, -19$ (f) $-\frac{3}{2}, -2$

2 (a) $x^2 - 4x + 5 = 0$ (b) $2x^2 + 6x + 1 = 0$ (c) $20x^2 - 5x - 12 = 0$
  (d) $2x^2 + x = 0$ (e) $3x^2 + 2 = 0$ (f) $x^2 - kx + k^2 = 0$
  (g) $x^2 + (k + 2)x + 3k^2 = 0$ (h) $qrx^2 - prx + q^3 = 0$

3 (a) T (b) F (c) T (d) T (e) T
4 (a) 3   (b) $\frac{25}{4}$ (c) $\frac{3}{5}$ (d) $\frac{19}{2}$ (e) $-\frac{11}{4}$
  (f) $\frac{15}{4}$ (g) $\frac{9}{4}$ (h) $-\frac{31}{4}$
5 (a) $x^2 - 5x + 5 = 0$ (b) $x^2 - 3x + 1 = 0$ (c) $x^2 - 7x + 1 = 0$
6 $3x^2 - 10x - 4 = 0$ 7 $2x^2 + 4x + 5 = 0$
8 $x^2 - 7x + 8 = 0$ 9 $x^2 + 9x + 11 = 0$
10 $k = 0$ 11 $m = -30$
12 (a) $\frac{9}{2}$ (b) 1   (c) $\frac{9}{4}$ (d) $\frac{17}{4}$
13 $53x^2 - 15x + 1 = 0$ 14 $52x^2 - 60x + 9 = 0$

Exercise 10.5

1 $x > \frac{4}{3}$ 2 $x > \frac{7}{3}$ 3 $x \geqslant \frac{3}{7}$
4 $x > -2$ 5 $x \geqslant \frac{15}{17}$ 6 $x < \frac{17}{7}$
7 $0 \leqslant x \leqslant 2$ 8 $x < -2$ and $x > 3$ 9 $-\frac{1}{2} < x < \frac{1}{3}$
10 $-1 < x < 3$ 11 $x < -\frac{1}{2}$ and $x \geqslant 2$ 12 $-1 < x < 2$
13 $x < -2$ and $x > 3$ 14 $-6 \leqslant x \leqslant 4$ 15 $x < -\frac{1}{2}$ and $x > 3$
16 all x 17 no solution   18 $x = 2$
19 all x 20 no solution

21 $x > 2 + \dfrac{1}{\sqrt{2}}$ and $x < 2 - \dfrac{1}{\sqrt{2}}$ 22 $-3 - \sqrt{\frac{2}{3}} \leqslant x \leqslant \sqrt{\frac{2}{3}} - 3$

23 $x \leqslant 0$ and $x > 1$ 24 $1 \leqslant x \leqslant \frac{3}{2}$
25 $-2 < x < 0$ and $x > 2$ 26 $x < -\frac{5}{2}$ and $x > -1$
27 $\frac{5}{2} \leqslant x \leqslant -1$ 28 $\frac{1}{2} - \sqrt{\frac{3}{20}} < x < \frac{1}{2} + \sqrt{\frac{3}{20}}$
29 $x < -1$ and $x > \frac{1}{3}$ 30 $-1 < x < 0$

Exercise 10.6

1 (a) $|x| < 4$ (b) $|x| < 2$ (c) $|x - 2| < 4$
  (d) $|x| > 3$ (e) $|2x - 2| < 5$ (f) $|3x + 2| < 2$
  (g) $|x + 1| > 3$ (h) $|x - 2| \leqslant \sqrt{5}$ (i) $|x - 2| > \sqrt{5}$
  (j) $|x + \frac{3}{2}| < \frac{\sqrt{5}}{2}$
2 (a) $-5 < x < 5$ (b) $-2 < x < 4$ (c) $-2 < x < \frac{1}{3}$
3 (a) $-5 < x < -1$ (b) $x > \frac{3}{2}$ and $x < -1$ (c) $1 < x < 4$
  (d) $x < -2$ and $x > \frac{8}{3}$
4 $1 < a < 4$ 11 $x < -2$ and $0 < x < 2$
5 $k \leqslant -2$ and $k \geqslant 3$ 12 $-6 < x < 0$ and $x > 6$
7 $0 < x < 1$ and $x > 2$ 13 $-7 < x < -1$ and $x > 4$
8 $-\frac{3}{2} < x < 1$ and $x > 4$ 14 $x < -3$ and $0 < x < \frac{1}{3}$
9 $x < -1$ and $\frac{3}{2} < x < 2$ 15 $]-\infty, -10] \cup [2, \infty[; k \leqslant -4; \frac{1}{2}, -1$
10 $x < -1$ and $x > 1$

Exercise 10.7

1 $\dfrac{2}{3(2 - x)} - \dfrac{1}{3(1 + x)}$ 6 $1 + \dfrac{5}{2(x - 3)} - \dfrac{1}{2(x + 3)}$

2 $\dfrac{6}{7(3x + 1)} + \dfrac{5}{7(x - 2)}$ 7 $\dfrac{5}{x} - \dfrac{5}{x - 1} + \dfrac{5}{(x - 1)^2}$

3 $\dfrac{1}{2(x - 1)} + \dfrac{3}{2(x + 3)}$ 8 $\dfrac{2}{3x} - \dfrac{2x}{3(x^2 + 3)}$

4 $\dfrac{3}{16(x + 2)} + \dfrac{3}{4(x - 2)^2} - \dfrac{3}{16(x - 2)}$ 9 $1 - \dfrac{1}{3(x + 1)} + \dfrac{1}{3(x - 2)}$

5 $\dfrac{7x}{2(x^2 + 2)} - \dfrac{3}{2x}$ 10 $\dfrac{7}{8(x + 2)} - \dfrac{7}{8x} + \dfrac{7}{4x^2} - \dfrac{3}{2x^3}$

11 $1 + \dfrac{6}{x-3} + \dfrac{9}{(x-3)^2}$

12 $\dfrac{2}{x+1} - \dfrac{2}{x-2} + \dfrac{1}{(x-2)^2}$

13 $\dfrac{2}{5(x+1)} + \dfrac{13x-8}{5(x^2+4)}$

14 $\dfrac{2}{1-x} + \dfrac{2x+4}{x^2+x+1}$

15 $\dfrac{3}{2(x+1)} + \dfrac{3}{2(x-1)} - \dfrac{2}{x}$

16 $\dfrac{1}{x-1} - \dfrac{3}{x-2} + \dfrac{2}{x-3}$

17 $\dfrac{2}{1+x^2} - \dfrac{4}{1-x} + \dfrac{2}{(1-x)^2}$

18 $1 + \dfrac{1}{2x} - \dfrac{x+4}{2(x^2+2)}$

19 $-\dfrac{1}{2(x+1)} + \dfrac{\sqrt{3}-1}{4(\sqrt{3}-x)} + \dfrac{1+\sqrt{3}}{4(\sqrt{3}+x)}$

20 $\dfrac{1}{8x} - \dfrac{x}{8(x^2+4)} - \dfrac{x}{2(x^2+4)^2}$

Exercise 10.8

1 $-4 - 2x - 10x^2; \quad |x| < \frac{1}{2}$

2 $x - 5x^2 + 19x^3; \quad |x| < \frac{1}{3}$

3 $-\dfrac{1}{6} + \dfrac{x}{36} + \dfrac{11x^2}{216}; \quad |x| < 2$

4 $-\dfrac{1}{2} + \dfrac{x}{4} - \dfrac{3x^2}{8}; \quad |x| < 1$

5 $\dfrac{2x}{3} + \dfrac{4x^2}{9} + \dfrac{14x^3}{27}; \quad |x| < 1$

6 $1 + 4x + 9x^2; \quad |x| < 1$

7 $\dfrac{1}{3} - \dfrac{x}{9} + \dfrac{19x^2}{27}; \quad |x| < 1$

8 $1 + 3x + 7x^2; \quad |x| < \frac{1}{2}$

9 $-1 + x - 4x^2; \quad |x| < \frac{1}{3}$

11 $\frac{1}{2}[1 - 3^{-(n+2)}]$

12 $(-2)^{n-1}; \quad |x| > 2$

13 $\dfrac{1}{2} - \dfrac{x}{2} + \dfrac{3x^2}{4}$

14 $\frac{1}{4}[1 + (-1)^n(6n+7)]$

Exercise 10.9

1 $\frac{1}{2}\ln\left|\dfrac{x-1}{x+1}\right| + c$

2 $\ln\frac{4}{3}$

3 $\frac{1}{4}[\ln\frac{4}{3} - \frac{1}{6}]$

4 $\frac{1}{2}\ln\frac{9}{5}$

5 $\frac{21}{121}\ln\frac{18}{7} + \frac{2}{33}$

6 $\frac{1}{2} + \frac{1}{12}\ln\frac{5}{3} + \frac{4}{3}\ln\frac{3}{4}$

7 $5\ln 3 - 8\ln 2$

8 $\dfrac{1}{2\sqrt{2}}\ln\left|\dfrac{\sqrt{2}+x}{\sqrt{2}-x}\right| + c$

9 $\dfrac{3}{4} - \dfrac{3}{2\sqrt{2}}\tan^{-1}\sqrt{2} + \dfrac{3}{2\sqrt{2}}\tan^{-1}\dfrac{1}{\sqrt{2}}$

10 $\frac{1}{4} - \frac{1}{4}\ln 2$

11 $4\ln\frac{2}{3} + \frac{5}{3}$

12 $4\ln|x+2| - \frac{1}{2}\ln|x^2+5| + \dfrac{2}{\sqrt{5}}\tan^{-1}\dfrac{x}{\sqrt{5}} + c$

Exercise 10.10

1 $\dfrac{1}{4}-\dfrac{1}{2(n+1)(n+2)}$ 6 $\frac{1}{2}$

4 $\dfrac{5}{12}-\dfrac{2n+5}{2(n+2)(n+3)}$ 7 $\dfrac{r+2}{(r+1)!}; 1-\dfrac{1}{(2n+1)!}; -1+\dfrac{1}{(2n+1)!}$

Miscellaneous Exercise 10B

1 $\left(1+\dfrac{2}{m^2},\dfrac{2}{m}\right)$ 2   6   3 $x^2-19x+9=0$

4 (a) $\frac{1}{2}\ln\dfrac{x^2}{2x+1}+c$ (b) $x^2-x+\ln(x+1)+c$

6 $(x-2)^2(x+1);$ $(-1,0),(2,0),(0,-4)$
7 $k=\frac{1}{2},\frac{9}{2};$ $x=-\frac{1}{3}$ and -3
8 $x=9,\frac{1}{27}$ 12 $-2,-3$

9 $a=-9,$ $b=24;$ $(x-2)^2(x-5)$ 13 $1+\dfrac{1}{3(2x-1)}-\dfrac{2}{3(x+1)};\frac{5}{2}$

10 (a) $\frac{5}{4}$ (b) $a<-1$ 14 (a) $-\frac{1}{3}\leqslant k\leqslant 1$ (b) $\dfrac{2\pm 3\sqrt{2}}{7}$

11 $]-2,-1[$ 15   3 and 75

16 $\dfrac{1}{1+x}+\dfrac{2}{(1-x)^2}-\dfrac{2}{1-x};$ $1+x+5x^2+\cdots$

17 (b) $8x^2-4(1-3k)x+(1-k)^3=0$
18 $(x+2)(2x-1)(x^2-x+2)$
19 $]-\infty,-4[\cup]-1,2[$
20 $p=6,$ $q=1;$ $(x-1)(3x-4)(2x+1)$
21 $(\alpha+\beta)(\alpha^2-\alpha\beta+\beta^2);-4;x^2-2x-4=0;$ $1\pm\sqrt{5}$ in either order

22 $\dfrac{1}{2x}-\dfrac{1}{2(x+2)};$ (a) $12[x^{-5}-(x+2)^{-5}]$ (b) $\frac{1}{2}\ln\frac{9}{5}$

23 $]2,4[$ 24 $\dfrac{1}{4(x+1)}+\dfrac{3}{4(1-3x)};$ $1+2x+7x^2+20x^3+\cdots;\frac{1}{4}((-1)^n+3^{n+1})$

25 (a) $\ln\frac{5}{2}$ (b) $2-\pi/2$
26 (a) $6\pm 2\sqrt{5}$ (b) $5kx^2-(k^2-2k+16)x+5k=0$
27 $]-\infty,-\frac{1}{3}[\cup]3,\infty[$

28 $\dfrac{1}{2}\left[\dfrac{1}{x+1}-\dfrac{1}{x+3}\right];\dfrac{n}{3(2n+3)}$

29 (a) $A=-1,$ $B=6,$ $C=-8$

    (b) $\frac{1}{2}\ln(x^2+4)+c;\frac{1}{2}\tan^{-1}\dfrac{x}{2}+c$ (c) $a=2,$ $b=1$

30 $p=3,$ $q=-5$
31 $[0,\infty[$
32 $6,1;34,198;$ coefficients are $2,6,17,33$

33 $\dfrac{2}{p},-p;3.2^{1/3}a$

34 $|n|>3$

Unit 11

Exercise 11.1

1   $-\sin 2x$

2   $3\tan^2 x \sec^2 x$

3   $55\cos 5x$

4   $\dfrac{2x}{x^2+1}$

5   $\cot x$

6   $6\sin 12x$

7   $3e^{3x+1}$

8   $3x^2 e^{x^3}$

9   $-2\tan 2x$

10   $-3\sin(3\theta - \pi/4)$

11   $\dfrac{3}{\sqrt{1-9x^2}}$

12   $\dfrac{2}{4+x^2}$

13   $3\cos 3\theta e^{\sin 3\theta}$

14   $1+\ln x$

15   $\dfrac{\ln x - 1}{(\ln x)^2}$

16   $2x\tan x + x^2 \sec^2 x$

17   $e^x(\cos x + \sin x)$

18   $\frac{3}{2}x^{1/2} - \frac{1}{2}x^{-3/2}$

19   $\tan^{-1} 2x + \dfrac{2x}{1+4x^2}$

20   $\dfrac{x\cos x + \sin x}{x\sin x}$

21   $-\dfrac{x}{2y}$

22   $\dfrac{3-y}{x+2y}$

23   $\dfrac{-y^2}{x^2(y+1)}$

24   $\dfrac{1}{t}, \; -\dfrac{1}{2at^3}$

25   2

26   $62\frac{5}{6}$ or $\frac{377}{6}$

27   $\frac{1}{4}$

28   $e^2 - e^{-2}$

29   $\frac{1}{2}\ln\frac{5}{3}$

30   $74\frac{2}{27}$

31   $\pi/4$

32   $\frac{1}{2}\ln 2$

33   1

34   $\frac{1}{4}$

35   $\frac{15}{8} + \ln 2$

36   $e^3 - e$

37   24.2

38   $2(e^{-1} - e^{-3})$

39   π

40   $\frac{1}{2}\ln 2$

41   $\dfrac{3}{2\sqrt{2}}\tan^{-1}\dfrac{x}{\sqrt{2}} + c$

42   $\dfrac{5}{\sqrt{3}}\sin^{-1}\sqrt{3x} + c$

Exercise 11.2

1   $\frac{1}{35}(5x-4)^7 + c$

2   $-\dfrac{1}{9(3x-2)^3} + c$

3   $(x^2-3)^{1/2} + c$

4   $\frac{1}{5}\sin^5 x + c$

5   $\sqrt{x} - \frac{3}{2}\ln(2\sqrt{x}+3) + c$

6   $\frac{1}{2}(\ln x)^2 + c$

7   $-\sqrt{1-x^2} + c$

8   $\frac{1}{6}$

9   $3(\sqrt{2}-1)$

10   $12\ln 4 - \frac{33}{2}$

11   $\frac{2}{3}(1+e^x)^{3/2} + c$

12   $2\sqrt{x-1} + 4\tan^{-1}\sqrt{x-1} + c$

13   $\frac{1}{2}e^{x^2+3} + c$

14   $\frac{77}{3}$

15   $\tan^{-1} e - \dfrac{\pi}{4}$

16   $\frac{1}{4}\ln\frac{32}{17}$

Exercise 11.3

1   $2\sqrt{x+1} + c$

2   $\dfrac{-1}{30(5x+2)^6} + c$

3   $\dfrac{9\pi}{4}$

4   $\frac{5}{24}$

5 $\frac{1}{4}\sin^{-1}4x + c$

6 $\dfrac{1}{4\sqrt{2}}$

7 $-\dfrac{1}{(x-2)} - \dfrac{2}{(x-2)^2} - \dfrac{4}{3(x-2)^3} + c$

8 $2 + \pi/\sqrt{3}$

9 $1 + 4\ln\frac{4}{3}$

10 $\sin x - \frac{1}{3}\sin^3 x + A$

11 $\dfrac{-1}{9}\left[\dfrac{1}{3(3x-1)^3} + \dfrac{1}{4(3x-1)^4}\right] + c$

12 $\frac{32}{3}$

13 $\frac{1}{9}(x-2)^9 + \frac{5}{8}(x-2)^8 + c$

14 $\frac{1}{5}\cos^5 x - \frac{1}{3}\cos^3 x + c$

15 $\frac{203}{480}$

16 $-\sqrt{25 - x^2} + c$

17 $\frac{1}{2}\ln 3$

Exercise 11.4

1 1

2 $\pi/12$

3 $\frac{2}{3}(1 - \cos\theta)^{3/2} + c$

4 $4\tan^{-1}\sqrt{2x - 1} + c$

5 $\pi/12$

6 $4\sqrt{3x^2 + 1} + c$

7 $2(e^x + 2)^{1/2} + c$

8 $\pi/2 + 1$

9 $\frac{1}{9}\ln(1 + 3x^3) + c$

10 1

11 $\dfrac{(9x^2 - 1)^{3/2}}{27} + c$

12 $\frac{8}{3}[\sqrt{2} - 1]$

13 $\pi/4$

14 $2 + \ln\dfrac{2}{e^2 + 1}$

15 $\dfrac{1}{8} + \dfrac{\pi}{16}$

16 $\pi/6$

17 π

Exercise 11.5

1 $x\sin x + \cos x + c$

2 $xe^x - e^x + c$

3 1

4 $\dfrac{x}{3}\sin 3x + \dfrac{\cos 3x}{9} + c$

5 $\dfrac{x^4}{4}\ln x - \dfrac{x^4}{16} + c$

6 $\frac{1}{4}(e^2 + 1)$

7 $\ln 12 - 1$

8 $x^2\sin x + 2x\cos x - 2\sin x + c$

9 $\frac{1}{3}[2 - \ln 3]$

10 $x\sin^{-1} x + (1 - x^2)^{1/2} + c$

11 1

12 $\dfrac{1}{2}\left[\dfrac{5\pi}{6} + 1 - \sqrt{3}\right]$

13 $\frac{1}{16}[\pi^2 - 4]$

14 $\dfrac{\pi}{3} - \dfrac{\sqrt{3}}{2}$

15 $\frac{1}{2}e^{-x}(\sin x - \cos x) + c$

16 $\frac{1}{2}e^{x^2} + c$

17 $(x-1)^2 e^x - 2(x-1)e^x + 2e^x + c$

18 $-\frac{1}{16}(1 - x^2)^8 + c$

19 $-e^{\cos x} + c$

20 $\frac{4}{9}[\frac{4}{3}(3x - 4)^{3/2} + \frac{1}{5}(3x - 4)^{5/2}]$

21 $\dfrac{x}{3}e^{3x+1} - \dfrac{e^{3x+1}}{9} + c$

22 $-\dfrac{\cos^6 x}{6} + c$

Exercise 11.6

1 (a) $\ln|1 + 2x| - \ln|3 - x| + c$ (b) $\dfrac{\pi}{3} + \dfrac{\sqrt{3}}{2}$

3 $\dfrac{2}{1 + 4x^2}$; (a) $\frac{1}{8}\ln(4x^2 + 1) + \tan^{-1}2x + c$ (b) $\tan^{-1}2x^2 + c$

4 (a) $\frac{1}{4}e^{2x}(6x+5)+c$ (b) $\frac{1}{64}(\pi+2)$

5 0.59

6 (a) 33.4   (b) $\frac{2}{3}$ (c) ln(1.4)

7 (a) $\frac{1}{2}\ln 2$ (b) $\frac{4}{15}$

8 (a) $\frac{1}{2}x^2\ln|x|-\frac{1}{4}x^2+c$ (b) $\frac{2}{3}(x-2)^{3/2}+4(x-2)^{1/2}+c$

9 (a) $\frac{4}{5}(\sin x)^{5/2}+c$

10 (a)(i) $(\pi-2)/8$ (ii) $\frac{16}{105}$ (b) $e(e-1)+\ln(e+1)$

Exercise 11.7

1 0.25   **3** 2.4   **5** 1.5%   **7** (a) 3.0005 (b) 0.2475

2 2π **4** 0.142   **6** (a) 32.2 (b) 31.4   (c) 0.05

8 $\dfrac{dT}{dx}=\dfrac{k}{2\sqrt{x}};\quad 0.1\%$

9 $\ln y = \ln\sin 3x - 2x;\ 1;\ 0.08$

10 $V=\frac{1}{6}\pi d^3; 9\%; 6\%$

Exercise 11.8 (Remember: general solutions are not unique)

1 $y^4 = 2x^2 + c$

2 $y = \dfrac{x^3}{3} + x + c$

3 $y = Ax$

4 $\sin y = \sin x + c$

5 $\ln y = e^x + c$

6 $x + \dfrac{1}{2y^2} = c$

7 $y^2 = c(x^2 - 1)$

8 $y = \tan^{-1}x + c$

9 $2\tan^{-1}y = \ln x + c$

10 $s^2 = t^2 + 6t + c$

11 $6x + \ln\cos y = c$

12 $\dfrac{y^3}{3} = xe^x - e^x + c$

13 $e^{-y} = \cos x + c$

14 $y^2 = (\ln x)^2 + c$

15 $r = \ln(\sin\theta + \cos\theta) + c$

16 $\tan y = 2\tan x + c$

17 $2y^{1/2} + e^{-x} = 7$

18 $y^4 = 3\left(\dfrac{x-2}{x+2}\right)$

19 $y = \dfrac{2x}{5-2x}$

20 $e^y = \frac{1}{4}[5 - (2x+1)e^{-2x}]$

21 $y = \dfrac{x^2+3}{3-x^2}$

22 $y = \sqrt{(2\ln x + 5 - x^2)}$

23 $y = e^{2x^2-2}$

Exercise 11.9

1 $t = \dfrac{1}{k}\ln(p/100); \dfrac{T\ln 1.5}{\ln 2}$

2 $t = \ln\left(1 + \dfrac{x^2}{4}\right) + \dfrac{3}{2}\tan^{-1}\dfrac{x}{2}$ **4** 93 days; 320 mg

5 $\dfrac{1}{V\beta}\ln 3; \dfrac{Ve^{v\beta t}}{\alpha(3t^{v\beta t})}$

Exercise 11.10

1 π **3** $21\frac{1}{3}$ **5** $e^3 - 1$

2 $(-1,0),(\frac{5}{2},0);\frac{407}{32}$ **4** $\frac{8}{3}$ **6** $\dfrac{1}{3}\left[\dfrac{1}{e} - \dfrac{1}{e^4}\right]$

7 (a) $\frac{14}{9}$ (b) 2 units^2 (c) $\pi \ln 4 \, \text{unit}^3$

8 0.048

9 $(0,0)$, $(\pi, \pi)(2\pi, 2\pi)$, $(3\pi, 3\pi)$, $(4\pi, 4\pi)$; $\frac{1}{6}[32\pi^2 + 15]$

10 $-e^{-x}(2\cos x)$

11 Min when $x = 3$;   max when $x = \frac{3}{5}$; 24.3 units^2

Miscellaneous Exercise 11B

2 $\dfrac{2}{2x-1} - \dfrac{x}{x^2+1}$; $\frac{1}{2}\ln\dfrac{(2x-1)^2}{(x^2+1)} + c$

3 $y + x + 3 = 0$

4 0.02

6 Max value $= 2$ at $x = \pi/6$;   Min value $= -2$ at $x = \dfrac{2\pi}{3}$

   $x = \dfrac{5\pi}{12}$ and $\dfrac{11\pi}{12}$; $\dfrac{\pi}{2}[\pi + \sqrt{3}]$

7 $\tan^{-1}e^x + \frac{1}{2}\ln(1 + e^{2x}) + c$ **11** (a) $-\dfrac{1}{2(1+e^{2x})} + c$ (b) $\dfrac{\pi}{3\sqrt{3}}$

8 $\dfrac{3}{4} + \dfrac{\pi}{3} - \dfrac{\sqrt{3}}{4}$ **12** (a) $\dfrac{\sqrt{3}}{8} + \dfrac{\pi}{12}$ (b) 1   (ii) $\ln 3$

9 $\dfrac{3}{1+3x} - \dfrac{1}{1+x}$; $y = \dfrac{x-1}{x+1}$ **13** $y = \ln(1+x^2) + e^{-x^2} - 1$, $p = \frac{1}{6}$, $q = -\frac{5}{24}$

14 Max $y = 2$;   Min $y = -2$; $x = \pi/3$ and π

15 (a) $\frac{61}{192}$ (b) $1 - \pi/4$ (c) π (d) $\ln\frac{1}{3}$

16 $\frac{1}{2}\tan^{-1}\dfrac{y}{2} = x + c$; $\dfrac{4}{\pi}\ln 2$ **20** $ye^2 + x = 4$; $1 - 1/e^2$

17 (a) $(10, 7)$ (b) $\delta y \approx \dfrac{4}{y+1}\delta x$ **21** (a) 10π (b)   3:2   (c) $3s$

18 $2x - y - 4 = 0$; $(4,4)$; $\frac{3}{2}$; $\frac{3}{4}$ **22** $y = x$; $\dfrac{\pi^4}{48} + \dfrac{\pi^2}{8}$

19 $4\cos x(e^{\sin x} - 2e^{1/2}\sin x)$ **23** $-\frac{3}{10}$

24 $\dfrac{mx}{\sqrt{1-m^2x^2}} + \sin^{-1}mx$; $x\sin^{-1}mx + \dfrac{1}{m}(1 - m^2x^2)^{1/2} + c$

Unit 12

Exercise 12.1

1 $(2,-1)$; $x = 2$ **7** $(0,-4)$; $x = 0$ **13** $(x-1)^2 + 4$

2 $(\frac{1}{2}, -\frac{1}{4})$; $x = \frac{1}{2}$ **8** $(3, \frac{1}{2})$; $x = 3$ **14** $(x+2)^2 - 9$

3 $(\frac{1}{2}, -\frac{9}{4})$; $x = \frac{1}{2}$ **9** $(\frac{1}{2}, -\frac{3}{4})$; $x = \frac{1}{2}$ **15** $(x - \frac{3}{2})^2 + \frac{7}{4}$

4 $(0,0)$; $x = 0$ **10** $(1, 1)$; $x = 1$ **16** $(x-1)^2 - 1$

5 $(0, 3)$; $x = 0$ **11** $(-\frac{3}{2}, \frac{3}{2})$; $x = -\frac{3}{2}$ **17** $(x - \frac{1}{2})^2 + \frac{1}{4}$

6 $(4, 0)$; $x = 4$ **12** $(-2, -3)$; $x = -2$ **18** $(x + \frac{1}{2}a)^2 + \frac{3}{4}a^2$

21 (a) $(x-6)^2 - 11$; -11 (b) $2(x+2)^2 - 8$; -8 (c) $-(x-2)^2 + 4$; 4

   (d) $4(x - \frac{3}{2})^2 + 2$

22 (a) $4(x^2) - 12$; $[-12, \infty[$ (b) $-3(x - \frac{1}{6})^2 + \frac{13}{12}$; $]-\infty, \frac{13}{12}]$

   (c) $5(x + \frac{9}{10})^2 - \frac{81}{20}$; $[-\frac{81}{20}, \infty[$ (d) $-6(x - \frac{1}{3})^2 - \frac{25}{3}$; $]-\infty, -\frac{25}{3}]$

24 (a) $[-3, \infty[$ (b) $[-4, \infty[$ (c) $[\frac{5}{2}, \infty[$ (d) $]-\infty, \frac{53}{4}]$

25 (a) $]0, 2[$ (b) $[-2, 2]$ (c) $[1, 5]$ (d) $]-\infty, 1[\cup]3, \infty[$

   (e) $]-1, \frac{1}{3}[$ (f) $]-\infty, -3]\cup[-\frac{1}{2}, \infty[$

26 $c > 12$

Exercise 12.4

2	1	**6**	$]-\infty, 1[$	**10**	$]-\infty, -6[\cup]-2, \infty[$	
3	-1	**7**	$]-\infty, -1[$	**11**	$]-5, -2[\cup]-2, -1[$	
4	$\frac{4}{3}, \frac{2}{3}$	**8**	$]-\infty, \frac{2}{3}[\cup]\frac{4}{5}, \infty[$	**13**	$]-2, 0[$	
5	$3, -4$	**9**	$]-\infty, -4[\cup]3, \infty[$	**14**	$-2 < a < \frac{4}{5}; -\frac{4}{5} < a < \frac{4}{5}$	

Exercise 12.5

1 $fg:x \to (x-3)^3; gf:x \to x^3 - 3; hf:x \to \sin x^3; hg:x \to \sin(x-3)fh:x \to \sin^3 x;$
$gh:x \to \sin x - 3, \cdot fff:x \to x^{27}; gg:x \to x - 6; g^{-1}; x \to x + 3$

2 $[0, 2\pi]$ **9** domain and range $[0, \infty[$

3 $(x-2)/(1-x)$ **10** g^{-1} does not exist

4 $[0, 27]$ **11** domain $[1, \infty[$, range $[0, \infty[$

5 $[-27, 0]$ **12** domain $[-1, \infty[$, range $[0, \infty[$

6 $[-1, 1]$ **13** k^{-1} does not exist

7 $[-1, 1]$ **14** domain $\mathbb{R}$, range $]0, \infty[$

8 (a) $f^{-1}(x) = x + 3$ **15** domain $]2, \infty[$, range $\mathbb{R}$
     (b) $g^{-1}(x) = e^x + 1$ **16** n^{-1} does not exist

17 (a) $\mathbb{R}, f(x) \neq 1$ (b) x (c) $\dfrac{x+3}{x-1}$

18 $f(g(x)) = \sin x, \quad 0 \leqslant x \leqslant \pi$

19 $f^{-1}(x) = e^{x+1} - 3, \mathbb{R},] - 3, \infty[$; reflection in $y = x$.

Exercise 12.6

1	c	**15**	All $x, x \neq 0, \pm\pi, \pm 2\pi, \ldots$
2	c		(b) All $x, x \neq \pm\pi/2, \pm 3\pi/2, \ldots$
3	an infinite discontinuity		(c) All $x > 0$
4	c	**16** (a) No (b) No (c) Yes
5	an infinite discontinuity	**17** $k = 1$
6	c	**18** $0, 1, 0, 0, 0$
7	an infinite discontinuity	**19** (a) 1 (b) $2\frac{1}{2}$
8	c	**20** (a) No (b) even (c) No
9	(b) odd (c) period 2π	**21** (a) Yes (b) neither (c) No
10	(b) odd (c) period π	**22** (a) Yes (b) odd (c) No
11	(b) even (c) period 2π	**23** (a) Yes (b) odd (c) Yes
12	(b) even (c) period 4π	**24** (a) Yes (b) even (c) Yes
13	(b) even (c) period π	**25** (a) No (b) No (c) Yes
14	(b) even (c) period π		

Exercise 12.7

1 (a) 2π (b) π (c) π (d) $2\pi/3$ (e) 2π (f) π
    (g) $\pi/2$ (h) 6π

2 (a) odd   (b) even   (c) odd   (d) neither   (e) odd   (f) even

3 $2 + \cos 2x; \pi; [1, 3]$ **9** (a) $f(x) = -f(-x)$; reflection in y-axis

4 $5; 65\frac{1}{3}$ (b) $f(x) = f(-x)$; half turn about 0

5 No; Yes; 10   domain $\mathbb{R}, x \neq k\pi, \quad k \in \mathbb{Z}$

10 (a) neither   (b) even

7 $\dfrac{1}{\sqrt{2}}; \dfrac{1}{\sqrt{2}}; 0$ **11** 3π

12 $\dfrac{2}{\ln 2}$

Miscellaneous Exercise 12B

1 (a) $[2, \infty[$ (b) $[-4, 5]$ (c) $[-\sqrt{2}, \sqrt{2}]$

2 $]-3, -1[\cup]1, 3[$ **5** $]-\infty, -3\frac{1}{2}[$ **8** $]0, \frac{4}{5}[$

3 $]-4, 4[$ **6** 1   **9** $\pi(4\pi + 3\sqrt{3})$

4 -1 and $\frac{5}{3}$ **7** $(-q, r); pr > 0;$
          $p = 4, \quad q = -\frac{3}{4}, \quad r = \frac{11}{4}$

10 $g^{-1}(x) = \dfrac{3x+5}{x-2}$; $\mathbb{R}, x \neq 2$

11 $]-\infty, -4[\cup]0, \infty[$

12 f is odd, g is even, f has period $\pi; \frac{1}{2}$

13 $]-4, 0[$

15 $]-\infty, \frac{9}{4}]$ no inverse; $]0, 1]$ inverse exists

16 (a) $[1 + \ln 2, \infty[$ (b) $\frac{1}{3}(e^{x-1} - 2)$ (c) $[1 + \ln 2, \infty[$

17 (b) $17\frac{1}{15}$

19 (a) $]-1, 0[\cup]1, \infty[$ (b) $]-\infty, -\sqrt{2}[\cup]1, \sqrt{2}[\cup]2, \infty[$
  (c) $]-\infty, \frac{1}{2}[\cup]2, \infty[$

20 $\pi, \pi, \pi; [0, 2]$ **21** $a = 5 - c;]\frac{1}{2}, 4\frac{1}{2}[$

Unit 13

Exercise 13.1

1 $4\mathbf{i} + 4\mathbf{k}$

2 $\begin{pmatrix} 3 \\ 3 \\ 13 \end{pmatrix}$

3 $\sqrt{14}$

4 $a = -1, \quad b = 2$

5 $a = \frac{4}{3}$

6 $\frac{1}{3}(2\mathbf{i} + 2\mathbf{j} - \mathbf{k})$

7 $\dfrac{4}{7}\begin{pmatrix} 6 \\ -3 \\ 2 \end{pmatrix}$

8 (a) $\begin{pmatrix} 2 \\ 3 \\ 1 \end{pmatrix}$ (b) $\begin{pmatrix} 4 \\ 5 \\ 2 \end{pmatrix}$ (c) $\begin{pmatrix} 2 \\ 2 \\ 1 \end{pmatrix}$ (d) $\begin{pmatrix} -2 \\ -2 \\ -1 \end{pmatrix}$

10 $\begin{pmatrix} 3 \\ 1 \\ 3 \end{pmatrix}$

11 $2\mathbf{i} + 5\mathbf{k}$

12 $2\mathbf{b} - \mathbf{a}, 4\mathbf{b} - 3\mathbf{a}, 3\mathbf{b} - 2\mathbf{a}$

13 $\mathbf{r} = 2\mathbf{i} - 2\mathbf{j} + \mathbf{k}$

14 (a) $\mathbf{p} - \mathbf{q}, -3\mathbf{p} + \mathbf{q}$ (b) $\frac{1}{2}(-3\mathbf{p} + \mathbf{q})$
  (c) $(k - \frac{3}{2})\mathbf{p} + (\frac{1}{2} - k)\mathbf{q}$ (d) $\frac{3}{2}$

Exercise 13.2

1 (a) 1   (b) 2

3 (a) $7, \frac{1}{3}\sqrt{7}$
  (b) $14, \sqrt{\frac{7}{19}}$

4 (a) $-\dfrac{1}{\sqrt{3}}$ (b) $\frac{1}{3}$

5 $\cos A\hat{B}C = \dfrac{\sqrt{14}}{70}$,
  $\overrightarrow{AC} = \mathbf{i} + 2\mathbf{j} + 6\mathbf{k}$,
  $B\hat{A}C = 51.2°$

6 $\dfrac{3}{\sqrt{2}}, \sqrt{\frac{3}{2}}$

7 4

8 $\frac{1}{2}|\mathbf{b} - \mathbf{a}| \times |\mathbf{c}|$

9 $-\frac{10}{27}$

10 $\mathbf{a} - \mathbf{b}, \quad \mathbf{a} + \mathbf{b}$

Exercise 13.3

1 $\mathbf{r} = \begin{pmatrix} 2 \\ 3 \\ 7 \end{pmatrix} + t\begin{pmatrix} 2 \\ 5 \\ -1 \end{pmatrix}$, $\mathbf{r} = \begin{pmatrix} 9 \\ 3 \\ 6 \end{pmatrix} + \lambda\begin{pmatrix} 1 \\ -2 \\ 0 \end{pmatrix}$,

  $\mathbf{r} = \begin{pmatrix} 2 \\ 3 \\ 7 \end{pmatrix} + \mu\begin{pmatrix} 7 \\ 0 \\ -1 \end{pmatrix}$ or $\begin{pmatrix} 9 \\ 3 \\ 6 \end{pmatrix} + \mu\begin{pmatrix} 7 \\ 0 \\ -1 \end{pmatrix}$, $\mathbf{r} = s\begin{pmatrix} 3 \\ 1 \\ 2 \end{pmatrix}$

2 $(3, 1, 9), (-6, 7, -6), (0, 3, 4), -\frac{3}{2}$

3 (a) $0°$(or $180°$)   (b) $42.8°$ (c) $111.5°$ (d) $90°$

4 $\begin{pmatrix} 1 \\ 2 \\ 0 \end{pmatrix}$ **7** $\begin{pmatrix} -2 \\ -2 \\ -6 \end{pmatrix}, \begin{pmatrix} -1 \\ -1 \\ -5 \end{pmatrix}, \mathbf{r} = \begin{pmatrix} 4 \\ 5 \\ 10 \end{pmatrix} + \lambda\begin{pmatrix} 1 \\ 1 \\ 5 \end{pmatrix}, \mathbf{r} = \begin{pmatrix} 1 \\ 2 \\ -1 \end{pmatrix} + \mu\begin{pmatrix} 1 \\ 1 \\ 3 \end{pmatrix}, (3, 4, 5)$

6 $-2, (-1, 3, 4)$

Exercise 13.4

1 (a) $\mathbf{r} \cdot \begin{pmatrix} 2 \\ 1 \\ -3 \end{pmatrix} = 0$ (b) $\mathbf{r} \cdot \begin{pmatrix} -2 \\ -2 \\ 1 \end{pmatrix} + 5$ (c) $\mathbf{r} \cdot \begin{pmatrix} 0 \\ 2 \\ -1 \end{pmatrix} = 2b - c$

2 $\dfrac{20}{\sqrt{29}}$

5 (a) 66.4°   (b) 36.6°
   (c) 65.1°

7 (4, 2, 3)

4 $\mathbf{r} \cdot \begin{pmatrix} 3 \\ -1 \\ 2 \end{pmatrix} = 12$

8 (a) $\mathbf{r} \cdot \begin{pmatrix} -2 \\ 2 \\ 1 \end{pmatrix} = 9$ $\theta = 90°$ (c) 9

Exercise 13.5

1 (a) $\dfrac{x-2}{3} = -y = \dfrac{z+1}{z}$ (b) $1 - x = \dfrac{y}{3} = -\dfrac{z}{4}$

2 (a) $2x + y + 3z = 5$ (b) $y = 1$ 3 $\mathbf{r} = \begin{pmatrix} 1 \\ 0 \\ 0 \end{pmatrix} + s\begin{pmatrix} 1 \\ 1 \\ 0 \end{pmatrix} + t\begin{pmatrix} 2 \\ 1 \\ 1 \end{pmatrix}$ but there are other possible answers

4 (a) $\mathbf{r} = \mathbf{i} - \mathbf{j} + s(\mathbf{i} + \mathbf{j} + \mathbf{k}) + t(\mathbf{i} - 2\mathbf{j} + 3\mathbf{k})$ (b) $\mathbf{r} \cdot (5\mathbf{i} - 2\mathbf{j} - 3\mathbf{k}) = 7$
   (c) $5x - 2y - 3z = 7$

5 (a) $\dfrac{x-4}{3} = \dfrac{y-1}{2}$, $z = 2$ (b) $y = 2$, $z = 0$

6 $\mathbf{r} = \mathbf{k}$; $\mathbf{i} - \mathbf{j} + \mathbf{k}$; $\mathbf{r} \cdot (\mathbf{i} - \mathbf{j} + \mathbf{k}) = 1$

7 (a) $3\mathbf{j} + 4\mathbf{k}$ (b) $\mathbf{r} \cdot (3\mathbf{j} + 4\mathbf{k}) = 0$ (c) $\dfrac{1}{\sqrt{26}}$

8 (a) $\mathbf{r} = a(-4\mathbf{i} + 4\mathbf{j} - \mathbf{k}) + ta(9\mathbf{i} - 6\mathbf{j} + 12\mathbf{k})$ (b) $\mathbf{r} \cdot (13\mathbf{i} + 17\mathbf{j} + 16\mathbf{k}) = 0$

   (c) $\dfrac{21}{\sqrt{35}\sqrt{33}}$; $\mathbf{r} \cdot (9\mathbf{i} - 6\mathbf{j} + 12\mathbf{k}) = 15a$;   (d) $3x - 2y + 4z = 5a$

   (e) $\dfrac{x+49}{3} = \dfrac{y-49}{-2} = \dfrac{z+9}{4}$

Miscellaneous Exercise 13B

1 $\sqrt{6}, \sqrt{2}, \pi/6$

6 $\frac{9}{2}$

2 3

7 6, 5

3 $7, 9, \frac{59}{63}, \sqrt{122}$

8 $\mathbf{r} = \begin{pmatrix} 6 \\ 4 \\ 2 \end{pmatrix} + t\begin{pmatrix} 1 \\ 3 \\ 4 \end{pmatrix}$, $(5, 1, -2)(2, -8, -14)$

4 $\dfrac{1}{\sqrt{54}}(2\mathbf{i} + \mathbf{j} - 7\mathbf{k})$

9 $\mathbf{r} = \frac{2}{3}(4\mathbf{i} - 6\mathbf{j} + 14\mathbf{k}) + t(\mathbf{i} - 2\mathbf{j} + 2\mathbf{k})$; $\frac{35}{39}$

5 $11\mathbf{i} - 11\mathbf{j} - \mathbf{k}$; $\dfrac{3}{\sqrt{35}}$

10 $\begin{pmatrix} -2q - 2p - 2 \\ q - p + 1 \\ q + p - 3 \end{pmatrix}$; $p = \frac{2}{5}$, $q = -\frac{3}{5}$; $\dfrac{8\sqrt{5}}{5}$

11 $\dfrac{1}{3\sqrt{3}}[5\mathbf{i}-\mathbf{j}+\mathbf{k}]$; 54.2° **12** $\mathbf{r}=\begin{pmatrix}4\\2\\6\end{pmatrix}+t\begin{pmatrix}1\\2\\1\end{pmatrix}$; (5, 4, 7); $\mathbf{r}=\begin{pmatrix}4\\2\\6\end{pmatrix}+\lambda\begin{pmatrix}5\\-2\\5\end{pmatrix}$

13 (b) -1 (c) $2\sqrt{17}$

14 $\mathbf{r}=\mathbf{a}+t(\mathbf{a}-\tfrac{1}{2}\mathbf{b})$, $\mathbf{r}=\mathbf{b}+5(\mathbf{b}-\tfrac{1}{2}\mathbf{a})$, $\tfrac{1}{3}(\mathbf{a}+\mathbf{b})$, $\mathbf{r}=\lambda(\mathbf{a}+\mathbf{b})$

16 $(0, -3, 0)$; $(3, -1, 7)$

17 $\mathbf{r}=\begin{pmatrix}1\\3\\1\end{pmatrix}+t\begin{pmatrix}3\\-2\\8\end{pmatrix}$ or $\mathbf{r}=\begin{pmatrix}4\\1\\-7\end{pmatrix}+t\begin{pmatrix}3\\-2\\-8\end{pmatrix}$; $\mathbf{r}.(2\mathbf{i}-\mathbf{j}+\mathbf{k})=0$

18 $2\mathbf{i}+\mathbf{k}$; $\mathbf{r}=a\mathbf{i}+a\mathbf{j}+2a\mathbf{k}+t(2\mathbf{i}+\mathbf{k})$

19 1:6; $\mathbf{r}=(\mathbf{i}-2\mathbf{j}+\mathbf{k})+t(13\mathbf{i}+4\mathbf{j}-5\mathbf{k})$; $-12\mathbf{i}-6\mathbf{j}+6\mathbf{k}$

20 (a) 30°   (c) $\begin{pmatrix}2-2p+q\\-3+p\\5-p+q\end{pmatrix}$; $p=1$, $q=-2$

21 $\tfrac{1}{2}(\mathbf{b}+\mathbf{c})$, $\tfrac{1}{2}(\mathbf{a}+\mathbf{c})$, $\tfrac{1}{2}(\mathbf{a}+\mathbf{b})$; $\mathbf{r}=\tfrac{1}{2}\mathbf{a}+\lambda(\mathbf{b}+\mathbf{c}-\mathbf{a})$;
$\mathbf{r}=\tfrac{1}{2}\mathbf{b}+\mu(\mathbf{a}+\mathbf{c}-\mathbf{b})$; $\tfrac{1}{4}(\mathbf{a}+\mathbf{b}+\mathbf{c})$
22 (a) $(-4, 11)$ (b) $(\tfrac{7}{3}, -\tfrac{2}{3}, \tfrac{4}{3})$

Unit 14

Exercise 14.1

1 (a) $2i$ (b) -1 (c) $-i$ (d) 1   (e) i (f) -1 (g) $\dfrac{i}{3}$

2 (a) $3-4i$ (b) $2+i$ (c) $-3i$ (d) $1+i\sqrt{3}$ (e) $2+\sqrt{3}-3i$

3 (a) $\pm i$ (b) $\pm 5i$ (c) $-\tfrac{1}{2}\pm\dfrac{i\sqrt{3}}{2}$ (d) $4\pm 3i$ (e) $-\tfrac{3}{4}\pm\dfrac{i\sqrt{7}}{4}$

4 (a) 2   (b) -3 (c) $2+3i$ (d) $2-3i$
5 (a) $x=y=0$ (b) $x=3$, $y=1$ (c) $x=-1$, $y=1$ (d) $x=3$, $y=4$

   (e) $x=0$, $y=\sqrt{2}$ (f) $x=2$, $y=-1$ (g) $x=2$, $y=6$

6 (a) $4+2i$ (b) 17   (c) $-7-24i$ (d) $\dfrac{2}{5}+\dfrac{i}{5}$

   (e) $\tfrac{6}{13}+\tfrac{9}{13}i$ (f) $\tfrac{1}{5}-\tfrac{2}{5}i$ (g) $\tfrac{2}{17}+\tfrac{9}{17}i$ (h) $2i$ (i) $-2-2i$

7 $\tfrac{3}{5}+\tfrac{4}{5}i$ **11** $a=\pm 1$
8 $\pm(5+2i)$ **12** (a) $x=2$, $y=-1$ (b) $x=3$, $y=2$
9 2; 2   **13** -1; 1
10 $z=2$ **15** $a=-2$, $b=4$

Exercise 14.2

3 $2-i$ **9** $1+i\sqrt{3}, 1-i\sqrt{3}, \tfrac{5}{2}+\dfrac{i\sqrt{3}}{2}, \tfrac{5}{2}-\dfrac{i\sqrt{3}}{2}$; 28

4 $-i$; 1   **10** $x=\lambda+\dfrac{1}{5\lambda}$, $y=\dfrac{1}{10\lambda}-\dfrac{\lambda}{2}$

7 $\pm(3-2i)$; $-5+12i$ **11** $\tfrac{1}{2}+\dfrac{i\sqrt{3}}{2}$

Exercise 14.3

1 (a) $(4, \pi/2)$ (b) $(\sqrt{2}, -\pi/4)$ (c) $(\sqrt{2}, -3\pi/4)$ (d) $(\sqrt{2}, 3\pi/4)$
   (e) $(2, -\pi/3)$ (f) $(3, 0.84^{c})$ (g) $(3, -2.3^{c})$

2 (a) $3(\cos \pi + i \sin \pi)$ (b) $2(\cos(-\pi/2) + i \sin(-\pi/2))$
   (c) $a\sqrt{2}(\cos \pi/4 + i \sin \pi/4)$ (d) $a\sqrt{2}(\cos(-\pi/4) + i \sin(-\pi/4))$
   (e) $4(\cos 0 + i \sin 0)$ (f) $4[\cos(2.42) + i \sin(2.42)]$

3 (a) $\frac{1}{2} + \frac{i\sqrt{3}}{2}$ (b) $2\sqrt{3} + 2i$ (c) $-3 + 3\sqrt{3}i$ (d) $-1 + i$ (e) $2i$
   (f) -3 (g) $2 - 2\sqrt{3}i$

4 (a) $(\sqrt{2}, -0.79^{c})$,   (b) $(\sqrt{5}, 1.11^{c})$ (c) $(5, 2.22^{c})$ (d) $(\sqrt{\frac{5}{2}}, 1.90^{c})$
   (e) $(\sqrt{10}, 0.32^{c})$

5 (a) $4, \pi/3$ (b) $2, -\pi/6$ (c) $64, \pi$ (d) $32, 7\pi/6$

6 $10, 96.9^{\circ}; \frac{5}{2}, -23.1^{\circ}$ 7 $7\pi/6$ (or $-5\pi/6$)

8 (a) $5\sqrt{5}, -\frac{1}{2}$ (b) $\frac{1}{\sqrt{5}}, -\frac{1}{2}$

Exercise 14.4

17 $|z - 2 - 3i| = 6$ 19 $|z - 3i| = 3$
18 $|z + 4 - 3i| = 1$ 20 $\arg(z - 1) = 3\pi/4$
21 $\arg(z - 2i) = 0$
22 $\arg(-2 + i - z) = 3\pi/4$
26 (a) $|z - 3| = |z + 6|$ (b) $|z - 3 + 5i| = |z + 2 - i|$
27 (a) $y = -\frac{1}{2}$ (b) $y = 2x$
32 (a) $y = x$ (included) (b) $y = 0$ (not included), $y = \sqrt{3}x - 2\sqrt{3}$ (included)

Miscellaneous Exercise 14B

2 $-4, 13$
3 (a) $(4, \pi/3)$ (b) $(\frac{1}{2}, -\pi/6)$
4 (a) $-\pi/2$ (b) $\pi/4$ (c) $\pi/12$
5 $x = 1, \quad y = 2$ and $x = -1, \quad y = -2$
6 $1 + i, 2 - i; z^2 - (4 + 4i)z + (2i + 8) = 0$
7 (a) $(\sqrt{2}, -\pi/4)$ (b) $(2, \pi/3)$ (c) $(2\sqrt{2}, \pi/12)$ (d) $(1/\sqrt{2}, -7\pi/12)$
8 $p = \sqrt{2}, \quad q = -2 - 2\sqrt{2}$ and $p = -\sqrt{2}, \quad q = -2 + 2\sqrt{2}$

9 $0, \pm\sqrt{3}, \pm\frac{1}{\sqrt{3}}$

10 $(1, 2\pi/3), (1, 4\pi/3)$ (a) $\sqrt{3}, -\pi/2$ (b) $-1, 0$

11 $5\sqrt{2}; \alpha = -2, \quad \beta = -1$

12 $1/\sqrt{2}, \frac{5\pi}{12}; n = 12; -\frac{1}{64}$

13 (a) $\pm(3 + 2i)$ (b) $\sqrt{2}\left(\cos\frac{3\pi}{4} + i\sin\frac{3\pi}{4}\right)$

14 $p = 3, \quad q = 1$ and $p = -3, q = -1$

15 $a = 1, \quad b = -2; \frac{3}{10} + \frac{21i}{10}$ 16 $(1, \pi/2), (1, \pi/4)$

17 $\frac{4}{3} + \frac{2\sqrt{2}}{3}i, \frac{2}{3} + \frac{4\sqrt{2}}{3}i$

18 $2\sqrt{2}(\cos \pi/4 + i\sin \pi/4); \sqrt{104}; \tan^{-1}\frac{1}{3}$

19 (a)(i) $2\left(\cos\dfrac{2\pi}{3} + i\sin\dfrac{2\pi}{3}\right)$ (ii) $2(\cos \pi/2 + i\sin \pi/2)$

    (b) 3 and -4 or -3 and 4

20 (b) $1 + 3i; -3; 30$

Unit 15

Exercise 15.2
14 $a = -\frac{3}{2}$

Exercise 15.3
1 $3ty + 2x = 3t^4 + 2t^2; \frac{1}{8}$ **2** $(26, 9)$
3 $-\frac{1}{2}; (-2, \frac{1}{2}), (1, -1)$
4 Q is $(4, 3)$ (a) $x + y = 7$ (b) $\frac{7}{24}$ (c) 3.5 sq. units
5 $(-\frac{14}{5}, -\frac{13}{5})$ and $(\frac{34}{5}, \frac{23}{5})$

Exercise 15.4

1 $(0, 0), 3$ **5** $(1, 2), \sqrt{5}$ **9** $(-\frac{1}{6}, 0), \sqrt{73}/6$
2 $(0, 0), \frac{1}{2}$ **6** Not a circle   **10** $(-\frac{3}{2}, 2), \frac{5}{2}$
3 Not a circle   **7** $(-1, \frac{1}{4}), \frac{9}{4}$ **11** $2y + x = 0$
4 $(1, -2), 2$ **8** not a circle   **12** $2y = x - 5; y + 2x = 0$
13 (a) 1 (b) $4x + 6y - 3 = 0$ (A circle of infinite radius !)
14 $x^2 + y^2 + 2x - 10y + 1 = 0$ and $x^2 + y^2 - 6x - 2y + 9 = 0$

15 $\dfrac{6 \pm \sqrt{6}}{4}; \frac{23}{25}$ **16** $7, 4; x = -2$

Exercise 15.5

2 $a = 10, b = 5$ **4** $\cos\theta y + \sin\theta x = a\sin\theta\cdot\cos\theta$ **5** $x = 2t, y = \dfrac{1}{t} - t^3$
3 (b) $2ay = a^2 + x^2$ $4x^2 + 4y^2 = a^2$

Miscellaneous Exercise 15B
1 5   **8** $(6a, -5a), 2\sqrt{15a}$

2 $5, (3, 4)$ **9** $\dfrac{16\sqrt{2}}{3}$

4 $x = -2, y = 1;$ $x = 0$ and $x = 2$
5 $\frac{8}{15}$ **10** $(-5, 3); 6$
6 $-\frac{3}{2} + 2i; \frac{5}{2}$ **12** $y + tx = 2t + t^3, ty = x + t^2$
                                              13 $(-1, 0)$ **15** $-1, 5; \frac{3}{4}; 0$ and $\frac{8}{9}$
7 $m^2x^2 - (2m^2 + 4)x + m^2 = 0; \left(1 + \dfrac{2}{m^2}, \dfrac{2}{m}\right)$ **15** $-1, 5; \frac{3}{4}; 0$ and $\frac{8}{9}$

17 $x = 0$ and $y = 1$ **18** (a) $y - 2at = -t(x - at^2)$ (b) $2y^2 = a(x - a)$
19 $(2, 3)$ and $(3, 0); y + 3x = 6\sqrt{2}; y + 3x + 6\sqrt{2} = 0$
20 (a) $(6, 7)$ (b) $\frac{16}{15}$ (c) $8\sqrt{2}$ (d) 80

21 $\left[\frac{1}{3}(p^2 + pq + q^2), \dfrac{pq}{2}(p + q)\right]; 27y^2 + 2 = 9x$

22 (a) $x = -\frac{1}{4}$ (b) $8y^2 + 1 = 4x$
23 $x^2 + y^2 - 10x - 2y + 1 = 0; 7$
24 7

Unit 16

Exercise 16.1

1　(a)　$(2x-3)x+4$　　　(b)　$((3x-2)x+1)x-3$　　　(c)　$(x^2-1)x^2+2$
　　(d)　$((4x^3-3)x+1)x$　　(e)　$((x-5)x+3)x$

2　(a)　18.6672　　(b)　2252.77　　(c)　-0.371283
3　There are many possible answers for the rough positions. One possible answer for
　each is: (a)　1 root in $]1,2[$　　(b)　1 root in $]0,\pi/2[$　　(c)　2 roots; $DC=0$ and
　in $]-\pi/2,0[$　　(d)　2 roots; one just <0, one just >3
4　$-1,0; 1, 2; 1.40$　　　7　$0; \infty$
5　1.306　　　　　　　　8　$\theta \approx 2^c$
6　2.24　　　　　　　　　9　Just $>-\pi/2$; Just <0; Just $<3\pi/2$; in $]\pi/2,\pi[$

Exercise 16.2

1　1.6875
2　3.65625
3　(a)　$x_2=2$, $x_3=1.66667$, $x_4=1.61905$, $x_5=1.61803$, $x_6=1.61803$;
　　　　1.61803　　(b)　-0.61803　　(c)　$x^2-x-1=0$
4　(b)　converges slowly　　(a)　does not converge.
5　-0.317　　　　7　1 and 2; 1.05
6　2.6053　　　　　8　2.93

Exercise 16.3

1　**5040**　　　3　60　　　　5　3360　　　7　1
2　6　　　　　　4　20160　　6　24　　　　8　10
9　(a)　20　　(b)　210　　(c)　840　　(d)　6
10　(a)　$n(n-1)$　　(b)　$(n-1)(n-2)(n-3)$　　(c)　$n!$
11　(a)　6　　(b)　27　　(c)　336　　(d)　512
12　12　　　　14　720　　　16　3360　　18　35　　　20　1　　　　　22　60; 40
13　20　　　　15　20160　　17　280　　19　4　　　21　362880

Exercise 16.4

1　(a)　256　　(b)　24　　(c)　12　　(d)　8　　2　5040　　3　13
4　(a)　22　　(b)　96　　5　2520　　6　110　　7　360　　(a)　180　　(b)　240
8　(i)　171　　(b)　640　　9　78

Exercise 16.5

1　(a)　256　　(b)　24　　(c)　12　　(d)　8
2　5040　　4　(a)　22　　(b)　96　　6　110　　　　　　8　(i)　171　　(b)　640
3　13　　　5　2520　　　　　　　　　7　360　(a)　180　　9　78
　　　　　　　　　　　　　　　　　　　　　　(b)　240

Exercise 16.6

1　1.51　　　4　1.71　　　7　2.96　　　10　1.15　　　13　0.199
2　5.96　　　5　3.98　　　8　2.28　　　11　-0.151　　14　1.37
3　0.168　　6　2.35　　　9　3.14　　　12　1.43

15　$\dfrac{2\pi}{3}(e^{-9}+4e^{-4}+2e^{-1}+2)$　　　16　35.43

Miscellaneous Exercise 16B

1　50400　　　　　8　(a)　120　　　14　234　　　　　　　　19　2.33; 18.6;
2　0.18　　　　　　　(b)　60　　　15　$(\ln 5)^2$; $1.4\,\text{units}^3$　　　$3\pi/4$; 3.1
3　22; 2.22　　　9　0.433　　　16　26^4 and 999;　　20　(a)　15500
4　224; 425　　10　50　　　　　26 × 25 and　　　　(b)　480
5　0.815; 0.815　11　$-\tfrac{7}{6}$; -1.18　　24 × 23 and 738　21　1.17
6　$\tfrac{4}{3}$; 1.25　　12　30　　　　17　3.59; 0.278
7　0.520　　　　13　0.879; 0.877　18　126

INDEX